Catalytic Selective Oxidation

ACS SYMPOSIUM SERIES **523**

Catalytic Selective Oxidation

S. Ted Oyama, EDITOR
Virginia Polytechnic Institute and State University

Joe W. Hightower, EDITOR
Rice University

Developed from a symposium sponsored
by the Division of Petroleum Chemistry, Inc.,
at the 204th National Meeting
of the American Chemical Society,
Washington, DC,
August 23–28, 1992

American Chemical Society, Washington, DC 1993

Library of Congress Cataloging-in-Publication Data

Catalytic selective oxidation: developed from a symposium / sponsored
 by the Division of Petroleum Chemistry, Inc., at the 204th
 National Meeting of the American Chemical Society, Washington,
 DC, August 23–28, 1992; S. Ted Oyama, editor, Joe W. Hightower,
 editor.

   p.   cm.—(ACS symposium series, ISSN 0097–6156; 523)

Includes bibliographical references and indexes.

ISBN 0–8412–2637–7

   1. Oxidation—Congresses. 2. Catalysis—Congresses. 3. Vanadium
catalysts—Congresses.

   I. Oyama, S. Ted, 1955–   . II. Hightower, Joe W. III. American
Chemical Society. Division of Petroleum Chemistry, Inc.
IV. American Chemical Society. Meeting (204th: 1992:
Washington, D.C.) V. Series.

QD63.O9C39   1993
660′.28443—dc20   92–45894
                                           CIP

Foreword

THE ACS SYMPOSIUM SERIES was first published in 1974 to provide a mechanism for publishing symposia quickly in book form. The purpose of this series is to publish comprehensive books developed from symposia, which are usually "snapshots in time" of the current research being done on a topic, plus some review material on the topic. For this reason, it is necessary that the papers be published as quickly as possible.

Before a symposium-based book is put under contract, the proposed table of contents is reviewed for appropriateness to the topic and for comprehensiveness of the collection. Some papers are excluded at this point, and others are added to round out the scope of the volume. In addition, a draft of each paper is peer-reviewed prior to final acceptance or rejection. This anonymous review process is supervised by the organizer(s) of the symposium, who become the editor(s) of the book. The authors then revise their papers according to the recommendations of both the reviewers and the editors, prepare camera-ready copy, and submit the final papers to the editors, who check that all necessary revisions have been made.

As a rule, only original research papers and original review papers are included in the volumes. Verbatim reproductions of previously published papers are not accepted.

M. Joan Comstock
Series Editor

Contents

REACTIVITY OF SINGLE CRYSTALS AND
WELL-DEFINED CRYSTAL FACES

CHARACTERIZATION OF OXIDATION CATALYSTS

Activation and Selective Oxidation of C_1–C_4 Alkanes

State-of-the-Art Engineering Concepts in Selective Oxidation

INDEXES

Preface

THE NEED TO CONSERVE RAW MATERIALS and enhance the yield of high-value products has led the petrochemical industry to rely increasingly on processes that involve catalytic selective oxidation. In turn, this reliance has led to the emergence of important new technologies in the past decade—processes that use selective catalytic oxidation now generate almost a quarter of all organic chemicals produced worldwide.

The six sections following the overview chapter deal with aspects of selective oxidation that range from theories and concepts to state-of-the-art engineering applications. Several chapters describe the synthesis, characterization, and performance of potentially attractive new catalytic materials. These catalysts range from single crystals with well-defined crystal faces to highly dispersed or amorphous solids. Most of the actual catalytic reactions studied involve the oxidation of hydrocarbons in the range from C_1 to C_4.

At least a third of the chapters in this book deal with the use of vanadium in one of its several oxidation states. New analytical techniques have become available that make it possible to characterize these materials more precisely and to determine the exact nature of the active centers. We appear to be on the threshold of exciting breakthroughs in fundamental research that should result in significant improvements to industrial oxidation processes. Our hope is that this book will provide insights and stimulate additional research that will achieve these goals.

Acknowledgments

The editors are grateful to the more than 80 authors who cooperated beautifully in completing this book in a timely fashion. We also deeply appreciate the financial support from the Petroleum Research Fund; Amoco Chemical, Arco Chemical, and BP Research Companies; the Center for Advanced Materials Processing at Clarkson University; Exxon Chemical and Exxon Research and Engineering Companies; and the Division of Petroleum Chemistry, Inc., of the American Chemical Society, which made the symposium and this book possible.

We express our appreciation to Anne Wilson, Peggy Smith, and Cathy Buzzell-Martin of the American Chemical Society for seeing the book through to publication. Finally, the artwork on the cover includes a picture of the Cu(1,1,1) face of Cu_2O from Chapter 10 by Schulz and Cox.

S. TED OYAMA
Virginia Polytechnic Institute
   and State University
Blacksburg, VA 24061–0211

JOE W. HIGHTOWER
Rice University
Houston, TX 77251–1892

December 18, 1992

Chapter 1

Research Challenges in Selective Oxidation

S. Ted Oyama[1], A. N. Desikan[2], and Joe W. Hightower[3]

[1]Chemical Engineering Department, Virginia Polytechnic Institute and State University, Blacksburg, VA 24061
[2]Department of Chemical Engineering, Clarkson University, Potsdam, NY 13699−5705
[3]Department of Chemical Engineering, Rice University, Houston, TX 77251−1892

Selective oxidation is an active field of research, as attested by the number of international conferences centered on the subject that have been held in recent years (1-4). Although to a large extent the activity has been driven by commercial applications, the field has also attracted attention because of its scientific challenge. Compared to catalysis on metals, progress has been slower on oxides in uncovering the relationship between catalyst structure and mechanism. This is because oxidation catalysis is complex: the reactions are often intricate networks, many times occurring on traces of non-crystalline surface phases, with poorly established kinetics. Nevertheless, improvements in instrumentation and techniques have allowed a deeper understanding of the inherent processes that are involved in oxidation. This overview will not simply summarize the subject of the present conference, but will highlight important areas of research which are currently at various levels of advancement. Both commercial and scientific issues will be addressed.

Commercial Perspective

Industrially, the scope of partial oxidation catalysis is wide, ranging from the large scale production of commodities to the synthesis of minute amounts of pharmaceuticals and fine chemicals. Rough estimates place the worth of world products that have undergone a catalytic oxidation step at US$ 20-40 billion (5). A compilation of market data for the United States in 1990 is provided in Fig. 1-4 (6). The total production of the top 50 industrial chemicals in the U.S. was 616 billion pounds of which catalytic organic chemicals production was 147 billion pounds (Fig. 1). Of these organic chemicals, counting those produced by heterogeneous oxidation, homogeneous oxidation, oxychlorination and dehydrogenation, about one half were produced by an oxidation step (Fig. 2). For world markets, statistics are less complete, but the available data is

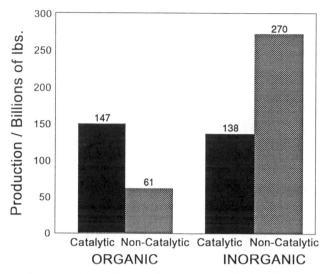

Figure 1. Survey of chemical production in the United States in 1991. (Reproduced from ref. 6. Copyright 1991 American Chemical Society.)

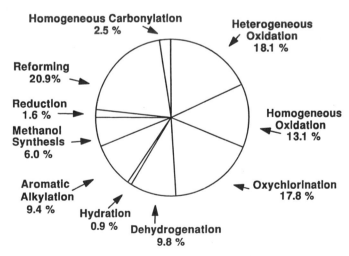

Figure 2. Catalytic production of major organic chemicals in the United States in 1991. (Reproduced from ref. 6. Copyright 1991 American Chemical Society.)

presented in Fig. 3. The figure shows for each region total production of chemicals (catalytic *and* noncatalytic) and the portion manufactured using catalytic selective oxidation. The values for U.S. production of organic chemicals in Fig. 1 are slightly less than those reported in Fig. 3, because the former represents the top 50 chemicals and the latter total chemicals. Considering catalysts alone, the value of oxidation catalysts produced for commerce in the U.S. was US$ 105 million, second only to polymerization catalysts (Fig. 4).

In industrial applications the achievement of higher activity and selectivity is of course desirable. However, beyond a certain point, they are not the driving forces for extensive research. For instance, current processes for epoxidation of ethylene to ethylene oxide on silver catalysts are so optimized that further increases in selectivity could upset the heat-balance of the process. Amoco's phthalic acid and maleic anhydride processes are similarly well energy-integrated (*7*). Rather than *incremental* improvements in performance, forces driving commercial research have been

1) Substitution of raw materials.
2) Formulation of alternative catalysts.
3) Reduction in the number of process steps.
4) Elimination of waste by-products.
5) Development of new processes.

Examples of the above topics abound, and a few are listed below.

1) In the raw materials area, the trend for substitution of paraffin for olefin or aromatic feedstocks is well known. Thus, processes are being developed using butane or pentane instead of benzene for maleic anhydride, and propane instead of propylene for acrylonitrile.

2) New catalysts are exemplified by the chlorine-free Pd^{2+}/Mo-V-P heteropolycompound systems developed to replace Wacker $PdCl_2$/$CuCl_2$ catalysts in olefin-to-ketone processes (*8*).

3) Reducing the number of steps in a chemical transformation can have a profound impact on its economics. Thus, there has been considerable interest in developing 1-step processes for propylene-to-acrylic acid or isobutylene-to-methacrylic acid, which currently involve the intermediate production of acrolein and methacrolein, respectively (*9*). Sometimes simplification of a process just involves the elimination of a purification step, so that the products of one reactor go straight to another without isolation of an intermediate.

4) Environmental concerns are stimulating the development of processes that reduce the amounts of waste side products. Considerable impact is expected in the fine chemicals area, where low volumes and high margins have until now allowed the use of stoichiometric reagents. Because in a multi-step synthesis the amounts of by-products salts can greatly exceed the quantity of final product (*10*), there is interest in developing catalytic routes that produce few side products. Another environmental area which will impact the development of future processes is the problem of release of carbon dioxide. Because in partial oxidation selectivity improvements are likely to be penalized by increased energy consumption, such improvements may not decrease *net* carbon oxides emmissions. Rather, than directly decreasing emissions, there is likely to be

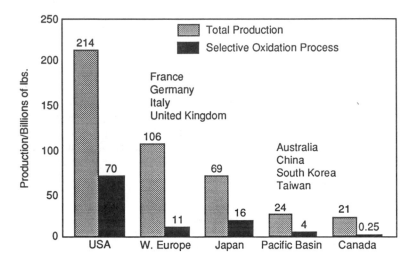

Figure 3. World organic chemical production in 1991. (Reproduced from ref. 6. Copyright 1991 American Chemical Society.)

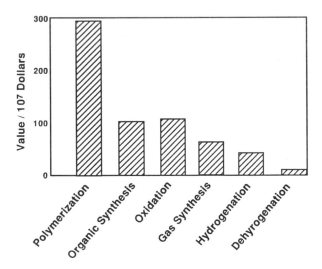

Figure 4. United States catalyst market in 1990. (Reproduced from ref. 6. Copyright 1991 American Chemical Society.)

interest in the use of CO and CO_2 as reactants. Here oxidative carbonylation (*11*) or carboxylation (*12,13*) may be important.

$$CH_3OH + CO + \tfrac{1}{2}O_2 \dashrightarrow O=C(OCH_3)_2 + H_2O$$

$$2ROH + 2CO + \tfrac{1}{2}O_2 \dashrightarrow ROOC\text{-}COOR + H_2O$$

Carbonylation

$$C_6H_5ONa + CO_2 \dashrightarrow (HO)C_6H_4COONa$$

$$C_2H_4 + CO_2 + \tfrac{1}{2}O_2 \dashrightarrow \begin{array}{c} H_2C\text{-}CH_2 \\ |\quad\ | \\ O\ \ O \\ \diagdown C \diagup \\ \| \\ O \end{array}$$

Carboxylation

5) The most desirable target in the development of new processes is so-called shut-down economics, where the price of one product is lower than the break-even point of another, even if that other product is manufactured in a plant that is fully depreciated. This is usually an elusive goal, especially for commodity chemicals, where great optimization has been carried out. However, there are a number of cases which for a variety of reasons present opportunities for the development of new processes involving the direct oxidation of a hydrocarbon (*11*). A few of them are listed below. Two cases are outstanding. The development of non-coproduct technologies (propylene oxide, phenol), and the monooxygenation of various reactants (methane to methanol, butane to tetrahydrofuran, propylene to propylene oxide, benzene to phenol).

Table 1
Potential Targets of New Selective Oxidation Processes

Product	Competing Technology
Methanol	Two-step, via CH_4 steam reforming
Formaldehyde	Three-step, via methanol
Ethylene	Cracking of naphtha
Propylene oxide	Co-product with t-butyl alcohol or styrene
Phenol	Co-product with acetone
1,4-Butanediol	Reppe acetylene chemistry
Tetrahydrofuran	Multi-step
Ethylene glycol	Hydration of ethylene oxide
Adipic acid	Multi-step
Isocyanates	Phosgene chemistry
Styrene	Co-product with propylene oxide
Methyl methacrylate	Two-step, via methacrolein
Methyl formate	Three-step, via methanol

Some of the expected future advances will be based on innovative engineering designs. For example, for maleic anhydride production from C-4 feedstocks new fluidized-bed technology offers advantages in the areas of heat generation and removal over fixed-bed operation (*14*). Molten nitrate salts have been reported as novel media for the selective epoxidation of propylene and other higher olefins without the generation of co-products (*15,16*).

$$CH_3\text{-}CH=CH_2 \quad \text{---> } \quad CH_3\text{-}CH\text{-}CH_2 \atop \diagdown O \diagup$$

As discussed in this volume, the use of membrane reactors (**Bernstein**, *et al.*), monoliths (**Hickman** and **Schmidt**), optimized catalyst distribution in pellets (**Gavriilidis** and **Varma**), and supercritical conditions (**Azzam** and **Lee**) are examples of engineering developments that may provide improvements over existing processes.

Scientific Perspective

This section gives a brief overview of selected issues in selective oxidation that are of scientific interest. No attempt is made to be comprehensive.

Elemental and Structural Characterization Many oxidation reactions occur on mixed oxides of complex composition, such as SbSn(Fe)O, VPO, FePO, heteropolycompounds, etc. Very often the active surfaces are not simple terminations of the three dimensional structure of the bulk phases. There is need to extensively apply structural characterization techniques to the study of catalysts, if possible in their working state.

Similarly, the surface composition of these complex catalysts need to be determined. This will not be a simple task. For example, for the multicomponent allylic oxidation catalyst,

$$M^{2+}\ Fe^{3+}\ Bi^{3+}\ Mo^{4+}\ A^+\ E^{2+}\ T\ O$$

where M^{2+} is a metal ion such as Ni^{2+} or Co^{2+}, A an alkali metal, E an alkaline earth metal and T is P, As or Sb, an XPS study showed that the surface composition changed under catalytic conditions (*17*). On Mg-Mn-O_x, used as an oxidative coupling catalyst, it has been deduced that the surface is partially reduced to a level different from that of the bulk (*18*). The quantification of the actual elemental constituents remains a difficult problem because of the possibility of nonhomogeneity problems (*19*).

The following table summarizes techniques useful for the study of oxide surfaces (*20*). Applications of a number of these are described in the papers by **Volta**, *et al.*, **Busca**, **Deo** and **Wachs**, **Okuhara**, *et al.* A transient technique is reported by **Rigas**, *et al.*

Studies of the *effect* of structure are much needed. Initial studies correlated selectivities to the relative numbers of different crystal faces in crystallites and indicated structure sensitivity (*21-23*). Such studies may need refinement (*24*), and one important direction is the use of single crystals. A number of papers in this volume deal with this topic (**Wickham**, *et al.*; **Aruga**, *et al.*; **Schulz** and **Cox**; **Chen**, *et al.*).

Another way of investigating structure is through the classical method on metals of varying catalyst particle size. The key to this method is to measure active catalyst surface areas in order to determine changes in turnover rates with ensemble size. In recent years several chemisorption techniques have been developed to titrate surface metal centers on oxides (*25*). In this volume **Rao** and **Narashimha** and **Reddy** report on the use of oxygen chemisorption to characterize supported vanadium oxide.

Table 2
Summary of Catalyst Characterization Techniques

Techniques	Depth of Analysis	Temperature Range	In situ Capability	Main Information
IR	0.1-1mm	LN$_2$-300 ^{o}C	possible	Functional groups Acid or basic sites
UV	0.1-1mm	LN$_2$-RT	possible	Oxidation state Environmental symmetry
Raman	0.1-1mm	LN$_2$-600 ^{o}C	possible	Functional groups Environmental symmetry
XRD	> 1mm	RT-1000 ^{o}C	possible	Crystallization Size of particles
TEM/STEM	< 100nm	LN$_2$-1000 ^{o}C	difficult	Morphology Crystallization
Mössbauer	> 0.01mm	LHe-400 ^{o}C	yes	Oxidation state Environmental symmetry
XPS/Auger	1-5 nm	LN$_2$-400 ^{o}C	difficult	Oxidation state Elemental composition
SIMS/ISS	0.1-0.3nm	RT	no	Topmost layer analysis
ESR	> 1mm	LHe-300 ^{o}C	possible	Paramagnetic species Environmental symmetry Surface properties
Magnetism	> 1mm	LHe-500 ^{o}C	yes	Magnetic properties
NMR	> 1mm	LN$_2$-500 ^{o}C	difficult	Functional groups Spatial nuclear environment
STM/AFM	0.001-1nm	LN$_2$-300 ^{o}C	difficult	Surface structure
XAS/EXAFS	0.001-1nm	LHe-500 ^{o}C	possible	Environmental symmetry Interatomic distances Nearest neighbors

New Materials There is increasing interest in the design of new catalysts by using principles of molecular architecture. This involves the engineering of such properties as structure, valence, electronegativity, or redox potential of components that make up a catalytic system. Examples of these are heteropolycompounds as described by **Hill**, *et al.*, supported metal oxides as described by **Deo** and **Wachs**, metalloporphyrins as described by **Meunier** and **Campestrini**, solution phase metal ions as described by **Partenheimer** and **Gipe**, mixed metal compounds as described by **Moser** and **Cnossen**, **Ponceblanc**, *et al.*, and **Smits**, *et al.*, noble metal catalysts by **van den Tillaart**, *et al.*, and **Mallat**, *et al.*, metal silicalites by **Bellussi**, *et al.*, and **Khouw**, *et al.*.

Types of Oxygen One of the accepted tenets of oxidation is that there are two types of oxygen, electrophilic and nucleophilic, responsible for total and partial oxidation, respectively (*2628*). Electrophilic oxygen comprise electron deficient adsorbed species such as superoxide O_2^-, peroxide O_2^{2-}, and oxide O^-, whereas nucleophilic oxygen includes saturated species such as terminal oxygen groups $M=O$, or μ-oxo bridging groups M-O-M, both with the oxygen atom in a nominal O^{2-} state.

There is credible evidence that this nucleophilic O^{2-} oxygen species is capable of carrying out selective oxidation, for example, from the observation that catalytic activity and selectivity persist at the same level even after gas phase oxygen is cut off. However, one cannot rule out the existence of an equilibrium between this nucleophilic species and another type of oxygen. In the case of electrophilic oxygen species, the evidence for their involvement in deep oxidation is even more tenuous. The species have been observed by electron spin resonance (esr) spectroscopy, but generally at subambient temperatures. At catalytic conditions the charge on the oxygen species has been determined by electrical capacitance methods, but this has been a deduction from total surface space charge measurements. As discussed by **Oyama**, *et al.*, there is need for conclusive spectroscopic and kinetic characterization of these species at reaction conditions. Surface potential measurements that can be carried out at catalytic conditions seem to offer a means of determining the charged nature of oxygen species on surfaces (*29,30*). Unfortunately, they have not been widely applied.

Activation of Alkanes The selective oxidation of these unreactive hydrocarbons continues to receive attention. Progress in this area is reported by **Matsumura**, *et al.*, **Driscoll**, *et al.*, **Bañares** and **Fierro**, **Erdohelyi**, *et al.*, **Owens**, *et al.*, and **Khouw**, *et al.*.

Kinetics There have been few comprehensive studies of the kinetics of selective oxidation reactions (*31,32*). Kinetic expressions are usually of the power-rate law type and are applicable within limited experimental ranges. Often at high temperature the rate expression is nearly first order in the hydrocarbon reactant, close to zero order in oxygen, and of low positive order in water vapor. Many times a Mars-van Krevelen redox type of mechanism is assumed to operate.

Precise rate expressions are difficult to obtain because of the existence of reaction networks in which the secondary reactions take place with ease. Sometimes, the relative rate constants in a reaction network are reported and presented as the mechanism of a reaction. For example, butane --> butene --> furan --> γ -butyllactone --> maleic anhydride. This is incorrect, as a different inaccessible experimentally. Theoretical determinations are more practical and have been applied successfully in the analysis of olefin epoxidation and allylic oxidation pathways (*39,40*).

Optimal State of Oxidation of a Surface   On metals with uniform surfaces it can be shown that the optimal coverage is $\theta = 1/2$. There has not been a similar theoretical determination for oxides. However, UV-Vis studies of spent oxide catalysts indicate that they are in a partially reduced state (*41*). As shown in Fig. 5 in ethylene oxidation on silver, selectivity increased with oxygen coverage (*42*). On allylic oxidation catalysts it has been suggested that the number of active oxygen atoms surrounding an adsorbed hydrocarbon should be limited in order to prevent total oxidation (*26*). This "site isolation" could be achieved by partial reduction of a metal oxide. Taking the surface to be a grid of oxygen atoms, a Monte Carlo calculation predicted that the maximum selectivity would be achieved at 60-70% reduction. More studies to determine surface oxidation state at catalytic conditions are highly desirable. This would be useful for deducing reaction mechanisms and for providing guidance for further catalyst design. Site isolation can also be achieved structurally as exemplified by the U-Sb-O system (*43*).

Role of Crystallographic Shear Planes   This area is tied to the subject of surface oxidation state. In redox kinetics it is believed that sites for hydrocarbon oxidation are different from that for reoxidation, and crystallographic shear planes (CSP) have been suggested to assist bulk oxygen movement between the sites (*44,45,46,47*). There have been a few studies of this phenomenon. In $WO_{2.95}$ and $WO_{2.9}$ it has been shown CSPs are involved in oxygen transfer but not the initial abstraction of hydrogen in propylene oxidation. In contrast $WO_{2.72}$, which has a tunnel structure without CSPs, cannot insert oxygen (*48*). In the Mo-O system the oxidation of $Mo_{18}O_{52}$ (001) to MoO_3 at 670 K and low pressures of oxygen, RHEED and SEM have shown that oxygen transport occurs by a vacancy exchange mechanism involving exclusively CSPs (*49*). On $Mo_{18}O_{52}$ oxygen chemisorbs on the CSP boundaries. LEED studies of rutile single crystals has shown the formation of (1x3), (1x5),and (1x7) superstructures on the (100) face that are similar to the CSP structures in the bulk (*50,51*). Although prevalent in the Mo, W, V, Nb, and Ti systems, in industrial Mo-Te-O and Sn-Sb-O systems CSPs are not observed (*52*). Clearly, further studies of the role of CSPs in selective oxidation are needed.

Novel Oxidants   In homogeneous catalysis the use of nontraditional oxidants such as potassium persulfate, iodozobenzene, cumene hydroperoxide, t-butyl hydroperoxide, NaOCl, and R_3NO has seen increasing use (*53*). In heterogeneous systems the employment of such oxidizing agents has not been as prevalent and there are opportunities for their use. An example is given by **Khouw**, *et al.* Among potentially useful oxidizing agents are hydrogen peroxide, ozone, nitrogen oxides, and halogens.

Theoretical Design of Catalysts   An area where work is needed in the field of selective oxidation is the theoretical design of catalysts. Although there has been some work in this area in the past thirty years (*54,55,56*) it has not

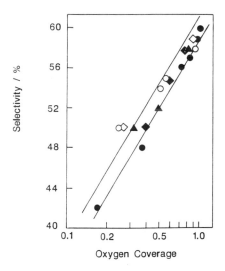

Figure 5. Dependence of selectivity on oxygen coverage in the oxidation of ethylene on various silver catalysts. (Reproduced with permission from ref. 42. Copyright 1984 Academic Press.)

set of steps may be occurring on the catalyst surface. Clearly, substantially more work in this area is needed.

Mechanisms There is a derth of knowledge about the mechanisms operative in selective oxidation reactions. The only exceptions are the reactions of ethylene to ethylene oxide on supported silver catalysts and of propylene to acrolein on bismuth molybdate type catalysts. For the latter, it is well established through isotopic labeling experiments that a symmetric allyl *radical* is an intermediate in the reaction and that its formation is rate-determining. Many studies simply extrapolate the results substantiated for this case to other reactions. New ideas on mechanisms are presented by **Oyama**, *et al.*, **Parmaliana**, *et al.*, and **Laszlo**.

It would be useful to classify reactions as to their type, heterolytic (ionic) or homolytic (free radical), radical cation, or radical anion. Below are a number of examples of these types of reaction.

Table 3
Classification of Oxidation Reactions

Reaction	Catalyst	Type	Ref.
Propylene ---> Acetone	MoO_3-SnO_2	Heterolytic	(33)
Methane ---> Ethane	Li-MgO	Homolytic	(34)
Toluene ---> Diphenylmethane	$[CoW_{12}O_{40}]^{5-}$	Radical cation	(35)
Benzene ---> Phenol	$VO(O_2)PIC(H_2O)_2$	Radical anion	(36)

PIC = picolinic acid

For many reactions the type of intermediate that is involved may be deduced from a study of a family of reactants. For example, by noting that in allylic oxidation the order of reactivity is isobutene > trans-2-butene > cis-2-butene > 1-butene one may deduce that an allyl radical or cation is an intermediate. For other oxidations, if the reaction rate order is primary > secondary > tertiary, then an anionic intermediate is implicated. However, care must be taken that the formation of the intermediate is involved in the rate-determining step and that there are no adsorption equilibrium effects. To rule out the latter, the reaction should be carried out at conditions of low coverage.

Surface Bond Energies Thermochemical data are very scant in the area of oxygen chemisorption (37). These data would be of great value for interpreting spectroscopic and kinetic data and for the analysis of reaction mechanisms. The vast majority of the available data are for low oxidation state systems (38). Although calorimetry offers a means for direct measurements, for analysis of reaction pathways it is necessary to have detailed values for many types of species (M-OH, MO-H, M-OR, M-R, M-O, M-H), and these are usually received major attention. Industrial catalyst development and optimization is still largely dependent on extensive empirical research. The molecular design of catalysts will depend strongly on advances in theoretical and computational chemistry. Methods are needed to relate selectivities, conversion levels, and kinetic behavior to catalyst structure. The prediction of the desired molecular structure will be followed by the rational synthesis of deliberately tailored practical catalysts. Characterization of these catalysts will play an important role. Indeed, careful experimental studies recently have shown the importance of surface structure in catalysts used in selective oxidation. Iwasawa, reviews the use of chemically and structurally controllable surface systems that differ from those produced by conventional impregnation methods (57). The advances presented in this volume are helping to pave the way for such molecular design of catalysts.

Acknowledgments

This overview was written by one of us (STO) in the laboratory of Dr. M. Haruta at the Government Industrial Research Institute (GIRIO), Osaka, Japan with sponsorship from the Research Institute of Innovative Technology for the Earth (RITE) through their Research Exchange Program.

Literature Cited

1. *New Developments in Selective Oxidation*, Proceedings of the European Workshop Meeting, Louvain-la-Neuve, Belgium, March 17-18, 1986; Delmon, B.; Ruiz, P. Eds.; *Catal. Today*, **1987**, 1, 1-366.
2. *Hydrocarbon Oxidation*, Proceedings of a Symposium held during the 193rd National ACS Meeting, New Orleans, August 30-September 4, 1987; Grasselli, R. K.; Brazdil, J. F. Eds.; *Catal. Today* **1988**, 3, 1-267.
3. *New Developments in Selective Oxidation*, Proceedings of an International Symposium, Rimini, Italy, September 18-22, 1989; Centi, G.; Trifiro, F. Eds.; Studies in Surface Science and Catalysis; Elsevier: Amsterdam, 1990; pp 1-891.
4. *New Developments in Selective Oxidation by Heterogeneous Catalysis*, Proceedings of the Third European Workshop Meeting on New Developments in Selective Oxidation by Heterogeneous Catalysis, Louvain-la-Neuve, Belgium, April 8-10, 1991; Delmon, B.; Ruiz, P. Eds.; Studies in Surface Science and Catalysis; Elsevier: Amsterdam, 1992; pp 1-486.
5. Delmon, B.; Ruiz, P. *Catal. Today* **1987**, 1, 1.
6. Desikan, A. N., compiled from C&EN, June 1991; Chemical Week, June 1991; Current Industrial Reports, 1990; U.S. International Trade Commission Report 2470, 1990.
7. Partenheimer, W., Amoco Chemical, Private communication.
8. Grate, J. H.; Hamm, D. R.; Mahajan, S. *Int. Patent Appl.* WO 91/13681, WO 91/13851, WO 91/13853, WO 91/13854, September 1991.
9. Ohara, T.; Hirai, M; Shimizu, N. *Hydrocarb. Process.* **1983**, 62, 73.

10. Sheldon, R. A. In *New Developments in Selective Oxidation*; Centi, G.; Trifiro, F., Eds.; Elsevier: Amsterdam, 1990; pp 1.
11. Pasquon, I. *Catal. Today* **1987**, 1, 297.
12. *U.S. Patent* 4,400,559, **1983**, Assigned to Halcon.
13. Dowden, D. A. In *Proceedings of the International School of Physics, Enrico Fermi*, Villa Marigola, Italy, June 20-July 8, 1988; Chiarotti, G. F.; Fumi, F.;Tosi, M. P. Eds.; Current Trends in the Physics of Materials; Elsevier: Amsterdam, 1990; p 591
14. Contractor, R. M.; Sleight, A. W. *Catal. Today* **1987**, 1, 587.
15. Pennington, B. T. *U.S. Patent* 4,785,123, **1988**, Assigned to the Olin Corporation.
16. Pennington, B. T.; Fullington, M. C. *U.S. Patent* 4,943,643, **1990**, Assigned to the Olin Corporation.
17. Prasada Rao, T. B. R.; Menon, P. G. *J. Catal.* **1978**, 51, 64.
18. Labinger, J. A.; Ott, K. C. *Catal. Lett.* **1990**, 4, 245.
19. Powell, C. J.; Seah, M. P. *J. Vac. Sci. Technol.* Mar/Apr **1990**, A8, 735.
20. Adapted from Vedrine, J. C. In *Surface Properties and Catalysis by Nonmetals*; Bonnelle,J. P.; Delmon, B.; Derouane, E. Eds.; Reidel: Dordrecht, 1983; pp 123.
21. Volta, J. C.; Tatibouet, J. M. *J. Catal.*, **1985**, 93, 467.
22. Volta, J. C.; Tatibouet, J. M.; Phichitkul, C.; Germain, J. E., Dechema, Ed., *Proc. 8th Int. Cong. Catal.*, IV, 451, Berlin, 1984.
23. Guerrero-Ruiz, A.; Abon, M.; Massardier, J.; Volta, J. C. *J. Chem. Soc., Chem. Commun.* **1987**, 1031.
24. Oyama, S. T. *Bull. Chem. Soc. Japan*, **1988**, 61, 2585.
25. Oyama, S. T.; Went, G. T.; Lewis, K. B.; Somorjai, G. A.; Bell, A. T. *J. Phys. Chem.* **1989**, 93, 6786.
26. Callahan, J. L.; Grasselli, R. K. *AIChE J.* **1963**, 9, 755.
27. Bielanski A.; Haber, J. *Cat. Rev.-Sci. Eng.*, **1979**, 19, 1.
28. Grasselli, R. K., In *Surface Properties and Catalysis by Non-Metals*; Bonnelle, J. P.; Delmon, B.; Derouane, E. Eds.; Reidel: Dordrecht, 1983; pp 273.
29. Barbaux, Y.; Bonnelle, J.-P.; Beaufils, J.-P. *J. Chem. Res (S).* **1979**, 48.
30. Barbaux, Y.; Elamrani, A.; Bonnelle, J.-P. *Catal. Today* **1987**, 1, 147.
31. *Rate Equations of Solid Catalyzed Reactions*; Mezaki, R.; Inoue, H. Eds.; University of Tokyo Press: Japan, 1991.
32. Kung, H. H. *J. Catal.* **1992**, 134, 691.
33. Buiten, J. *J. Catal.* **1968**, 10, 188.
34. Lee, J. S.; Oyama, S. T. *Catal. Rev. Sci. Eng.* **1988**, 30, 249.
35. Chester, A. W. *J. Org. Chem.* **1970**, 35, 1797.
36. Bonchio, M.; Conte, V.; Di Furia, F.; Modena, G. *J. Org. Chem.* **1989**, 54, 4368.
37. Roberts, M. W. *Chem. Rev. Soc.* **1989**, 18, 451.
38. *Bonding Energetics in Organometallic Compounds*; Marks, T. J. Ed.; ACS Symp. Ser. 428; American Chemical Society: Washington, D. C., 1990; pp. 1-305
39. Rappé, A. K.; Goddard, W. A., III *J. Am. Chem. Soc.* **1982**, 104, 3287.
40. Rappé, A. K.; Goddard, W. A., III *J. Am. Chem. Soc.* **1980**, 102, 5114.
41. Vedrine, J. C.; Coudurier, G.; Forissier, M.; Volta, J. C. *Catal. Today* **1987**, 1, 261.

42. Akella, L. M.; Lee, H. H. *J. Catal.* **1984**, 86, 456.
43. Grasselli, R. K.; Suresh, D. D. *J. Catal.*, **1972**, 25, 273.
44. O'Keefe, M. *Fast Ion Transport in Solids*; North Holland: Amsterdam, 1973; p. 233.
45. Anderson, J. S. In *Problems of Nonstoichiometry*; A. Rabenau, Ed.; North Holland: Amsterdam, 1970; Chap. 1, pp 1-85.
46. Krylov, O. V. In *Khimija poverkhnoti okisnykh katalizatorov*, (Chemistry of the Surface of Oxide Catalysts); Rozovskij, A. Ja. Ed.; Nauka: Moskow, 1979; pp 33-52.
47. Grasselli, R. K.; Bradzil, J. F.; Burrington, J. D. *J. Appl. Catal.*, **1986**, 25, 335.
48. de Rossi, S.; Jacono, M. W.; Pepe, F.; Schiavello, M.; Tilley, R. J. D. *Z. Phys. Chem.*, **1982**, 130, 109.
49. Floquet, H.; Betrand, O. *Surf. Sci.*, **1988**, 198, 449.
50. Chung, Y. W.; Lo, W. J.; Somorjai, G. A. *Surf. Sci.* **1977**, 64, 588.
51. Firment, L. E. *Surf. Sci.* **1981**, 116, 205.
52. Gai, P. L.; Boyes, E. D.; Bart, J. C. J. *Philos. Mag. A*, **1987**, 45, 531.
53. Sheldon, R. A. *Catal. Today* **1987**, 1, 351.
54. Dowden, D. A. *Chem. Eng. Progr. Symp.*, **1967**, 63, 90.
55. Trimm, D. L. *Design of Industrial Catalysts*; Elsevier: Amsterdam, 1980.
56. van Santen, R. A. *Chem. Eng. Sci.*, **1990**, 45, 2001.
57. Iwasawa, Y. *Adv. Catal.*, **1988**, 35, 187.

RECEIVED October 30, 1992

THEORIES AND CONCEPTS
IN SELECTIVE OXIDATION

Chapter 2

Adsorbate Bonding and Selectivity in Partial Oxidation

S. Ted Oyama[1], A. N. Desikan[2], and W. Zhang[3]

[1]Chemical Engineering Department, Virginia Polytechnic Institute and
State University, Blacksburg, VA 24061
[2]Department of Chemical Engineering and [3]Center for Advanced Materials
Processing, Clarkson University, Potsdam, NY 13699–5705

In this paper selectivity in partial oxidation reactions is related
to the manner in which hydrocarbon intermediates (R) are bound
to surface metal centers on oxides. When the bonding is through
oxygen atoms (M-O-R) selective oxidation products are favored,
and when the bonding is directly between metal and hydrocarbon
(M-R), total oxidation is preferred. Results are presented for two
redox systems: ethane oxidation on supported vanadium oxide
and propylene oxidation on supported molybdenum oxide. The
catalysts and adsorbates are studied by laser Raman spectroscopy,
reaction kinetics, and temperature-programmed reaction. Thermochemical calculations confirm that the M-R intermediates are
more stable than the M-O-R intermediates. The longer surface
residence time of the M-R complexes, coupled to their lack of
ready decomposition pathways, is responsible for their total
oxidation.

In the investigation of hydrocarbon partial oxidation reactions the study of the
factors that determine selectivity has been of paramount importance. In the past
thirty years considerable work relevant to this topic has been carried out.
However, there is yet no unified hypothesis to address this problem. In this
paper we suggest that the primary reaction pathway in redox type reactions on
oxides is determined by the structure of the adsorbed intermediate. When the
hydrocarbon intermediate (R) is bonded through a metal oxygen bond (M-O-R)
partial oxidation products are likely, but when the intermediate is bonded
through a direct metal-carbon bond (M-R) total oxidation products are favored.
Results on two redox systems are presented: ethane oxidation on vanadium
oxide and propylene oxidation on molybdenum oxide.

A number of other explanations for selectivity have been suggested in the
past. Sachtler and De Boer (1) attempted to correlate the difference in

0097–6156/93/0523–0016$06.00/0
© 1993 American Chemical Society

selectivity among different catalysts for the oxidation of propylene to the reducibility of the catalysts. They hypothesized that, in general, the higher the reducibility of the catalysts, the higher would be the conversion and lower the selectivity. However, there were marked differences in the catalytic activity for samples of very similar reducibilities. They attributed these discrepancies to the difference in active site density in these catalysts. Reducibility should be related to heat of formation, but a number of workers were unsuccessful in correlating catalytic activity of oxides in hydrocarbon oxidation and heat of formation of the oxides (*2-4*). Subsequently, Sachtler, *et.al*, correlated oxidation activity and selectivity to the *differential* ($\partial\Delta H/\partial x$) increase of the heat of oxygen release (ΔH) with increasing reduction level (x) (*5*).

Bielanski and Haber (*6*) explained selectivity by postulating that lattice oxygen is responsible for partial oxidation, while adsorbed ionic or radical oxygen species cause total oxidation. Among the evidence that suggests that the nature of the oxygen species influences selectivity is the finding on bismuth molybdate that propylene is converted to acrolein in good yield, even when the supply of gas-phase oxygen is interrupted for a significant time (>1 turnover) (*7,8*). This is consistent with bulk (lattice oxygen) being responsible for selective oxidation (*9,10*). Conversely, it is found that oxides such as Co, Ni, Cr, Mn, that tend to adsorb oxygen in ionic forms, produce mainly CO_x (*6*). Sancier, *et al.*, (*11*), however, have presented the contrary view that *both* adsorbed and lattice oxygen participate in propylene oxidation on silica-supported bismuth molybdate. There is good evidence that lattice oxygen does promote selective oxidation, but support for the role of the ionic oxygen species is more tenuous. Those species have not been directly observed on oxide surfaces at catalytic conditions (*12*).

Sachtler (*13*) addressed the question of the effect of bonding on oxidation selectivity but concluded at the time that there was insufficient data for a definitive answer. Davydov and coworkers have done extensive work in relating the structure of adsorbed hydrocarbons and oxidation selectivity (*14,15*). By means of thermal desorption of propylene they identified two forms of propylene adsorbed on the catalyst surface: (i) a weakly bound, reversibly adsorbed form with a peak maximum in the desorption spectrum of 380 K and (ii) a strongly bound species which only desorbed with decomposition beginning at temperatures greater than 470 K. They found that propylene primarily adsorbed irreversibly on catalysts for complete oxidation (CuO, Cr_2O_3) and largely reversibly on catalysts for selective oxidation (Cu_2O, MoO_3). Based on extensive IR spectroscopic measurements Davydov, *et al.* have postulated the presence of two type of adsorbed intermediates i) an oxygen bonded intermediate which leads to selective oxidation and ii) a carbonate-carboxylate intermediate which leads to total oxidation (Scheme 1). In a recent review Sokolovskii (*16*) suggests a general set of requirements for total and selective oxidation catalysts, but does not consider the effect of bonding.

In this paper we present a molecular view of the origin of selectivity differences based on bonding of intermediates. The suggestion is that hydrocarbon intermediates bonded to metal centers through oxygen atoms

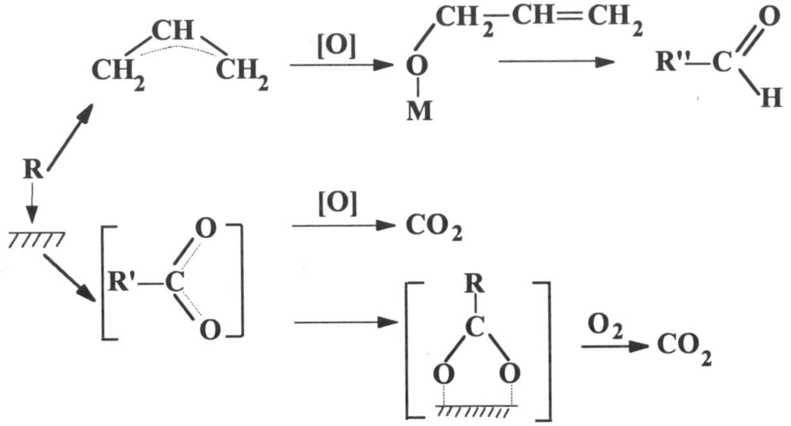

Scheme 1

(M-O-R) result in selective oxidation products, whereas intermediates bonded directly through metal carbon bonds (M-R) produce total oxidation products. We discuss kinetic, thermodynamic and transient data that indicate that the nature of the bonding of the intermediate is directly related to its further reaction via selective and unselective pathways.

Experimental

The catalysts employed in this study consisted of V_2O_5/SiO_2 and MoO_3/SiO_2 prepared by incipient wetness impregnation of the support (Cabosil L90) with aqueous solutions of vanadyl oxalate (Aldrich, 99.99%) and ammonium molybdate (Aldrich, 99%), respectively. The catalysts were dried at 623 K for 6 h and calcined at 773 K for 6 h.

Catalytic activity was measured in a 14 mm ID quartz packed-bed reactor, at atmospheric pressure. In ethane oxidation studies on V_2O_5/SiO_2 the partial pressures were, $P_{CH_3CH_3} = 13$ kPa, $P_{O_2} = 28$ kPa, $P_{H_2O} = 10$ kPa, $P_{He} = 50$ kPa. Product analysis was carried out by gas chromatography. For the temperature programmed studies on MoO_3/SiO_2, 60 µmol of allyl alcohol or allyl iodide were dosed on the samples (0.5g). After the system was purged, a flow of 33 µmol s^{-1} of He was established, and the product distribution was followed by mass spectrometry as the temperature was ramped at 0.17 Ks^{-1}.

Laser Raman spectra were obtained with the 647.1 nm line of a krypton ion laser (Lexel, Model 95). The radiation intensity at the sample was 300 mW. The scattered radiation was passed through a double monochromator

(Spex II, Model 1403) with 1800 grooves mm[-1] holographic gratings coupled to a photomultiplier (RCA) with associated electronics (EG&G Princeton Applied Research, Model 1112).

For the *in situ* Raman studies the MoO_3/SiO_2 samples were pressed into wafers at 21 MPa. The pressed wafers were placed in a rotatable sample holder that was enclosed in a quartz cell equipped with an oven. The samples were oxidized at 773 K for 1 hour in O_2 to minimize sample fluorescence, then were cooled to 343 K in flowing helium. The adsorbates were introduced through an injection port of the cell at 343 K with flowing helium, and the spectra were recorded at the same temperature.

The reference samples were allyl alcohol (Aldrich, 98%), allyl iodide (Aldrich, 98%), and acrolein (Aldrich, 97%). They were placed in quartz tubes, and their spectra were recorded at room temperature.

Results

Ethane Oxidation on Supported Vanadium Oxide. Figure 1 shows the rates of production of the major products of ethane oxidation over a series of silica-supported vanadium oxide catalysts. As was described earlier, the structure of the catalyst changed considerably with the active-phase loading (*17*). The low loading samples (0.3 - 1.4%) were shown to consist primarily of $O=VO_3$ monomeric units, while the high loading catalysts (3.5 - 9.8%) were composed of V_2O_5 crystallites.

The rates of production are reported as turnover rates based on the number of V metal atoms as titrated by the adsorption of oxygen (*17*). These rates represent the number of product molecules produced per site per unit time and thus are measures of the actual product yield (conversion x selectivity) of the catalytic site. Conversions were less than 10% so that conclusions are derived for primary oxidation products.

Temperature-Programmed Reaction of Adsorbed Allylic Species.
Temperature programmed reaction (TPR) experiments were carried out by adsorbing allyl alcohol and allyl iodide on a 9.0 wt% MoO_3/SiO_2 sample and monitoring the evolved products by mass spectrometry. The Raman spectra of the pure liquid reference compounds are shown in Fig. 2. They agree well with those reported earlier (*18-20*).

The spectra of the adsorbed intermediates prior to reaction are shown in Fig. 3. The spectra show considerable differences from the solution reference spectra. The peaks for both surface species are also slightly different from each other, indicating differences in their structures. TPR traces of the main products are reported in Fig. 4. The complete TPD data were given earlier (*21*). In the case of allyl alcohol, there is formation of acrolein at relative low temperatures. Only at high temperatures, 700 K, is there production of CO_x. In the case of allyl iodide the TPR spectra are considerably different. There is no production of acrolein, instead, only CO_x is formed at high temperatures, > 700 K.

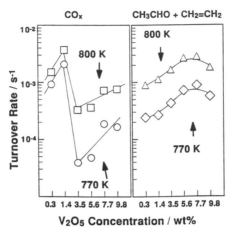

Figure 1. Turnover rates in the oxidation of ethane on V_2O_5/SiO_2 catalysts.

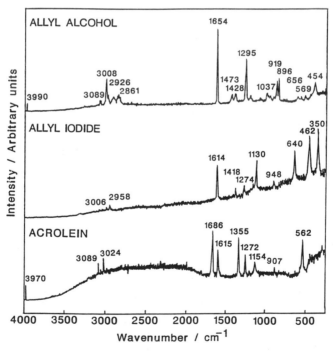

Figure 2. Laser Raman spectra of pure liquid reference compounds.

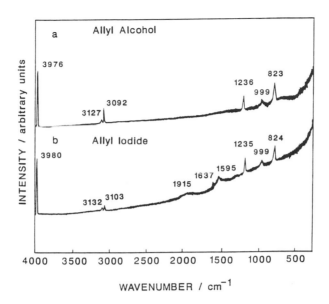

Figure 3. Laser Raman spectra of adsorbed intermediates on MoO_3/SiO_2 prior to reaction.

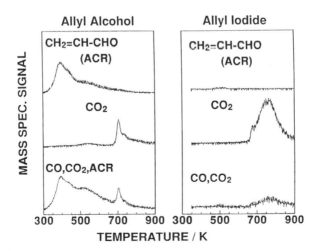

Figure 4. Temperature programmed reaction traces of adsorbed allyl alcohol and allyl iodide on MoO_3/SiO_2.

Discussion

Reaction Product Analysis. In the ethane oxidation reaction (Fig. 1), the turnover rate for CO_x production depends greatly on the catalyst structure. The rate is considerably higher for the 0.3 and 1.4 wt% V_2O_5 samples than for the higher loading samples. The sudden fall in CO_x production rate (factor of x30) with loading is indicative that the reaction is structure-sensitive (22). In contrast, the relative invariance of the CH_3CHO and CH_2CH_2 production rates with catalyst structure suggest that these C_2 producing reactions are structure-insensitive.

An important conclusion that may be derived from these data are that the intermediates responsible for CO_x or C_2 products are probably different. If a common intermediate existed,

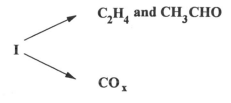

$$I \nearrow C_2H_4 \text{ and } CH_3CHO$$
$$I \searrow CO_x$$

Scheme 2

a *decrease* in the production of one of the products should lead to an *increase* in the formation of the other. This is not observed. It is concluded that there are two independent paths for the production of the two product types. A likely possibility is,

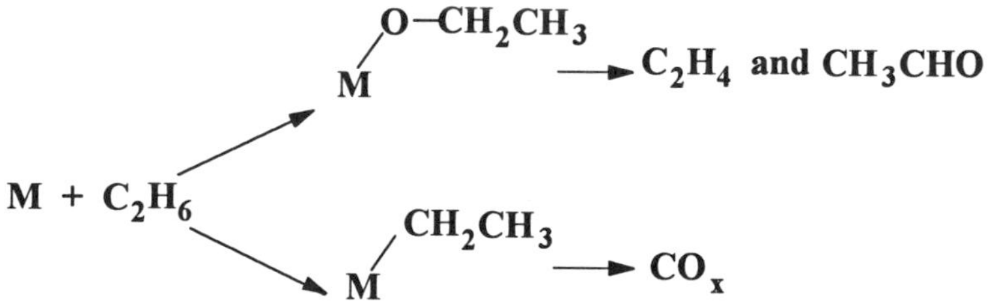

$$M + C_2H_6 \nearrow \overset{O-CH_2CH_3}{M} \longrightarrow C_2H_4 \text{ and } CH_3CHO$$
$$M + C_2H_6 \searrow \overset{CH_2CH_3}{M} \longrightarrow CO_x$$

Scheme 3

The rates of formation of the final products depend on the rate constants for decomposition of the respective intermediates *and* their relative concentrations. These concentrations will be affected by the structure of the catalyst, and this is why there is a dependence on active phase loading. In fact, it is likely that the M-O-R intermediate is favored on the crystallites because of steric constraints. The metal center is octahedrally coordinated and not susceptible to direct M-R bonding as is the more open $O=VO_3$ four coordinated species.

The two intermediates depicted above differ fundamentally from each other. The CO_x-producing intermediate has a direct metal-carbon (M-R) bond whereas the C_2-producing intermediate has a metal-oxygen-carbon (M-O-R) bond. From known organic decomposition pathways, the formation of selective oxidation products from the M-O-R intermediate is likely. An α-H elimination produces acetaldehyde and a β-H elimination produces ethylene.

In the M-R bonded intermediate an α-elimination is not possible, whereas a β-elimination produces a metal hydride. This is energetically unfavorable compared to the oxy-bonded intermediate partly because of the lower M-H bond strength compared to MO-H (see section on thermodynamics). Thus, because of kinetics (lack of decomposition pathways) and thermodynamics (energetics), the metal bound M-R intermediate is less reactive on the surface than the M-O-R intermediate.

Temperature Programmed Reaction. Examination of another redox system, propylene oxidation on MoO_3, provides further insight. It is well accepted that propylene oxidation on molybdenum-based catalysts proceeds through formation of allylic intermediates. From isotopic studies it has been demonstrated that formation of the allylic intermediate is rate-determining (H/D effect), and that a symmetric allylic species is formed (^{14}C labelling).

Determination of the mode of bonding of the allylic intermediate formed would provide direct evidence that an M-O-C species produces acrolein. This is difficult because formation of the allylic intermediate is rate-determining and its subsequent reaction is fast. Thus, the surface concentration of the intermediate is small and not amenable to standard spectroscopic observation.

In order to circumvent this problem the rate-determining step was bypassed by using more reactive reagents, allyl alcohol and allyl iodide. These allylic probes were expected to adsorb on the molybdenum oxide surface to provide, respectively, M-O-C and M-C bonded intermediates. These studies were carried out with a sample of 9.0 wt% MoO_3/SiO_2, which Raman spectroscopy and x-ray diffraction showed to consist of fine (~ 5nm) crystallites of MoO_3 (*23*).

The laser Raman spectra of the adsorbed allyl alcohol and allyl iodide are shown in Fig. 3a and 3b. It can be seen that the spectra for both species are slightly different, although the grouping of peaks are related. The assignment of these peaks are summarized in Table 1.

Unfortunately, there are no bands that can be clearly identified with M-C or M-O-C vibrations. These modes may be difficult to observe by Raman spectroscopy because the bonds are only weakly polarized. In addition it is believed that the vibrations of light atoms bonded to a metal center are broadened by coupling to the support (*24*). Nevertheless, the differences in the spectra of the two species suggest that the proposed structures are formed. For the case of allyl alcohol, isotopic substitution experiments on supported molybdates have shown that the original oxygen atom is retained in the acrolein product (11), consistent with the scheme above.

Table 1. Adsorbed Species Assignment

Frequency (cm^{-1})	Assignment	Ref.
3960-3990	νOH	25
3132-3127	ν=CH2, νCH=	15,14
3103- 3092	νCH	15
1915(br)	νCO or νCO$_2$	26
1637(vw)	ν_{as} C-O	14
1595	ν C=C in σ-complex	15,14
1235	γ CH=, δCH=	14,27,28
999	νMo=O	29
823	νMo-O-Mo	29

Our view that M-O-R intermediates result in partial oxidation products is compatible with the evidence that links lattice oxygen and selectivity. Our view is that lattice oxygen *per se* does not result in selectivity. Indeed, it has been shown to produce CO$_x$ (*11*), possibly from deep oxidation of M-R intermediates. We suggest it is the formation of M-O-R intermediates from surface lattice oxygen species (O^{2-}) that leads to selective oxidation products. Conversely, it is *possible* that adsorbed electrophilic oxygen does carry out deep oxidation. Our view is that this would occur primarily on M-R species because of their long residence time on the surface. It should be stressed that these ideas apply for primary selectivities, without consideration of oxidation of products in secondary reactions. Also we are concerned with single monodentate bonding to the surface. Bidentate bonding such as occurs with carboxylates (Scheme 1) may result in CO$_x$, despite the presence of M-O-C bonds. Similarly we exclude metallacyclic structures (Scheme 4a), and π-allyl structures (Scheme 4b), which have been suggested in epoxidation, and allylic oxidation, respectively. These intermediates have M-C bonds but yield selective oxidation products.

(a)   (b)

Scheme 4

Thermodynamics . The lack of reactivity of the M-C intermediate may be further related to its thermodynamic stability. The work of Rappé and Goddard is illustrative (*30*). For example, for the interaction of ethanol with high valent metal oxo centers (Cr or Mo) two structures are possible that differ only in that they contain M-C or M-O-C bonds. Otherwise they contain the same number and types of atoms.

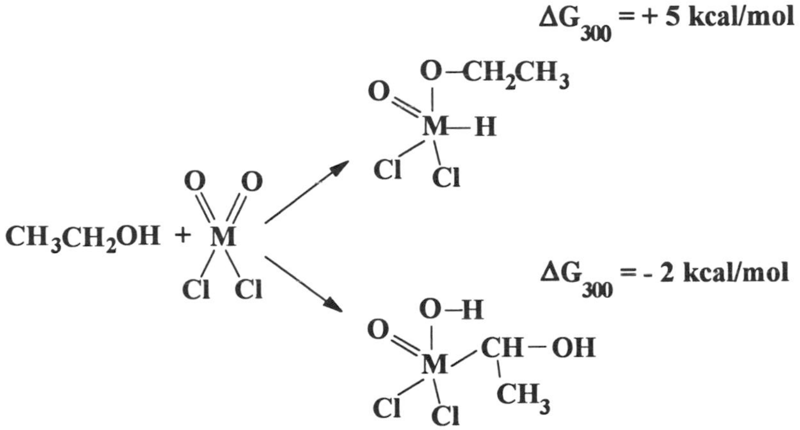

$$\Delta G_{300} = + 5 \text{ kcal/mol}$$

$$\Delta G_{300} = - 2 \text{ kcal/mol}$$

Scheme 5
Formation of M-O-R and M-R Intermediates

Valence bond calculations indicate that the M-C intermediate is more stable than the M-O-C intermediate by 7 kcal/mol. This reinforces the ideas presented earlier. The lack of ready decomposition pathways in conjunction to their intrinsic stability renders M-C intermediates susceptible to oxidative dehydrogenation. They stay on the surface until they are completely oxidized to CO_x.

Further support for the relative stability of the M-R intermediates over the M-O-R species is given by calculations (*31*) of the free energy of their transformation (Scheme 6).

The standard free energy of transformation (ΔG_{300}) of the M-C (ethyl) species to form ethylene is +17 kcal/mol. Compared to this, ΔG_{300}'s of the M-O-C (ethoxide) species are +5.4 and -1.1 kcal/mol to form ethylene and acetaldehyde, respectively.

The results indicate that the β-H elimination of the M-R intermediate is more difficult than either the α-H or the β-H elimination of the M-O-R species. Thus, the M-O-R species can readily decompose to yield stable hydrocarbon products, while the M-R intermediate is relatively stable. Its long lifetime on the surface makes it susceptible to irreversible oxidation to CO_x.

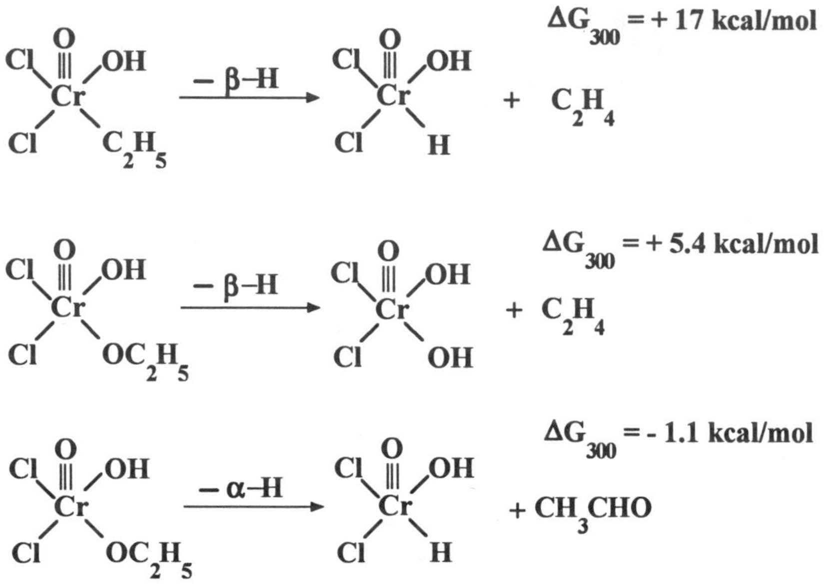

$$\Delta G_{300} = +17 \text{ kcal/mol}$$

Cl–Cr(=O)(OH)(Cl)(C$_2$H$_5$) $\xrightarrow{-\beta\text{-H}}$ Cl–Cr(=O)(OH)(Cl)(H) $+$ C$_2$H$_4$

$$\Delta G_{300} = +5.4 \text{ kcal/mol}$$

Cl–Cr(=O)(OH)(Cl)(OC$_2$H$_5$) $\xrightarrow{-\beta\text{-H}}$ Cl–Cr(=O)(OH)(Cl)(OH) $+$ C$_2$H$_4$

$$\Delta G_{300} = -1.1 \text{ kcal/mol}$$

Cl–Cr(=O)(OH)(Cl)(OC$_2$H$_5$) $\xrightarrow{-\alpha\text{-H}}$ Cl–Cr(=O)(OH)(Cl)(H) $+$ CH$_3$CHO

Scheme 6
Decomposition of M-O-R and M-R Intermediates

Involvement of Adsorbed Oxygen Species in Catalytic Oxidation. Various types of oxygen species have been suggested to be associated with the surface of metal oxides. Fully reduced atomic oxygen species (O^{2-}) with coordination varying from 1-4 corresponding to M=O (terminal oxo groups), M-O-M (bridging μ-oxo groups), M$_3$O and M$_4$O (capping oxygens) may be considered as surface lattice oxygens whose role is simply to maintain surface stoichiometry. Ionic or radical species such as ozonide (O_3^-), superoxide (O_2^-), peroxide (O_2^{2-}) and oxide (O^-), may also be formed, associated with reduced surface metal ions, Lewis acid sites, or oxygen vacancies. Even subsurface oxygen has been suggested to occur on perovskite type oxides with properties (generally unselective) different from those of lattice oxygen (*32*).

At low temperatures the O^-, O_2^-, and the absorbed O_3^- species can be regarded as well characterized. Under controlled conditions, O^- and O_2^- exists on MgO (*33,34*), MoO$_3$/SiO$_2$ and MoO$_3$/Al$_2$O$_3$(*35-37*), V$_2$O$_5$/SiO$_2$, V$_2$O$_5$/Al$_2$O$_3$ (*36,37*) and ZnO (*38*). O_3^- has been found to exist on MgO (*33,39,40*), TiO$_2$/SiO$_2$ (*41*) and V$_2$O$_5$/SiO$_2$ (*42*). In analogy to their gas phase properties, these species have been considered electrophilic (*6*). However, it has been pointed out that such properties may change upon complexation with a metal center (*43*).

There is evidence from surface potential measurements at catalytic conditions (673 K) of charging of the surface of oxides. Although different oxygen species have been suggested (O^-, O^{2-}), it is not clear whether these actual species exist or whether an average surface charge is being measured (*44,45*).

The main body of direct evidence on the nature of different oxygen species have come from experiments which have been designed to stabilize the various species in well-defined environments (77 K or 300 K). This is far removed from the situation in many catalytic reactions which occur at considerably higher temperatures. However Shvets, *et. al*, (*36, 37*) have identified O_2^- and O^- ions at 573 K on V_2O_5/SiO_2 in an O_2 atmosphere, whereas Yoshida, *et. al*, (*46*) observed for the same system that heating for 15 min at 423K caused a decrease in O_2^- concentration by 80%. In a extensive study Takita, *et. al*, (*47*) and Iwamoto, *et. al*, (*48*) on MgO have determined the thermal stabilites of the oxygen ions (O^-, O_2^-, and O_3^-). They have concluded that these ions are stable up to 373 K but that their concentrations decrease rapidly between 450 K and 573 K. The concentrations decreased to almost zero at 473 K. Thus, at temperatures typical of reactions these electrophilic species are not detected.

On ZnO the signal from O_2^- adsorbed at 298 K disappears entirely after heating to 493 K (*49*). On MoO_3/Al_2O_3 the signal attributed to O_2^- stabilized on an anion vacancy persisted only to 423 K (*50,51*). Similarly, on MoO_3/SiO_2 oxygen and hydrogen adsorb as O_2^- and OH radicals at 77 K. They react above 77 K to produce O^- and HO_2, which subsequently decompose on warming to room temperature (*52*). IR studies at room temperature of oxygen adsorption on heat-treated TiO_2 specimens revealed absorption bands at 1630, 1650, and 1680 cm^{-1}, attributed to neutral molecular oxygen. The bands disappeared after heating to 373 K. Again, there was no evidence for charged species, O_2^- or O_2^{2-} (*53*).

Based on the existence of the reported species oxygen chemisorption has been suggested to occur in the sequence:

$$O_2 \rightarrow O_2^- \rightarrow O_2^{2-} \rightarrow O_2^- \rightarrow O^{2-}$$

In fact, this sequence of transformation has not been observed in any one system. A continuous gradation of such species may exist which will depend on the specific environment at the adsorption site (*12*). These species may also easily interconvert at the temperatures of catalytic reactions.

As mentioned earlier, Haber, *et. al*, have suggested that in the partial oxidation of hydrocarbons nucleophilic lattice oxygen (O^{2-}) is responsible for selective oxidation, while electrophilic ionic or radical oxygen species (O_2^-, O_2^{2-}, and O^-) cause deep oxidation to CO_x (*6*). This concept may be an over-simplification, since it is likely that a range of oxygen species exists at catalytic conditions.

Furthermore, there are a number of reactions where adsorbed oxygen species rather than lattice oxygen ions are thought to be the principal oxidizing agent. Tagawa, *et. al*, have concluded that O^- forms on the surface and abstracts the β-hydrogen from the adsorbed complex in the dehydrogenation of ethyl benzene (*54*) . Szakács, *et. al.*, suggest that a mobile O^- radical is responsible for both partial and total oxidation of 1-butene and n-butene to maleic anhydride on V-P-O_x (55). Akimoto, et. al, have given evidence that O_2^- ions are involved in the oxidation of butadiene to maleic anhydride over supported molybdena catalysts (56). Yoshida, et. al, have reported the reaction of O_2^- on V_2O_5 with propylene and benzene to form aldehydes while the lattice oxygen shows little reactivity below 423 K (57).

Thus, the labelling of nucleophilic oxygen as promoting selective oxidation and electrophilic as favoring total oxidation may be too simplistic. As suggested here, details of the bonding of intermediates to the sample need also to be considered

Conclusions

1) On supported vanadium oxide, oxidation of ethane results in two distinct surface intermediates that lead respectively to acetaldehyde and CO_x.
2) On supported MoO_3, thermal desorption of adsorbed allyl alcohol produces acrolein, whereas desorption of allyl iodide produces CO_x.
3) Thermodynamic calculations indicate that hydrocarbon intermediates (R) bound to surface metal centers through oxygen atoms (M-O-R) are more stable than intermediates bound directly through metal carbon bonds (M-R).
4) The availability of organic decomposition pathways suggests that M-O-R intermediates produce selective oxidation products, whereas M-R intermediates produce CO_x.

Acknowledgment

This paper was written with support from the New York State Science and Technology Foundation and the Director, Division for Chemical and Thermal Systems of the National Science Foundation under Grant CTS-8909981.

Literature Cited

1. Sachtler,W. M. H.; De Boer, N. D. *Proc. Int. Cong. Catal. 3rd.* Amsterdam, 1964, *1*, 252, 1965.
2. Roiter,V. A.; Golodets, G. I. *Proc. Int. Cong. Catal. 4th*, Moscow, 1968, *1*, 365, 1970.
3. Germain, J. E.; Laugier, R. *Bull. Soc. Chim. Fr.* **1972**, 541.
4. Moro-oka, Y.; Morikawa, Y; Ozaki, A. *J. Catal.* **1967**, *7*,23.
5. Sachtler, W. M. H.; Dorgelo, G. H.; Fahrenfort, J.; Voorhoeve, R. J. H. *Rec. Trav. Chim. Pays-Bas.* **1970**, *89*, 460.
6. Bielanski, A.; Haber, J. *Cat. Rev.-Sci. Eng.* **1979**, *19*, 1.
7. Mitchell, A. G.; Lyne, M. P.;Scott, K. F.; Phillips, C. S. G. *J. Chem. Soc. Faraday Trans. I.* **1981**, *77*, 2417.
8. Snyder, T. P.; Hill, Jr., C. G. *Cat. Rev.-Sci. Eng.* **1989**, *31*, 43.
9. Keulks, G. W. *J. Catal.* **1970**, *19*, 232.

10. Wragg, R. D.; Ashmore, P. G.; Hockey, J. A. *J. Catal.* **1971**, *22*, 49.
11. Sancier, K. M.; Wentrcek, P. R.; Wise, H. *J. Catal.* **1975**, *39*, 141.
12. Che, M.; Tench, A. J. *Adv. Catal.* **1983**, *32*, 1.
13. Sachtler, W. M. H. *Cat. Rev.-Sci. Eng.* **1970**, *4*, 27.
14. Davydov, A. A.; Mikaltechenko, V. G.; Sokolovskii, V. D.; Boreskov, G. K. *J. Catal.* **1978**, *55*, 299.
15. Davydov, A. A.; Yeferemov, A. A.; Mikaltechenko, V. G.; Sokolovskii, V. D. *J. Catal.* **1979**, *58*, 1.
16. Sokolovskii, V. D. *Cat. Rev.-Sci. Eng.* **1990**, 32, 1.
17. Oyama, S. T.; Went, G. T.; Lewis, K. B.; Bell, A. T.; Somorjai, G. A. *J. Phys. Chem.* **1989**, *93*, 6786.
18. Silvi, B.; Perchard, J.P. *Spectrochim Acta, Part A.* **1976**, *32*, 11.
19. McLachlan, R.D.; Nyquist, R.A. *Spectrochim Acta, Part A.* **1968**, *24*, 103.
20. Harris, R. K. *Spectrochim Acta, Part A.* **1968**, *20*, 1129.
21. Desikan, A. N.; Zhang, W.; Oyama,S. T. *Div. Pet. Chem. ACS* **1992**, *37*, 1069.
22. Oyama, S. T. *J. Catal.* **1991**, *128*, 210.
23. Desikan, A. N.; Huang, L.; Oyama, S. T. *J. Phys. Chem.* **1991**, *95*, 10050.
24. Morrow, B. A. In *Vibrational Spectroscopy for Adsorbed Species*; Bell, A. T.; Hair, M. L. Eds.; ACS Symposium Series 137; American Chemical Society: Washington, DC, 1980; pp 119.
25. Boehm, H.P.; Knözinger, H. In *Catalysis: Science and Technology*; Anderson, J. R.; Boudart, M. Eds.; Springer - Verlag: New York, 1983, 4, 49.
26. Little, L. H. *Infrared Spectra of Adsorbed Species*; Academic: London, New York, 1966.
27. Harrison, P. G.; Maunders, B. *J. Chem. Soc., Faraday Trans. 1.* **1985**, *81*, 1329.
28. Dent, A. L.; Kokes, R. J. *J. Amer. Chem. Soc.* **1970**, *92*, 6709.
29. Griffith, W. P.; Lesniak, P. J. B. *J. Chem. Soc. A.* **1969**, 1066.
30. Rappé, A. K.; Goddard, W. A. *J. Amer. Chem. Soc.* **1982**, *104*, 3287.
31. These free energies were calculated from the data presented in ref. 30, and from tabulations in O'Neal, H. E.; Benson, S. W. "Thermochemistry of Free Radicals" in *Free Radicals*, Kochi, J. K. Ed.; p. 275, Wiley, New York, 1973, and in *Perry's Chemical Engineering Handbook*, Sixth Edition, Perry, R. H.; Green, D. W.; Maloney, J. O. Eds.; p.3-147, McGraw-Hill, New York, 1984. The free energy of hydration of Cl_2CrO_2 to $Cl_2CrO(OH)_2$ was estimated by correlating bond dissociation and heat of formation data for a series of compounds to free energies reported in ref. 30.
32. Seiyama, T.; Yamazoe, N.; Eguchi, K., *Ind. Eng. Chem. Prod. Res. Dev.* **1985**, *24*, 19.
33. Lunsford, J. H. *Cat. Rev. -Sci. Eng*. **1973**, *8*, 115.
34. Tench, A. J.; Holroyd, P. J. *J Chem. Soc., Chem. Comm.* **1968**, 471.
35. Ben Tarrit, Y.; Lunsford,J. H. *Chem. Phys. Lett.* **1973**, *19*, 348.
36. Shvets ,V. A.; Kazansky,V. B. *J. Catal.* **1972**, *25*, 123.

37. Shvets, V. A.; Vorotinzev, V. M.; Kazansky, V. B. *J. Catal.* **1969**, *15*, 214.
38. Wong, N. B.; Ben Tarrit, Y.; Lunsford,J. H. *J. Chem. Phys.* **1974**, *60*, 2148.
39. Wong, N. B.; Lunsford, J. H. *J. Chem. Phys.* **1972**, *56*, 2664.
40. Tench, A. J. *J. Chem. Soc. Faraday. Trans. I.* **1972**, *68*, 1181.
41. Nikisha, V. V.; Surin, S. A.; Shelimov, B. N.; Kazansky, V. B. *React. Kinet. Catal. Lett.* **1974**, *1*, 141.
42. Kazanzky, V. B.; Shvets, V. A.; Kon, M. Y.; Nikisha, V. V.; Shelimov, B. N. In *Catalysis* ; Hightower, J. Ed.; North-Holland: Amsterdam, 1973, pp 1423.
43. Vaska, L. *Acc. Chem. Res.* **1976**, *9*, 175.
44. Barbaux, Y.; Bonnelle, J. -P.; Beaufils, J. -P. *J. Chem, Res*, **1979**, *5*, 48.
45. Barbaux, U.; Elamrani, A.; Bonnelle, J. -P. *Catal. Today.* **1987**, *1*, 147.
46. Yoshida, S.; Matsuzaki, T.; Kasiwazaki, T.; Mori, K.; Tarama,K. *Bull. Chem. Soc. Jpn.* **1974**, *47*, 1564.
47. Takita , Y.; Lunsford, J. H. *J. Phys. Chem.* **1979**, *83*, 663.
48. Iwamoto, M.; Lunsford, J. H. *J. Phys. Chem.* **1980**, *84*, 3079.
49. Horiguchi, H.; Setaka, M.; Sancier, K. M.; Kwan, T. *Proc. Int. Cong. Catal. 4th.* Moscow, 1968, 1, 81, 1971.
50. Ishii, Y.; Matsuura, J. *Nippon Kagaku Zasshi* **1968**, *89*, 553.
51. Ishii, Y.; Matsuura, J. *Nippon Kagaku Zasshi* **1971**, *92*, 302.
52. Balistreri, S.; Howe C. R. F. in *Magnetic Resonance in Collloid and Interface Science*; Fraissard, J. P.; Resing, H. A. Eds.;Reidel: 1980, pp. 489.
53. Davydov. A. A. *Kinet. Katal.* **1979**, *20*, 1506.
54. Tagawa, T.; Hattori, T.; Murakami, Y. *J. Catal.* **1982**, *75*, 66.
55. Szakács, S.; Wolf, H.; Mink, G.; Bertóti, I.; Wostneck, N.; Locke, B.; Seeboth, H. *Catal. Today* **1987**, *1*, 27.
56. Akimoto, M.; Echigoya, E. *J. Catal.* **1974**, *35*, 278.
57. Yoshida, S.; Matsuzaki, T.; Ishida, S.; Tarama, K. *Proc. Int. Cong. Catal. 5th.*, Miami Beach, Florida, 1972, *21*, 1049, 1973.

RECEIVED October 30, 1992

Chapter 3

Surface Oxide–Support Interactions in the Molecular Design of Supported Metal Oxide Selective Oxidation Catalysts

Goutam Deo and Israel E. Wachs

Department of Chemical Engineering, Zettlemoyer Center for Surface Studies, Lehigh University, Bethlehem, PA 18015

A series of metal oxides were deposited on the surface of different oxide supports to study the surface oxide - support interactions. The dehydrated Raman spectra of the supported metal oxide catalysts reveal the presence and structure of the supported metal oxide phases. The same surface metal oxide species were found on the different oxide supports for each of the supported metal oxide systems. The reactivity of the surface metal oxide species, however, depends on the specific oxide support $(TiO_2 \sim ZrO_2 > Nb_2O_5 > Al_2O_3 \sim SiO_2)$. For a given oxide support, the reactivity depends on the specific surface metal oxide species (e.g. $VO_x > MoO_y$). The redox activation energy for all the surface metal oxide phases lie in the range of 18-22 kcal/mole. The similar activation energies suggests that the number of active sites and/or the activity per site is responsible for the difference in reactivity. The redox TON for the methanol oxidation reaction correlates with the reduction temperature during TPR experiments, which suggests that the bridging M-O-Support bond controls the activity during redox reactions.

Supported metal oxide catalysts are formed when one metal oxide component (i.e., Re_2O_7, CrO_3, MoO_3, WO_3, V_2O_5, Nb_2O_5, etc.), the supported metal oxide phase, is deposited on a second metal oxide substrate (i.e., Al_2O_3, TiO_2, SiO_2, etc.), the oxide support [1]. The supported metal oxide phase is present on the oxide support as a surface metal oxide species. The reactivity of these

0097–6156/93/0523–0031$06.00/0

supported metal oxide catalysts have been intensively investigated over the past decade in numerous catalytic applications and the main emphasis has been to relate the reactivity with the structure of the surface metal oxide species [1,2]. In determining the structure of these surface metal oxide species Raman spectroscopy has proven to be indispensable because of the ability of Raman spectroscopy to discriminate between different metal oxide species that may simultaneously be present in the catalyst. The reactivity studies have demonstrated that these surface metal oxide species are the active sites for many catalytic reactions [3]. The combined structural and reactivity information currently available about these oxide catalysts is beginning to allow us to develop an understanding of the surface oxide support interactions and to apply this knowledge for the molecular design of supported metal oxide catalysts.

The molecular design of supported metal oxide catalysts requires that we specify the synthesis method, oxide support, catalyst composition, calcination temperature, location of the surface metal oxide species, as well as its reactivity. Consequently, the influence of each of the above parameters upon the structure and catalytic properties of supported metal oxide catalysts needs to be examined. The present study primarily focuses on the molecular design aspects of supported vanadium oxide catalysts because these catalysts constitute a very important class of heterogeneous oxide catalysts. However, comparison with other supported metal oxide systems (MoO_3, Re_2O_7, and CrO_3) will also be made.

Experimental

The oxide supports employed in the present study were: TiO_2 (Degussa, ~55 m^2/g), Al_2O_3 (Harshaw, ~180 m^2/g), SiO_2 (Cabot, ~300 m^2/g), ZrO_2 (Degussa, ~39 m^2/g) and Nb_2O_5 (Niobium Products Co., ~50 m^2/g). Many different synthesis methods have been used to prepare supported metal oxide catalysts. In the case of supported vanadium oxide catalysts, the catalysts were prepared by vapor phase grafting with $VOCl_3$, nonaqueous impregnation (vanadium alkoxides), aqueous impregnation (vanadium oxalate), as well as spontaneous dispersion with crystalline V_2O_5 [4]. No drastic reduction of surface area of the catalysts was observed.

Structural characterization of the surface metal oxide species was obtained by laser Raman spectroscopy under ambient and dehydrated conditions. The laser Raman spectroscope consists of a Spectra Physics Ar^+ laser producing 1-100 mW of power measured at the sample. The scattered radiation was focused into a Spex Triplemate spectrometer coupled to a Princeton Applied Research OMA III optical multichannel analyzer. About 100-200 mg of the pure catalysts were pelletized and used for obtaining the Raman spectra in the dehydrated mode. For ambient spectra 5-20 mg of catalysts was placed on a KBr backing.

The supported metal oxide catalysts were examined for their reactivity in the methanol oxidation reaction. The reactor was operated using milligram amounts of catalysts that provide differential reaction conditions by keeping conversions below 10%. A methanol/oxygen/helium mixture of ~6/13/81 (mole %) at 1 atm pressure was used as the reactant gas for all the data presented. The analysis was performed with an online gas chromatograph (GC) (HP 5840A) containing two columns (Poropak R and Carbosieve SII) and two detectors (FID and TCD). Reaction data at 230 °C are presented in the form of turnover number (TON) - defined as the number of moles of methanol converted per mole of vanadium per second. The reaction data for some catalysts were also obtained at 200, 230, and 240 °C to calculate the activation energy and check for diffusional limitations in the reactor. No mass and heat transfer limitations were observed.

Results and Discussion

The vanadium oxide species is formed on the surface of the oxide support during the preparation of supported vanadium oxide catalysts. This is evident by the consumption of surface hydroxyls (OH) [5] and the structural transformation of the supported metal oxide phase that takes place during hydration-dehydration studies and chemisorption of reactant gas molecules [6]. Recently, a number of studies have shown that the structure of the surface vanadium oxide species depends on the specific conditions that they are observed under. For example, under ambient conditions the surface of the oxide supports possesses a thin layer of moisture which provides an aqueous environment of a certain pH at point of zero charge (pH at pzc) for the surface vanadium oxide species and controls the structure of the vanadium oxide phase [7]. Under reaction conditions (300-500 °C), moisture desorbs from the surface of the oxide support and the vanadium oxide species is forced to directly interact with the oxide support which results in a different structure [8]. These structural transformations taking place during hydration and dehydration conditions of the oxide support suggest that the correlation of the structure - reactivity data should be performed with the structural data obtained under dehydration conditions, and correlating such structural information with the reactivity data.

A series of ~1% V_2O_5 catalysts was prepared by non aqueous impregnation of vanadium tri-isopropoxide oxide (final calcination in oxygen at 450/500 °C) in order to investigate the influence of different oxide supports upon the dehydrated molecular structure and reactivity of the surface vanadium oxide species. Under ambient conditions Raman bands due to orthovanadate, pyrovanadate, metavanadate, and decavanadate species are observed. Assignments of the Raman bands to the different species are made elsewhere [7,9]. At these surface coverages a single surface vanadium oxide species is predominantly present on the different oxide supports as is evident by the presence of a dominant 1015-1039 cm^{-1} band in the dehydrated Raman spectra and potential complication due to additional surface vanadium oxide species are eliminated. The dehydrated Raman band due to the V=O bond was found to vary from 1015-1039 cm^{-1} as a function of the different oxide supports as shown in Figure 1. The slight difference in band position is due to slightly different V=O bond lengths of the isolated surface vanadium oxide species on the different oxide supports and correlates with the different oxygen coordination of the oxide supports. The Raman spectra reveal that essentially the same surface vanadium oxide species is present on all the different oxide supports. This surface vanadium oxide species is described as a distorted four coordinated species possessing a single terminal bond (V=O) and three bonds to the support (V-O-S). The same conclusion is reached from solid state ^{51}V NMR studies of these catalysts [9]. Low surface coverages, ~1% metal oxide, of supported molybdenum oxide (MoO$_3$) [10], rhenium oxide (Re$_2$O$_7$) [11,12], and chromium oxide (CrO$_3$) [11,12] also indicate the presence of a single surface metal oxide species. The similar Raman band positions of the supported vanadium-oxygen (V=O) stretching frequency during dehydrated conditions are given in Table I. In addition, the structural transformation taking place due to dehydration is observed by comparing columns 2 and 3 of Table I. Thus, at low coverages the dehydrated surface vanadium oxide and related metal oxide molecular structures (Re$_2$O$_7$, CrO$_3$, and MoO$_3$) are independent of the specific oxide support.

Table I. Dehydrated Raman band position for the V=O terminal stretching vibrations for 1% V$_2$O$_5$ on different oxide supports along with the highest ambient Raman band position

Oxide Support	V=O band (cm^{-1}) dehydrated conditions	Highest band (cm^{-1}) ambient conditions
SiO$_2$	1039	1000
Nb$_2$O$_5$	1031	970-980
TiO$_2$	1027	940-950
ZrO$_2$	1024	960
Al$_2$O$_3$	1015	920-930

The reactivity of the surface vanadium oxide species (~1% V_2O_5) on the different oxide supports was probed by the methanol oxidation reaction. The methanol oxidation reaction is very sensitive to the nature of surface sites present. Surface redox sites form formaldehyde, methyl formate, and dimethoxy methane as the reaction products. Formaldehyde being formed as the first oxidation product from the methoxy intermediate. Subsequent reactions of the methoxy intermediate produces methyl formate and dimethoxy methane. Surface acid sites, Lewis as well as Bronsted, result in the formation of dimethyl ether. Surface basic sites yield CO/CO_2 as the reaction products [9]. For all the supported vanadium oxide catalysts, with the exception of alumina, the surface vanadia redox sites produced formaldehyde almost exclusively. On alumina, only a trace of formaldehyde was formed because the surface acid sites produced dimethyl ether. Thus, for the V_2O_5/Al_2O_3 system the formaldehyde produced was taken as representative of the reactivity of the surface vanadia redox sites. The reactivity of the surface vanadia species on different oxide supports was found to depend dramatically on the specific oxide support as shown in Table II. As the support was changed from silica to zirconia the turnover number (TON) for the surface vanadium oxide species was found to increase by three orders of magnitude. Similar trends were also observed for supported molybdenum oxide [10], rhenium oxide [12], and chromium oxide [12].

Table II. The TON and selectivity to formaldehyde for the methanol oxidation reaction on various 1% supported vanadium oxide catalysts

Oxide Support	TON (sec^{-1})	Selectivity to HCHO (%)
SiO_2	$2.0*10^{-3}$	~80
Al_2O_3	$2.0*10^{-2}$	<1
Nb_2O_5	$7.0*10^{-1}$	~94
TiO_2	$1.8*10^0$	~98
ZrO_2	$2.3*10^0$	~96

The reactivity of the supported vanadium oxide catalysts for other oxidation reactions also show similar trends as the oxide support is varied from titania to silica [13]. The activity and selectivity for partial oxidation products of vanadium oxide supported on titania being higher than vanadium oxide supported on silica. The oxidation activity of the supported vanadium oxide catalysts is related to the ability to donate oxygen to form the required oxidation products. The origin of this support effect is either due to differences in the terminal V=O bond or the bridging V-O-Support bond. Many publications have proposed that the terminal bond is responsible for catalysis and its activity is directly related to the V=O bond strength [14]. However, comparison of the Raman position of the V=O bond (shorter bond corresponds to higher Raman position) from Table I and TON value for the different supported vanadium oxide catalysts from Table II shows that no relationship exists between the bond strength $^{\circ}V=O$ and activity. A more elaborate analysis of the TON versus bond strength for various supported metal oxide catalysts are performed elsewhere [15]. A more plausible conclusion is that the reactivity is related to the bridging V-O-Support bond since the oxide support has a very significant effect on the reactivity. The trend in reactivity with specific oxide support appears to be related to the surface reducibility of the oxide supports. The more reducible oxide supports (TiO_2, ZrO_2, and Nb_2O_5) always exhibit very high TON while the irreducible oxide supports (Al_2O_3 and SiO_2) always exhibit very low TON [16]. In addition, the importance of the M-O-Support bond during the methanol oxidation reaction suggests that the reactivity should also be a function of the specific supported metal oxide species. Indeed, supported molybdenum oxide catalysts are about one order of magnitude less reactive than supported vanadia catalysts.

Additional information about the reactivity was obtained by determining the kinetic parameters during methanol oxidation for vanadia, molybdena, rhenia, and chromia on different oxide supports. For all these systems the activation energy is approximately the same, 18-22 kcal/mol. The activation energy corresponds to that expected for the breaking of the C-H bond of a surface methoxide intermediate, CH_3O_{ads}, and should be independent of the specific catalyst [17]. The pre-exponential factors, however, vary by orders of magnitude as the oxide support is varied. The similar structures of the supported metal oxide catalysts suggests that the difference in pre-exponential factors is related to either the number of active sites or the activity per site. Irrespective of the exact parameter that affects the pre-exponential factor, it is shown here that the oxide support has a large influence on the reactivity of the surface metal oxide species.

To investigate the effect of the synthesis method on the structure-reactivity relationship of the supported metal oxide catalysts, a series of V_2O_5/TiO_2 catalysts were synthesized by equilibrium adsorption, vanadium oxalate, vanadium alkoxides and vanadium oxychloride grafting [14]. The dehydrated Raman spectra of all these catalysts exhibit a sharp band at ~1030 cm^{-1} characteristic of the isolated surface vanadium oxide species described previously. Reactivity studies with methanol oxidation exhibited similar turnover numbers. Thus, the synthesis method does not affect the final structure or reactivity of the surface vanadium oxide species on titania. Similar conclusions were also found for molybdenum oxide supported on titania [14], silica [18], and alumina [18]. Consequently, the preparation method is not a critical parameter to consider for the design of supported metal oxide catalysts. However, care must be taken to form the surface metal oxide phase without destroying the oxide support or introducing extraneous impurities during preparation.

To study the effect of loading, various amounts of vanadia were deposited on the titania support. The Raman spectra of titania supported vanadia catalysts as a function of vanadia loading reveal the presence of three different vanadia species on the TiO_2 support under dehydration conditions as shown in Figure 2. At low loadings (~1% V_2O_5), a single sharp band is present at ~1027 cm^{-1} which is assigned to an isolated tetrahedral coordinated surface vanadium oxide species containing one terminal V=O bond and three bridging V-O-Ti bonds [19]. At intermediate loadings (2-5% V_2O_5), a second band is present at 920-930 cm^{-1} which has been assigned to a polymerized, tetrahedral coordinated surface vanadium oxide species [19] in addition to the first band at ~1030 cm^{-1}. At high loadings (>6% V_2O_5), a third sharp band is present at 994 cm^{-1} due to crystalline V_2O_5 which indicates that the close-packed surface vanadium oxide monolayer has been formed and all the reactive surface hydroxyls consumed. The first two bands are also present at high loadings. Similar trends in Raman bands are also observed for vanadium oxide supported zirconia, niobia, alumina, and silica and the formation of molecularly dispersed vanadium oxide species always preceeds the formation of crystalline V_2O_5. Thus, the catalyst composition is a critical parameter since it influences the formation of different vanadium oxide structures.

The reactivity of the titania supported vanadium oxide catalysts was probed by the methanol oxidation reaction. The oxidation of methanol over the titania supported vanadia catalysts exclusively yielded formaldehyde, 95%+, as the reaction product. The titania support in the absence of surface vanadia yielded dimethyl ether and trace amounts of CO_2. The almost complete formation of formaldehyde demonstrates that the reactivity of the titania supported vanadia catalysts is due to the surface vanadia redox sites. The turnover numbers of the various vanadia titania catalysts are presented in Table III as a function of the vanadia loading. The TON increases somewhat with initial surface vanadium oxide coverage, and decreases slightly at surface coverages approaching and exceeding monolayer coverage. Note that the TON of bulk V_2O_5 [16] is two orders of magnitude less than the titania supported vanadia catalysts indicating that crystalline V_2O_5 is

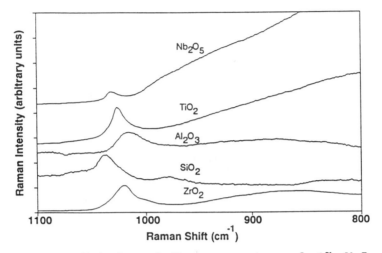

Figure 1.   Dehydrated Raman spectra of 1% V_2O_5 on different oxide supports.

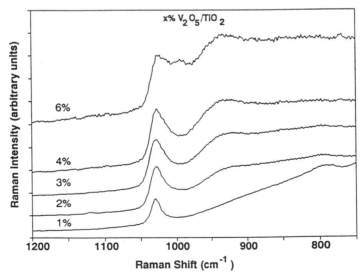

Figure 2.   Dehydrated Raman spectra of vanadia-titania catalysts as a function of vanadia loading.

significantly less active than surface vanadia species. The slight increase in TON with increasing surface coverage is not related to the presence of the polymerized surface vanadia species (Raman band at 920-930 cm^{-1}) as the Raman associated with the polymerized species increases continuously up to monolayer coverages and the turnover numbers are comparable. Similar coverage effects were also observed for vanadium oxide supported on zirconia, alumina, and niobia catalysts, as well as for supported rhenium oxide [11], molybdenum oxide [10] and chromium oxide catalysts [12]. Thus, the reactivity of the surface vanadium oxide species essentially does not depend on the surface coverage.

Table III. The TON of V_2O_5/TiO_2 catalysts as a function of loading

wt.% V_2O_5/TiO_2	T.O.N. (sec^{-1})
0.5	1.0
1.0	2.0
2.0	2.7
3.0	1.6
4.0	1.5
5.0	1.3
6.0	1.1
7.0	1.2
bulk V_2O_5	0.022

Calcination Temp

    The nature of the supported vanadium oxide phase is influenced by the calcination temperature. Moderate calcination temperatures, 350-500 °C, are required to decompose the metal oxide precursors (oxalates, alkoxides, oxychlorides, etc.) to form the surface vanadium oxide species [20]. Insufficient calcination temperatures do not completely decompose the precursors and, consequently, the precursors do not react with the surface hydroxyls to form the surface metal oxide species. However, high calcination temperatures, greater than 600 °C, can result in shrinking of the surface area of the oxide support and decreasing the available surface area for the surface metal oxide species. Consequently, high calcination temperatures increase the surface coverage of the metal oxide species and, in severe cases, destroy the surface metal oxide phase and form crystalline V_2O_5 or solid state solutions [7,20]. Thus, calcination temperature is an important parameter that controls the activation and deactivation of supported metal oxide catalysts. However, supported metal oxide catalysts are typically prepared by calcining at 400-500 °C which would eliminate problems with catalyst activation and deactivation.

Promoters

Typically, the properties of supported metal oxide catalysts are modified by the addition of promoters. To examine the influence of promoters, a series of promoters (tungsten oxide, niobium oxide, silica, potassium oxide and phosphorous oxide) were added to a 1% V_2O_5/TiO_2 catalyst. The influence of the different promoters upon the structure of the surface vanadium oxide species was examined with Raman spectroscopy and the results are presented in Table IV. The addition of monolayer amounts of tungsten oxide and niobium oxide to 1% V_2O_5/TiO_2 catalysts show similar effects observed as the vanadium oxide loading is increased, namely, the appearance of the Raman band at 920-930 cm^{-1}. Monolayer amounts of silica on previously prepared 1% V_2O_5/TiO_2 shows the presence of a single band at 1024 cm^{-1}. In any case the major Raman band is at 1024-1031 cm^{-1} indicating that the vanadium oxide species before and after addition of tungsten oxide, niobium oxide, and silica are structurally similar. However, the addition of potassium oxide and phosphorous oxide had a significant effect on the structure of the surface vanadia species. The addition of potassium oxide decreased the position of the Raman band (corresponding to an increase in V=O bond length), and the addition of phosphorous oxide resulted in the formation of crystalline $VOPO_4$ (major Raman bands at 1038 and 928 cm^{-1} in the 700-1200 cm^{-1} region). The non-interacting promoters (oxides of W, Nb, and Si) did not affect the activity or selectivity of the 1% V_2O_5/TiO_2 catalyst. However, the interacting promoters (potassium oxide and phosphorous oxide) significantly reduced the TON. Thus, promoters that preferentially coordinate to the oxide support (tungsten oxide, niobium oxide and silica) do not affect the structure or reactivity of the surface vanadium oxide species, whereas, promoters that coordinate with the surface vanadia site (phosphorous and potassium) influence the structure and reactivity of the surface vanadia species.

Table IV. Raman band position for promoters on
1% V_2O_5/TiO_2

Promoters on 1% V_2O_5/TiO_2	Raman band positions (700-1200 cm^{-1}) region
None	1027
6% Nb_2O_5	1031, 928
3% SiO_2	1024
7% WO_3	1031, 925
0.3% K_2O	1023, 997
0.7% K_2O	1009, 980
5% P_2O_5	1035, 925

Conclusions

The above discussion demonstrates that it is possible to molecularly design supported metal oxide catalysts with knowledge of the surface oxide - support interactions made possible by the assistance of characterization methods such as Raman spectroscopy and the methanol oxidation reaction. The formation and location of the surface metal oxide species are controlled by the surface hydroxyl chemistry, and the surface metal oxide species are located in the outermost layer of the catalysts as an overlayer. The specific oxide support is a critical parameter since it dramatically affects the reactivity of the surface metal oxide species, even though the surface metal oxide structure is independent of the specific oxide support. The catalyst composition is a critical parameter since it affects the presence of different metal oxide species (isolated surface species, polymerized surface species, and crystalline phases), and the reactivity, TON, is similar for surface metal oxide coverage. The preparation method is not a critical parameter since it does not influence the structure or reactivity of the surface metal oxide species. Calcination temperature is an important parameter that controls activation and deactivation of supported metal oxide catalysts, but calcination temperature is not critical if moderate temperatures, 350-450 °C, are used. Additives that interact with the oxide support do not influence the structure and reactivity of the surface metal oxide species. However, additives that interact with the surface metal oxide do influence the properties of the surface metal oxide species. In summary, the critical parameters that affect the catalytic properties are the specific oxide support, catalyst composition (type of additives) and surface metal oxide coverage.

Acknowledgments

We would like to thank Dr. D.S.Kim, H. Hu, and M. A. Vuurman for their helpful discussions. The financial support by NSF grant # CTS-9006258 is gratefully acknowledged.

Literature Cited

1. (a) Dixit, L.; Gerrard, D. L.; Bowley, H. *Appl. Spectrosc. Rev.* 1986, 22 189.
    (b) Bartlett, J. R.; Cooney, R. P. In *Spectroscopy of Inorganic-based Materials*; Clark, R.J.H. and Hester, R.E., Eds; Wiley: New York, 1987; pp 187.
    (c) Hardcastle, F.D.; Wachs, I. E. In *Proc. 9th Intern. Congr. Catal.*; Phillips, M.S. and Ternan, M., Eds., Chemical Institute of Canada, Ontario, 1988, Vol. 4; pp 1449.
2. Wachs, I. E. *Chem. Eng. Sc.* 1990, 45(8), 2561.

3. Bond, G. C.; Tahir, S. F. *Appl. Catal.* <u>1991</u>, 71, 1.
4. Machej, T.; Haber, J.; Turek, A. M.; Wachs, I. E. *Appl. Catal.*<u>1991</u>, 70, 115.
5. (a) Segawa, K.; Hall, W.K. *J. Catal.* <u>1982</u>, 77, 221.
   (b) Kim, D. S.; Kurusu, Y.; Segawa, K.; Wachs, I. E. In *Proc. 9th Inter. Congr. Catal.*; Phillips, M.S. and Ternan, M., Eds., Chemical Institute Canada, Ontario, 1988, Vol. 4, pp 1460.
(c) Turek, A. M.; Wachs, I. E.; Decanio, E. *J. Phys. Chem* <u>1992</u>, 96, 5000.
6. (a) Went, G. T.; Oyama, S. T.; Bell, A. T. *J. Phys. Chem.* <u>1990</u>, 94, 4240.
   (b) Vuurman, M. A.; Wachs, I. E.; Hirt, A. M. *J. Phys. Chem.* <u>1991</u>, 95, 9928.
7. Deo, G.; Wachs, I. E. *J. Phys. Chem.* <u>1991</u>, 95, 5889.
8. (a) Anpo, M.; Tanahashi, I.; Kubokawa, Y. *J. Phys. Chem.* <u>1980</u>, 84, 3440.
   (b) Cristiani, C.; Forzatti, P.; Busca, G. *J. Catal.* <u>1991</u>, 116, 586.
9. (a) Eckert, H.; Wachs, I. E. *J. Phys. Chem.* <u>1989</u>, 93, 6796.
   (b) Eckert, H.; Deo, G.; Wachs, I. E., unpublished results.
10. Hu, H.; Wachs, I. E., in preparation.
11. Vuurman, M. A.; Wachs, I. E.; Stufkens, D.J.; Oskam, A. *J. Mol. Catal.*, <u>1992</u>, in press.
12. Kim, D. S.; Wachs, I. E. *J. Catal.* <u>1992</u>, in press.
13. (a) Wainwright, M. S.; Foster, N. R. *Catal. Rev.* <u>1979</u>, 19, 211.
   (b) Hauffe, K.; Raveling, H. *Ber. Bunsenges. phy. Chem.* <u>1980</u>, 84, 912.
14. Klissurski, D.; Abadzhijeva, N. *React. Kinet. Catal. Lett.* <u>1975</u>, 2, 431.
15. Wachs, I. E.; Deo, G.; Kim, D. S.; Vuurman, M. A.; Hu, H. In *Preprints of the Proceedings of the 10th Internation Congress of Catalysis*, 1992, 72.
16. Deo, G.; Wachs, I.E. *J. Catal.* <u>1991</u>, 129, 307.
17. (a) Yang, T. S.; Lunsford, J. H. *J. Catal.* <u>1987</u>, 103, 55.
   (b) Fareneth, W. E.; Ohuchi, F.; Staley, R. H.; Chowdhry, U.; Sleight, A. W. *J. Phys. Chem.* <u>1985</u>, 89, 2493.
18. (a) Williams, C. C.; Ekerdt, J. G.; Jehng, J.-M.; Hardcastle, F. D.; Turek, A. M.; Wachs, I. E. *J. Phys. Chem.* <u>1991</u>, 95, 8781.
   (b) Williams, C. C.; Ekerdt, J. G.; Jehng, J.-M.; Hardcastle, F. D.; Wachs, I. E. *J. Phys. Chem.* <u>1991</u>, 95, 8791.
19. Wachs, I. E. *J. Catal.* <u>1990</u>, 124, 570.
20. Saleh, R. Y.; Wachs, I. E.; Chan, S. S.; Chersich, C. C. *J. Catal.* <u>1986</u>, 98, 102.

RECEIVED October 30, 1992

Chapter 4

Silica-Supported MoO$_3$ and V$_2$O$_5$ Catalysts in Partial Oxidation of Methane to Formaldehyde

Factors Controlling Reactivity

A. Parmaliana[1], V. Sokolovskii[2], D. Miceli[3], F. Arena[1], and N. Giordano[3]

[1]Dipartimento di Chimica Industriale, Università degli Studi di Messina, Salita Sperone 31, c.p. 29, I–98168 Messina, Italy
[2]Department of Chemistry, University of Witwatersrand, Johannesburg, P.O. Wits 2050, South Africa
[3]Istituto CNR–TAE, Salita Santa Lucia 39, I–98126 Santa Lucia, Messina, Italy

The partial oxidation of methane to formaldehyde with molecular O$_2$ has been investigated on bare SiO$_2$, 4%MoO$_3$/SiO$_2$ and 5%V$_2$O$_5$/SiO$_2$ catalysts at 550-650°C in batch, pulse and continuous flow reactors at $1.7 \cdot 10^2$ kPa. The HCHO productivity ($g_{HCHO} \cdot kg_{cat}^{-1} \cdot h^{-1}$) results in the order 4%MoO$_3$/SiO$_2$ < SiO$_2$ < 5%V$_2$O$_5$/SiO$_2$. On the basis of a series of experiments performed by continuous scanning of the reaction mixture with a quadrupole M.S., the participation of lattice oxygen in the main reaction pathway has been ruled out. The acidic properties of the catalysts have been compared by ZPC and NH$_3$-TPD measurements. A straight correlation between the density of reduced sites, evaluated in steady-state conditions by O$_2$ chemisorption, and the reaction rate has been disclosed. MoO$_3$ and V$_2$O$_5$ dopants modify the catalytic properties of SiO$_2$ by affecting the process of oxygen activation on the catalyst surface.

Although during the last decade the catalytic partial oxidation of methane to formaldehyde has attracted a great deal of research interest, a scarce attention has been devoted to rationalising both the selection of the catalysts and the factors controlling their reactivities. Several kinetic studies on SiO$_2$ *(1)* and silica supported MoO$_3$ *(2,3)*, V$_2$O$_5$ *(4,5)* or heteropolyacids *(6)* involving the use of either oxygen or nitrous oxide as oxidant have been performed, but very few have attempted to correlate catalytic activity with the physico-chemical properties of the solids. Barbaux et al. *(7)*, studying the title reaction on a series of differently loaded MoO$_3$/SiO$_2$ catalysts, observed that the yield and selectivity to HCHO increase with the amount of silicomolybdic acid in the catalyst. The promoting effect of MoO$_3$, up to a surface coverage of 0.1 monolayers, on the reactivity of bare silica

0097–6156/93/0523–0043$06.00/0

catalysts has been claimed by Spencer *(3)*. Recently, Hodnett and his group *(8)* have reported the inability of Fe, Ag, Cr, Na and Co ions in improving the activity of MoO_3 and V_2O_5 catalysts. For unpromoted V_2O_5/SiO_2 catalysts, they found that V_2O_5 loading in the range 1.8-7.2 wt% gave the best performance. Such a higher activity of the medium loaded V_2O_5 catalysts has been attributed to their capability to readily undergo a reduction-reoxidation process in the reaction conditions. However, the results achieved by the different research groups, either in terms of selectivity or yield to formaldehyde, appear to be controversial and not greatly encouraging *(9)*. In addition, several authors have even questioned the catalytic nature of the partial oxidation of methane to HCHO *(1,10-11)*. Indeed, Burch and his group *(10)* have posed serious doubts on the convenience to use solid catalysts stating that the yields of HCHO obtained from the catalytic reaction are lower than that obtained from the purely homogeneous gas-phase reaction. Nevertheless, Garibyan and Margolis *(11)*, in reviewing the mechanism of methane oxidation over SiO_2 catalysts, proposed a heterogeneous-homogeneous reaction scheme which predicts the generation of HCHO on the catalyst surface only. On this account, we have previously pointed out *(12)* the unique suitability of the SiO_2 surface in catalysing the title reaction assessing also the influence of the preparation methods, Na loading, thermal and mechanical pretreatments on its reactivity.

The surface and the structure features of MoO_3 *(13)* and V_2O_5 *(14)* supported on silica have been extensively studied. The influence of the MoO_3 loading on its dispersion and reactivity has been assessed *(13)*. In particular, on medium loaded MoO_3 catalysts (4.2-11.2 wt% MoO_3) the formation of polymeric molybdate species with octahedral coordination of the Mo ions has been detected *(13)*. On the contrary, it has been claimed that V ions form surface compounds with a lower coordination number in medium loaded V_2O_5/SiO_2 catalysts *(14)*. In fact, the presence of both oligomers with tetrahedral coordination around a V ion and ribbons with V ions in the centre of a square pyramid has been identified *(14)*. Similarly, Centi et al. *(15)* have elucidated the coordination and nature of V sites in V-silicalite catalysts providing a further evidence of the presence of V^{4+} ions in a tetrahedral environment which give rise to the formation of very reactive O_2^- species. Although the reduction-oxidation mechanism of hydrocarbons on unsupported *(16)* and silica supported V_2O_5 catalysts *(17)* as well as the reactivity of the different oxygen adsorbed species on such systems has been largely recognized *(18)*, still it lacks a valid attempt aimed to evaluate the amount of active sites present on the catalyst surface in steady-state conditions during the partial oxidation of methane or higher hydrocarbons *(19-20)*. Oyama et al. *(21)* have recently developed an oxygen chemisorption method to characterise unsupported and silica supported MoO_3 *(22)* and V_2O_5 *(21)* observing an interesting effect of structure on the reaction pathway of ethane partial oxidation on V_2O_5/SiO_2 catalysts *(20)*.

In this paper we present a comparative study of the catalytic properties of bare SiO_2, 4% MoO_3/SiO_2 and 5% V_2O_5/SiO_2 catalysts in the partial oxidation of methane to formaldehyde. The role of the acidic properties and reduced sites, evaluated in steady state conditions, is addressed. A co-operative reaction pathway, involving the presence of two kinds of active sites capable of methane activation and oxygen binding, is proposed.

Experimental

Catalyst. A pure precipitated SiO$_2$ sample (grade Si 4-5P, AKZO product; BET S.A., 395 m$^2\cdot$g^{-1}; Pore Volume, 1.14 cm$^3\cdot$g^{-1}) has been used "as received", while 4% MoO$_3$/SiO$_2$ (BET S.A., 187 m$^2\cdot$g^{-1}) and 5% V$_2$O$_5$/SiO$_2$ (BET S.A., 231 m$^2\cdot$g^{-1}) catalysts have been prepared by incipient wetness impregnation of the "Si4-5P" SiO$_2$ with a basic solution (pH$=$11) of ammonium heptamolybdate and ammonium metavanadate respectively *(23)*. The impregnated samples were dried at 90°C for 24h and then calcined at 600°C for 16h.

Catalyst Testing.
Batch reactor. Methane partial oxidation experiments were performed using a specifically designed batch reactor provided with an external recycle pump and a liquid product condenser placed downstream of the reactor and kept at -15°C. The quartz tube reactor measured 4 mm in diameter (i.d.) and 90 mm in length. All the products were analysed with a Hewlett-Packard 5890A gas-chromatograph equipped with a Thermal Conductivity Detector (TCD). The reaction mixture consisted of 18 mmol of methane, 9 mmol of oxygen (CH$_4$/O$_2$=2), 18 mmol of nitrogen as standard for GC analysis and 53 mmol of helium as diluent. All the runs were carried out with 0.05 g of catalyst (particle size, 16-25 mesh), at 1.7$\cdot$10^2 kPa and in the T range 550-650°C with a flow-rate of 1.0$\cdot$10^3 Ncm^{3}min^{-1}.
Pulse-Flow reactor. A conventional flow apparatus operating in both continuous and pulse mode has been used. Catalyst samples (0.05 g) were heated "in situ" up to reaction temperature (550-650°C) in a He carrier flow (50 Ncm^{3}min^{-1}) and then 5.5 Ncm^3 CH$_4$/O$_2$/He (P$_{CH4}$:P$_{O2}$:P$_{He}$ = 2:1:7) pulses were injected onto the sample until steady state conditions were reached. The reactor stream was analysed by an on-line connected "Thermolab" *(Fisons Instruments)* Quadrupole Mass Spectrometer (QMS).

Catalyst Characterization.
Oxygen chemisorption measurements were performed in the above flow apparatus using He as carrier gas (30 Ncm$^3\cdot$min^{-1}). Prior to chemisorption measurements, catalyst samples (0.25 - 1.00 g) were treated "in situ" for 15 min in a flow of CH$_4$/O$_2$/He (P$_{CH4}$: P$_{O2}$: P$_{He}$ = 2 : 1 : 7) reaction mixture at 550-650°C. O$_2$ uptakes were determined in a pulse mode (V$_{O2\ pulse}$=1.7 μmol) at the same temperature of the pretreatment by using a TCD connected to a DP 700 Data Processor *(Carlo Erba Instruments)*. The number of reduced sites was calculated by assuming the chemisorption stoichiometry O$_2$/"reduced site" of 1/2.
NH$_3$-TPD measurements were performed in the same apparatus. Prior to NH$_3$ saturation, the samples (0.05 g) were treated at 600°C in O$_2$/He or CH$_4$/O$_2$/He flowing mixtures for half an hour. After saturation at 150°C, the samples were purged at the same temperature in the He carrier flow (30 Ncm$^3\cdot$min^{-1}) for 1h.

The TPD process was run with a heating rate of $12°C·min^{-1}$, the NH_3 desorption being monitored and quantified by the QMS.

Zero Point Charge (ZPC) measurements. Potentiometric titration of samples (3.0 g) was carried out in an aqueous suspension (500 ml electrolytic KNO_3 solution) according to the procedure reported by Parks (24).

Results and Discussion

Catalytic Activity. The world-wide interest focused in the catalytic partial oxidation of methane to formaldehyde has led to a great variety of conflicting results (9). The main reason of such discrepancies lies in the lack of a generally valid rule for evaluating and comparing the proposed catalytic systems. In effect, this reaction involves a very complex pathway since the desired partial oxidation product, HCHO, exhibits a limited thermal stability at $T > 400°C$ and can be oxidized to CO_x more easily than CH_4 itself. Hence, a suitable reactor device and appropriate operating conditions result to be of fundamental importance in order to attain reliable data unaffected by experimental artefacts.

The batch reactor, above described, permits both to operate at quasi-zero conversion per pass and to evaluate the catalytic activity at finite values of the reagents conversion. A typical test performed on SiO_2 catalyst at 600°C is presented in Figure 1. It is remarkable how in our approach the product selectivity is unaffected by the methane conversion. A special care was taken to avoid oxygen-limiting conditions and, hence, methane conversion data obtained for oxygen conversions below 20% only have been used for the calculation of reaction rates.

The product selectivity is strongly affected by the flow rate, reactor geometry (i.e., internal diameter and "heated" zone) and weight of catalyst. On this account, the space time yield to HCHO - or HCHO productivity - ($g_{HCHO}·kg_{cat}^{-1}·h^{-1}$) appears to be the more definite parameter to evaluate the reactivity of the partial oxidation catalysts.

In an earlier paper (12) we have observed that the SiO_2 surface presents a significant activity in the partial oxidation of methane, whereas from the literature it emerges that MoO_3 (3) and V_2O_5 (5) are effective promoters of its activity. Then, in order to carefully evaluate such a promoting effect, a comparative investigation of the catalytic pattern of unpromoted SiO_2 and 4% MoO_3 and 5% V_2O_5 promoted silica systems in the methane partial oxidation with molecular oxygen has been performed. The reactivity data of such catalytic systems, in the T_R range 550-650°C, are summarized in Table I in terms of overall reaction rate, product selectivity and HCHO productivity. The contribution of the gas-phase reactions has been also quantified by performing a series of experiments with the empty reactor. The related results are also included in Table I. On the whole these data indicate that MoO_3 significantly depresses the reactivity of the bare SiO_2 catalyst, while the V_2O_5 greatly enhances the functionality of the SiO_2 surface towards the production of HCHO.

The opposite influence exerted by the MoO_3 and V_2O_5 on the SiO_2 reactivity has been fully confirmed by evaluating the catalytic behaviours of 0.2-10.0 wt% MoO_3 and V_2O_5 loaded SiO_2 catalysts. In particular, by raising the MoO_3 loading from 0.2 to 4.0 wt% the HCHO productivity gradually decreased, while for vanadia dopant, the HCHO yield increased with loading in the range 0.2-5.0 wt%. A further increase in the loading of the oxides did not modify the reactivities of the medium loaded (4% MoO_3, 5% V_2O_5) supported oxide catalysts *(25)*.

Table I. Activity and Selectivity of SiO_2, 4% MoO_3/SiO_2 and 5% V_2O_5/SiO_2 Catalysts in Methane Partial Oxidation. Batch reactor data [a]

Catalyst	T_R (°C)	Reaction Rate ($10^7 mol_{CH4} \cdot g_{cat}^{-1} \cdot s^{-1}$)	Selectivity (mol%) CO₂	CO	HCHO	HCHO Productivity ($g \cdot kg_{cat}^{-1} \cdot h^{-1}$)
	550	0.02	-	-	-	-
Empty Reactor	600	0.04	1	36	63	0.3
	650	0.15	2	46	52	0.8
	550	4.0	10	11	79	34.0
SiO_2	600	14.4	7	18	75	116.0
	650	44.3	14	23	63	304.0
	550	1.0	30	-	70	7.6
4% MoO_3/SiO_2	600	3.1	30	-	70	22.7
	650	10.5	28	2	70	76.1
	550	15.3	37	2	61	101.0
5% V_2O_5/SiO_2	600	60.8	38	14	48	318.0
	650	210.6	14	51	35	793.0

[a] For reaction conditions see Figure 1.

The validity of the data obtained with the batch reactor has been confirmed by continuous flow reactor tests. In Table II are shown comparative tests in the reactivity of the unpromoted SiO_2 catalyst by using batch and flow reactors at 600 and 650°C. The good agreement of these tests confirms the adequacy of our batch reactor approach.

On the basis of the above data obtained with the empty reactor and of the results of other series of experiments performed with differently sized empty and SiC or quartz filled reactors as well as with an empty stainless-steel reactor, we observe that the contribution of the gas-phase reaction in the T range investigated is absolutely negligible.

Table II. Methane Partial Oxidation over SiO$_2$ Catalyst in Batch and Continuous Flow Reactors

Experiment	T_R (°C)	Reaction Rate ($10^{-7}mol_{CH4} \cdot g_{cat}^{-1} \cdot s^{-1}$)	Selectivity (mol%)			HCHO Productivity ($g \cdot kg_{cat}^{-1} \cdot h^{-1}$)
			CO_2	CO	HCHO	
Batch Reactor	600	13.2	13	50	37	52
	650	38.4	15	59	26	108
Flow Reactor [b]	600	10.4	1	61	38	43
	650	41.6	13	57	30	166

[b] Reactor size: $\phi_{i.d.}$, 10 mm; l, 100 mm.
Reaction conditions: P_R, $1.1 \cdot 10^2$ kPa; flow rate, 1000 Ncm$^3 \cdot$min^{-1}; W_{cat}, 1.0 g.
Reaction mixture composition (mol%):CH$_4$, 18.4; O$_2$, 9.2; N$_2$, 18.4 and He,54.0.

Moreover, as neither the concept of surface initiated homogeneous-heterogeneous reaction (11) can be invoked to explain our results, it can be stated that the methane partial oxidation reaction proceeds via a surface catalysed process which likely involves specific catalyst requirements. However, by comparing the HCHO productivity of the different catalytic systems previously proposed (9) with that of our 5% V$_2$O$_5$/SiO$_2$ catalyst, it emerges that our findings constitute a relevant advancement in this area (23).

Catalyst Properties and Reaction Mechanism.   Apart from the generally accepted theory of the pure homogeneous gas phase oxidation of methane, based on a chain reaction of free radicals (26), three distinct theoretical approaches have been adopted to formulate the mechanism of the partial oxidation of methane to formaldehyde on oxide catalysts. One consists in a classical Langmuir-Hinshelwood model, where all the reactions take place on the surface (27). It is assumed that the reacting molecules interact simultaneously with the catalyst surface and the reaction between such activated species gives rise to the formation of the reaction products without the participation of oxygen from the oxide lattice (28). Dowden et al. (27) pointed out that the dissociation of methane and the activation (i.e., interaction with surface activated oxygen species) require quite different sites. This led to the suggestion of using bifunctional catalysts (29,30). By contrast, Lunsford's group (31) claimed a stepwise mechanism involving the reduction of the oxide surface by the reacting CH$_4$ and its reoxidation by the oxidant N$_2$O. This cyclic mechanism invokes the formation of active O$^-$ ions on the surface and their interaction with CH$_4$ molecules to form methyl radicals which rapidly react with the surface yielding methoxyde complexes and then HCHO and/or CH$_3$OH. Therefore, the oxide lattice seems to supply the oxygen atoms of the oxygenated molecules. A third approach deals with the proposition of an heterogeneous-homogeneous mechanism *(11)*. Even if the HCHO is assumed to be formed on the surface, it essentially entails the generation of radicals on the surface and formation of products in gas phase. This last theory still presents some undefined points providing on the whole semiquantitative analyses of such processes *(11)*.

Therefore, two fundamental issues still remain under debate: *i)* the direct participation of lattice oxygen in the formation of selective oxidation products and *ii)* the role of the acidic properties of the catalysts on their reactivities *(32)*. In order either to answer the above questions or to understand the antithetic behaviours of the silica supported MoO_3 and V_2O_5 catalysts, basic insights into the reaction mechanism have been gained by comparing the catalytic behaviour of pure SiO_2, $4\%MoO_3/SiO_3$ and $5\%V_2O_5/SiO_2$ catalysts in the presence and absence of molecular O_2. These studies have been performed in a pulse microreactor by continuous scanning of the reaction mixture with QMS technique. In Figures 2-4 are shown the rates of formation of HCHO, CO, CO_2 in presence of O_2 in the T_R range 550-650°C. These data further support the superior activity and efficiency in HCHO production of the $5\%V_2O_5/SiO_2$ catalyst and the apparent poisoning effect exerted by the MoO_3 on the reactivity of the SiO_2 surface.

The amount of the products formed over the studied catalysts, in the presence and absence of molecular O_2, are listed in Table III. It is evident that the formation of the oxidation products is associated with the gas phase oxygen supply. Then, as the reaction rates in the mixture of reactant and in separate steps differ *(19)*, these data exclude the participation of lattice oxygen in the partial oxidation of methane via a two step redox mechanism as main reaction pathway proving the occurrence of a "concerted mechanism".

Since the acid-base interactions of reagents (or products) with the catalyst could in principle affect the reaction pathway *(32)*, the acidic properties of bare SiO_2, $4\%MoO_3/SiO_2$ and $5\%V_2O_5/SiO_2$ have been comparatively evaluated by ZPC and NH_3-TPD measurements. The results are reported in Table IV in terms of ZPC value, amount of NH_3 uptake and temperature of NH_3 desorption peak maximum (T_M). The bare SiO_2 surface, after treatment at 600°C either in O_2/He or $CH_4/O_2/He$ atmosphere, is unable to adsorb NH_3. However, it is evident that both $4\%MoO_3/SiO_2$ and $5\%V_2O_5/SiO_2$ are considerably more acidic that the bare SiO_2 sample, with $4\%MoO_3/SiO_2$ resulting the most acidic system. These evidences do not find any correspondence in the reactivity scale, hence the change in the acidic properties cannot be the reason for the difference in the catalytic activity of the studied systems.

On the other hand, it is generally accepted that the redox properties of the selective oxidation catalysts control the oxygen activation as well as the surface stabilization of the oxygen activated species and their reactivity *(19)*. In particular, the stabilization of active oxygen forms requires the presence of reduced sites on the surface. In fact, the peculiar behaviour of Mo, V and Fe oxides in selective oxidation reactions is strictly linked with the stabilization of reduced states *(19)*.

This point has stimulated a growing interest in providing correlation between the degree of reduction *(32)* or the extent of reduced sites *(20)* and the reactivity in

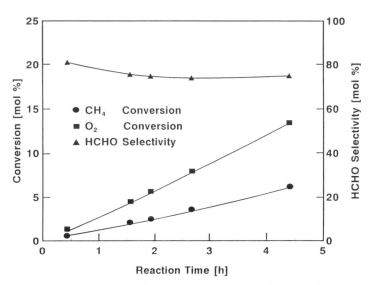

Figure 1. Batch reactor test of methane partial oxidation on SiO_2 catalyst at 600°C.

Reaction conditions: P_R, $1.7 \cdot 10^2$ kPa; recycle flow rate, 1000 Ncm$^3 \cdot$min^{-1}; W_{cat}, 0.05 g; reaction mixture composition (mmol): CH_4, 18; O_2, 9; N_2, 18 and He, 53.

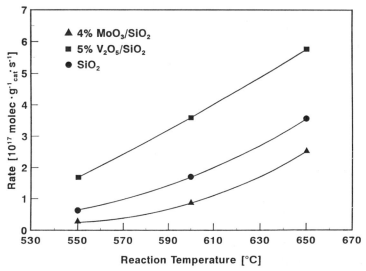

Figure 2. Methane Partial Oxidation. Rate of HCHO formation on unpromoted SiO_2, $5\%V_2O_5/SiO_2$ and $4\%MoO_3/SiO_2$ catalysts.

Reaction conditions: P_R, $1.2 \cdot 10^2$ kPa; He carrier flow, 50 Ncm$^3 \cdot$min^{-1}; W_{cat}, 0.05 g; pulse, 5.5 Ncm^3 of CH_4, O_2 and He mixture ($CH_4/O_2/He=2/1/7$).

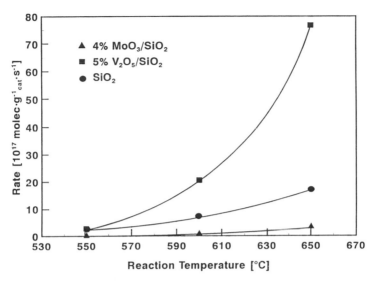

Figure 3. Methane Partial Oxidation. Rate of CO formation on unpromoted SiO$_2$, 5%V$_2$O$_5$/SiO$_2$ and 4%MoO$_3$/SiO$_2$ catalysts.
For reaction conditions see Figure 2.

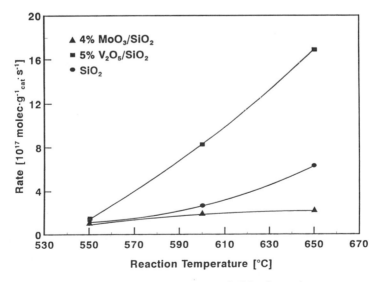

Figure 4. Methane Partial Oxidation. Rate of CO$_2$ formation on unpromoted SiO$_2$, 5%V$_2$O$_5$/SiO$_2$ and 4%MoO$_3$/SiO$_2$ and catalysts.
For reaction conditions see Figure 2.

Table III. Influence of the Gas Phase Oxygen on the Products Formation in
Methane Partial Oxidation over SiO_2, 4% MoO_3/SiO_2 and 5% V_2O_5/SiO_2
Catalysts. Pulse Reactor Data [a]

Catalyst	T_R (°C)	Reagents	Product Amount (10^{17} molec/pulse)		
			HCHO	CO	CO_2
	550	$CH_4 + O_2$	0.21	0.87	0.41
		CH_4	-	-	-
SiO_2	600	$CH_4 + O_2$	0.56	2.30	0.87
		CH_4	-	-	-
	650	$CH_4 + O_2$	1.18	5.60	2.11
		CH_4	-	0.25	-
	550	$CH_4 + O_2$	0.10	0.16	0.36
		CH_4	-	-	-
4% MoO_3/SiO_2	600	$CH_4 + O_2$	0.30	0.35	0.65
		CH_4	-	-	-
	650	$CH_4 + O_2$	0.85	1.20	0.75
		CH_4	-	-	-
	550	$CH_4 + O_2$	0.56	0.82	0.46
		CH_4	-	-	-
5% V_2O_5/SiO_2	600	$CH_4 + O_2$	1.20	6.68	2.76
		CH_4	-	0.45	-
	650	$CH_4 + O_2$	1.92	25.50	5.60
		CH_4	0.10	2.40	1.10

[a] W_{cat}, 0.05g; He carrier flow, 50 $Ncm^3 \cdot min^{-1}$; pulse, 5.5 Ncm^3 of (i) CH_4 and He mixture ($CH_4/He=2/7$) and (ii) CH_4, O_2 and He mixture ($CH_4/O_2/He=2/1/7$).

selective oxidation of light alkanes. On this account, we have estimated the density of reduced sites (number of reduced sites per g of catalyst, s_r / g_{cat}) by oxygen chemisorption after treating the catalyst sample under reaction conditions. It is worth noting that this approach allows the probing of the surface redox properties of "working catalysts" under steady state conditions. Care was taken in avoiding a bulk reduction of the oxide catalyst (21). Separate experiments have indicated that the extent of O_2 uptake depends upon the chemisorption temperature. Therefore the estimation of reduced active sites after treating the catalyst in atmospheres different from the reaction mixture as well as the evaluation of the O_2 chemisorption at temperatures different from that of reaction could lead to a very erratic assessment of the amount of reduced sites. As shown in Figure 5, the density of reduced sites on 5% V_2O_5/SiO_2 catalyst is 4-10 times higher than that of bare SiO_2 and 4% MoO_3/SiO_2 catalysts. Then, according to Le Bars et al. (32), we have observed that the reaction mixture is effective in reducing to some extent the catalytic surface. The presence of reduced sites signals the capability of the catalyst surface in activating the gas phase oxygen *(19)*.

Table IV. Acidity Characterization of SiO_2, 4% MoO_3/SiO_2
and 5% V_2O_5/SiO_2 Catalysts

Sample	*ZPC*	*Treatment at 600°C*	$\mu mol_{NH3} \cdot g_{cat}^{-1}$	$T_M(°C)$
			NH₃-TPD	
SiO_2	4.7	O_2	-	-
		$CH_4 + O_2$	-	-
$4\%MoO_3/SiO_2$	3.0	O_2	61.6	198
		$CH_4 + O_2$	59.3	194
$5\%V_2O_5/SiO_2$	3.1	O_2	20.1	185
		$CH_4 + O_2$	10.3	188

The rates of oxygen conversion on the studied catalysts in the T range 550-650°C are shown in Figure 6. This parameter corresponds to the sum of the rates of products formation or in other words to the overall rate of methane conversion. These values, even if obtained with a transient technique (pulse reactor data), are in satisfactory agreement with those obtained with the batch and continuous reactors (see Tables I and II). However, by comparing the data presented in Figure 5 and 6, it emerges that the trends of oxygen conversion (Fig. 6) are similar to those of the density of the reduced sites (Fig. 5). This indicates that the reactivity of SiO_2, $4\%MoO_3$ and $5\%V_2O_5$ promoted SiO_2 catalysts is mainly controlled by the amount of reduced sites. Assuming that the methane partial oxidation occurs according to a concerted mechanism, it follows that an active catalyst must explicate concomitantly two actions: *i)* the activation of the CH_4 molecule and *ii)* the activation of the gas phase O_2. Then, taking into account the above experimental evidences and the literature data *(33)* dealing with the suitability of the SiO_2 surface in activating CH_4 molecules, the following reaction scheme seems adequate to describe the partial oxidation of methane on the bare SiO_2 surface:

$$CH_4 \; + \; L \; \rightleftharpoons \; (CH_4)^* \qquad (1)$$

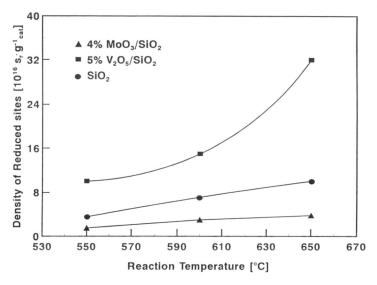

Figure 5. Density of reduced sites of SiO_2, $5\%V_2O_5/SiO_2$ and $4\%MoO_3/SiO_2$ catalysts in steady state conditions.

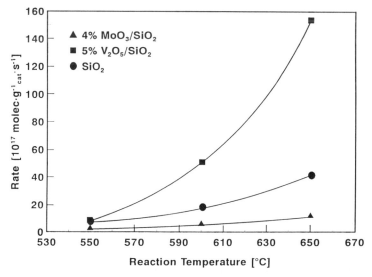

Figure 6. Rate of oxygen conversion on SiO_2, $5\%V_2O_5/SiO_2$ and $4\%MoO_3/SiO_2$ catalysts.
For reaction conditions see Figure 2.

$$O_2 \;+\; R \;\rightleftharpoons\; 2(O)^* \qquad\qquad (2)$$
$$CH_4^* \;+\; O^* \;\longrightarrow\; P^* \qquad\qquad (3)$$
$$P^* \qquad\qquad\;\; \longrightarrow\; P + L + R \qquad (4)$$

where $(CH_4)^*$ and $(O)^*$, indicate CH_4 and O_2 surface activated species respectively, L and R represent specific SiO_2 centres for CH_4 activation and reduced site respectively, while P refers to HCHO and CO_2. It can be also envisaged that the acidic properties of silica are quite sufficient in promoting the CH_4 activation, while its redox properties are rather weak. Therefore, for the bare SiO_2 catalyst it can be argued that the rate determining step is the oxygen activation. This statement has been confirmed by performing additional experiments on the catalytic behaviour of several commercial SiO_2 samples *(12)* prepared by different methods (pyrolysis, sol-gel, precipitation and extrusion). In fact, a linear relationship between the density of reduced sites and the reactivity of such SiO_2 samples has been found *(34)*.

The higher activity of the $5\%V_2O_5/SiO_2$ catalyst can be explained in terms of higher density of reduced site and of stabilization of reduced state of vanadium ions *(8, 14-15, 19, 21, 32)*. Such features enhance the rate of oxygen activation and likely favour the stabilization of active oxygen forms providing the formation of selective oxidation products *(15, 18-19, 27)*. On the contrary, MoO_3 lowers the "degree of reduction" of the SiO_2 surface and correspondingly the rate of oxygen activation (Figure 6) and the activity of the catalyst (Table I). In order to explain this evidence, it can be inferred that on the silica surface molybdenum ions exist in an octahedral coordination *(13)* which assists the stabilization of the highest oxidation state (Mo^{6+}).

Conclusions

- The partial oxidation of methane to formaldehyde with molecular O_2 over unpromoted SiO_2 and $4\%MoO_3$ or $5\%V_2O_5$ promoted SiO_2 catalysts, in the T range 550-650°C, proceeds via a concerted mechanism.
- The functionality of the SiO_2 surface towards the formation of HCHO is significantly promoted by V_2O_5, while it is depressed by the MoO_3.
-The stronger acidity of $4\%MoO_3/SiO_2$ and $5\%V_2O_5/SiO_2$ catalysts with respect to the unpromoted SiO_2 support seems to exert no direct influence on the reaction pathway of the methane partial oxidation.
- A direct relationship exists between the density of reduced sites and the reactivity of such oxide catalysts in the partial oxidation of methane to formaldehyde.

Acknowledgments

The financial support of this work by the Consiglio Nazionale delle Ricerche (Roma) "Progetto Finalizzato Chimica Fine II" is gratefully acknowledged.

Literature cited

1 Kastanas, G.N.; Tsigdinos, G.A.; Schwank, J. *Appl. Catal.* **1988**, *44*, 33
2 Khan, M.M.; Somorjai, G.A. *J. Catal.* **1985**, *91*, 263
3 Spencer, N.D. *J. Catal.* **1988**, *109*, 187
4 Zhen, K.J.; Khan, M.M.; Mak,C.H.; Lewis, K.B.; Somorjai, G.A. *J.Catal.* **1985**, *94*, 501
5 Spencer, N.D.; Pereira, C.J. *J. Catal.* **1989**, *116*, 399
6 Kasztelan, S.; Moffat, J.B. *J. Catal.* **1987**, *106*, 512
7 Barbaux, Y.; Elamrani, A.R.; Payen, E.; Gengembre, L.; Bonnelle, J.P.; Grzybowska, B. *Appl. Catal.* **1988**, *44*, 117
8 Kennedy,M.; Sexton,A.; Kartheuser,B.; Mac Giolla Coda,E.; Mc Monagle, J.B.; Hodnett,B.K. *Catal. Today* **1992**, *13*, 447
9 Brown, M.J.; Parkyns, N.D. *Catal .Today* 1991, *8*, 305
10 Baldwin, T.R.; Burch, R.; Squire, G.D.; Tsang, S.C. *Appl. Catal.* **1991**, *74*, 137
11 Garibyan, T.A.; Margolis, L. YA. *Catal. Rev.- Sci. Eng.* **1989-90**, *31* (4),355
12 Parmaliana, A.; Frusteri, F.; Miceli, D.; Mezzapica, A.; Scurrell, M.S.; Giordano, N. *Appl. Catal.* **1991**, *78*, L7
13 Liu, T.; Forissier, M.; Coudurier, G.; Védrine, J.C. *J. Chem. Soc., Faraday Trans.* I **1989**, *85* (7), 1607
14 Schraml-Marth, M.; Wokaun, A.; Pohl, M.; Krauss, H.L. *J.Chem. Soc., Faraday Trans.* **1991**, *87* (16), 2635
15 Centi,G.; Perathoner,S.; Trifirò, F.; Aboukais,A.; Aïssi,C.F.; Guelton,M. *J. Phys.Chem.*1992, *96*, 2617
16 Shaprinskaya,T.M.; Korneichuk,G.P.; Stasevich,V.P. *Kinet.Katal.* **1970**, *11*, 139
17 Erdöhelyi,A.; Solymosi,F. *J.Catal.* **1991**, *129* 497
18 Yoshida,S.; Matsuzaki,T.; Ishida,S.; Tarama,K. *Proceedings of the Fifth International Congress on Catalysis*, Preprint paper no 76, Amsterdam, **1972** (North Holland Publishing Co., Amsterdam)
19 Sokolovskii, V.D. *Catal. Rev.-Sci. Eng.* **1990**, *32* (1-2), 1
20 Oyama, S.T.; Somorjai, G.A. *J. Phys. Chem.* **1990**, *94*, 5022
21 Oyama, S.T.; Went, G.T.; Lewis, K.B.; Bell, A.T.; Somorjai, G.A. *J. Phys. Chem.* **1989**, *93*, 6786
22 Desikan, A.N.; Huang, L.; Oyama, S.T. *J. Phys. Chem.* **1991**, *95*, 10050
23 Parmaliana, A.; Frusteri, F.; Miceli, D.; Mezzapica, A.; Arena, F.; Giordano, N. Pending Italian Patent Application, June **1992**.
24 Parks, G.A. *J.Phys. Chem.* **1962**, *66*, 967
25 Miceli, D.; Arena, F.; Frusteri, F.; Parmaliana, A. *J.Chem.Soc., Chem.Comm.* **1992** (submitted).
26 Pitchai, R.; Klier, K. *Catal. Rev-Sci. Eng.* **1986**, *28* (1), 13
27 Dowden, D.A.; Schnell, C.R.; Walker, G.T. *Proceedings of the Fourth International Congress on Catalysis,* Moscow **1968**, 201
28 Kazanskii, V.B. *Kinet.Katal.* **1973**, *14* (1), 95
29 Stroud, H.J.F. U.K. Patent 1,398,385, **1975**

30 Otsuka, K.; Hatano, M. *J. Catal.* **1987**, *108*, 252

31 Liu, H.F.; Liu, R.S.; Liew, K.Y.; Johnson, R.E.; Lunsford, J.H. *J.Am. Chem.Soc.* **1984**, *106*, 4117

32 Le Bars, J.; Vedrine, J.C.; Auroux, A.; Pommier, B.; Pajonk, G.M. *J. Phys. Chem.* **1992**, *96*, 2217

33 Low, M.J.D. *J. Catal.* **1974**, *32*, 103

34 Parmaliana, A.; Sokolovskii, V.; Miceli, D.; Arena, F.; Giordano, N. in preparation.

RECEIVED October 30, 1992

Chapter 5

Oxygenation Reactions Catalyzed by Supported Sulfonated Metalloporphyrins

Bernard Meunier and Sandro Campestrini

Laboratoire de Chimie de Coordination du Centre National de la Recherche Scientifique, 205 route de Narbonne, 31077 Toulouse Cédex, France

A new trend in the field of oxidations catalyzed by metalloporphyrin complexes is the use of these biomimetic catalysts on various supports: ion-exchange resins, silica, alumina, zeolites or clays. Efficient supported metalloporphyrin catalysts have been developed for the oxidation of peroxidase-substrates, the epoxidation of olefins or the hydroxylation of alkanes.

After a decade on the modeling of cytochrome P-450 with soluble metalloporphyrin complexes associated with different oxygen atom donors (1,2) (iodosylbenzene, hypochlorite, organic or inorganic peroxides, molecular oxygen in the presence of a reducing agent), a new trend in this field is the development of supported metalloporphyrin catalysts. The use of supported catalysts has the following advantages: (i) facile catalyst recovery and (ii) physical separation of active sites by dispersion on the support to avoid self-destruction of the catalyst. Metalloporphyrins can be attached to polymers by covalent links (3), but this approach requires a multi-step synthesis of the porphyrin ligand and/or a chemical modification of the polymer. The metalloporphyrin can also be held by basic residues (imidazole or pyridine) acting as axial ligands, but in this latter case, the fixation is reversible (4).

Another approach is to immobilize metalloporphyrins on (i) inorganic materials (5) (silica, alumina, zeolites (6) or clays) in order to have inert supports for catalytic oxidation reactions or (ii) on organic polymer like ion-exchange resins (7). We recently used sulfonated metalloporphyrins supported on cationic ion-exchange resins in the modeling of ligninase, a peroxidase involved in the oxidative degradation of lignin in wood (8). These peroxidase models are also highly efficient in the oxidation of recalcitrant pollutants (9) like DDT or lindane and they can be used to study the *in vitro* metabolism of drugs (10). Two main advantages of these resin-metalloporphyrin

0097–6156/93/0523–0058$06.00/0

catalysts are that they can be used in aqueous solutions and, because of the large number of available ion-exchange resins, it is possible to design highly sophisticated catalysts based on the understanding at the molecular level of interactions between the metalloporphyrin and the support. For example, in the case of polyvinylpyridine polymer, a "proximal effect" (*i.e.* the strong coordination of a basic ligand to the metalloporphyrin) is provided by pyridine units of the polymer whereas the remaining residues, after protonation or methylation, are involved in electrostatic interactions with sulfonato groups of the porphyrin ligand (11).

We recently reported the use of robust sulfonated iron and manganese porphyrins supported on polyvinylpyridinium polymers in olefin epoxidation and alkane hydroxylation reactions (12). Sulfonated derivatives of *meso*-tetramesitylporphyrin, TMPS, *meso*-tetrakis(2,6-dichlorophenyl)porphyrin, TDCPPS, *meso*-tetramesityl-β-octabromoporphyrin Br$_8$TMPS and *meso*-tetrakis(3-chloro-2,4,6-trimethylphenyl)-β-octachloroporphyrin, Cl$_{12}$TMPS (see Scheme I for structures). TDCPPS and Cl$_{12}$TMPS are a mixture of atropoisomers and were used as such. The manganese and iron derivatives of the non-sulfonated version of Br$_8$TMPS and Cl$_{12}$TMPS have been used as soluble catalysts in oxygenation reactions performed in liquid phase (2e,13).

Preparation of supported metalloporphyrin on polyvinylpyridine polymers. Free polyvinylpyridine is obtained from an alkaline treatment of poly[4-vinylpyridinium (toluene-4-sulfonate)] crosslinked with 2% of divinylbenzene. This polymer contains 3.5 mmol of pyridine units per g of resin. Three different types of supported catalysts have been prepared: (i) by direct attachment of the manganese complex to the polymer by coordination of one pyridine unit to the complex, (ii) by further protonation of the remaining pyridine sites and creation of electrostatic interactions between polymer pyridiniums and the sulfonato groups of the metalloporphyrin and (iii) by methylation of the free pyridine units (instead of protonation) (see Scheme II). For the last two cases, [(M-Porp-S)-PVPH$^+$][TsO$^-$] and [(M-Porp-S)-PVPMe$^+$][TsO$^-$], there is an additional electrostatic interactions between the pyridinium units of the polymer and the sulfonato groups of the metalloporphyrin, assuring a second strong binding mode of the catalyst to the support. The axial ligation by pyridine in these PVP supported metalloporphyrins has been confirmed by UV-visible spectroscopy (12a).

These PVP polymers provide a "proximal effect" without addition of free pyridine in the reaction mixture. Different studies have shown that only one pyridine per manganese catalyst is sufficient to enhance the rate of the catalytic oxygen atom transfer from the high-valent metal-oxo species to the organic substrate. The advantage of PVP polymer over a cationic Amberlite resin (see Scheme II for structures) have been recently illustrated in the modeling of ligninase (11).

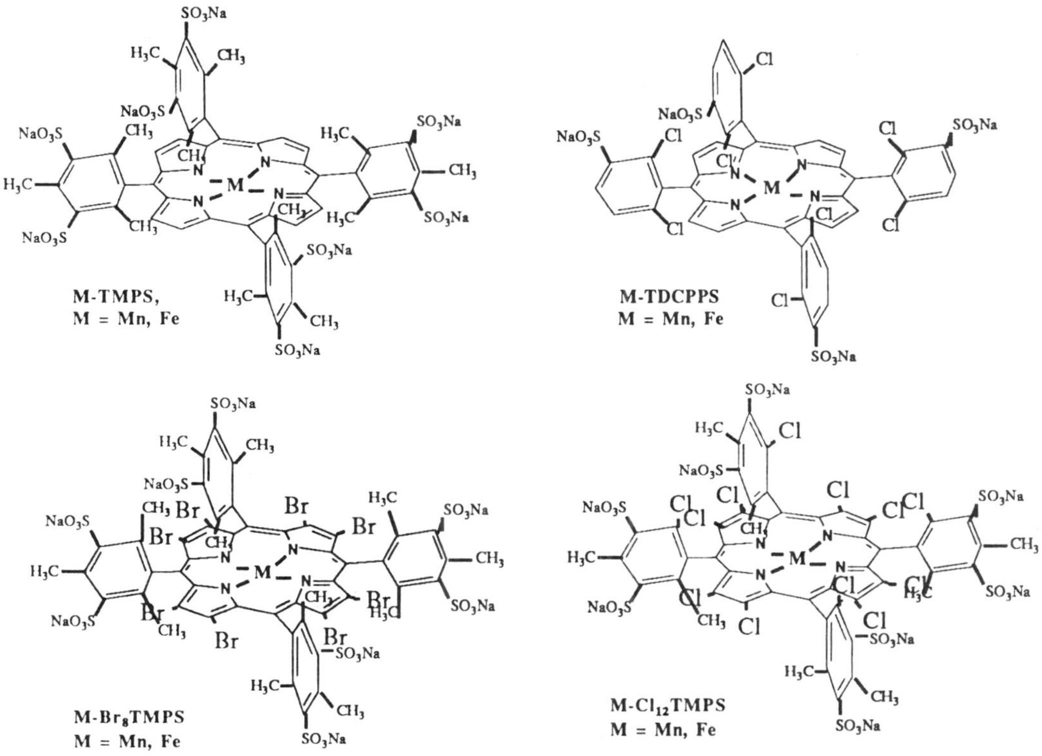

M-TMPS, M = Mn, Fe

M-TDCPPS M = Mn, Fe

M-Br$_8$TMPS M = Mn, Fe

M-Cl$_{12}$TMPS M = Mn, Fe

Scheme I. Structures of sulfonated robust metalloporphyrin complexes. The axial ligand of these water-soluble metalloporphyrins depend on the pH value, at acidic pH an hydroxo ligand occupies the axial position whereas it is a water molecule at basic pH values.

Cyclooctene epoxidation catalyzed by supported sulfonated metalloporphyrins.

We investigated the behavior of PhIO as oxygen donor with these catalysts supported on PVP. The results of the PhIO epoxidation of cyclooctene catalyzed by [MnTMPS-PVPMe$^+$][TsO$^-$], [MnTDCPPS-PVPMe$^+$][TsO$^-$], [MnBr$_8$TMPSPVPMe$^+$] [TsO$^-$] and [MnCl$_{12}$TMPS-PVPMe$^+$] [TsO$^-$] are reported in Figure 1. No epoxide formation is detected when PhIO is used with PVP or methylated PVP (PhIO is not decomposed by these polymers). But for a low loading of methylated PVP by the four sulfonated manganese porphyrin complexes and a molar ratio catalyst /substrate equal to 1.3%, an efficient catalytic epoxidation of cyclooctene is observed: 90% of the olefin is converted by [MnCl$_{12}$TMPS-PVPMe$^+$][TsO$^-$] in 7 h at room temperature (the turnover rate of this reaction is 18 cycles/h, based on the first 50% of olefin conversion). The two best catalysts are halogenated on the pyrrolic β-positions, the less active being the PVP-supported MnTDCPPS, just below the activity of the analogue catalyst based on tetramesitylporphyrin.

Supported iron porphyrins are less reactive than the corresponding manganese derivatives in the PhIO epoxidation of cyclooctene. 80% of olefin conversion was reached with MnBr$_8$TMPS-PVP in 2 h, whereas only 10% was obtained with FeBr$_8$TMPS-PVP in 6 h.

Recycling experiments with MnCl$_{12}$TMPS immobilized on methylated PVP. Recycling experiments were performed by re-using four times the same MnCl$_{12}$TMPS catalyst supported on methylated PVP in the PhIO epoxidation of cyclooctene. For [MnCl$_{12}$TMPS-PVPMe$^+$][TsO$^-$], it should be noted that the first two cycles are exactly the same, only a small activity decrease is observed for the third run. With 1 μmol of MnCl$_{12}$TMPS immobilized on [PVPMe$^+$][TsO$^-$] the total epoxide production after 4 runs is 1970 μmol. The overall selectivity based on PhIO is 65% .

Adamantane hydroxylation by PhIO catalyzed by MnCl$_{12}$TMPS supported on PVP polymers. First of all, the different forms of PVP supported MnCl$_{12}$TMPS were compared to the same manganese porphyrin immobilized on Amberlite-IRA-900 (Table I). The distribution of adamantane derivatives were analyzed after 2 h and 7 h of reaction at room temperature. [MnCl$_{12}$TMPS-PVPMe$^+$][TsO$^-$] gave the highest selectivity ratio for adamantanols versus adamantanone: 53, indicating that the hydroxylation reaction is actually the main oxidation reaction with this catalyst. With the same catalyst, the product distribution did not change after 2 h at room temperature, 312 μmoles of adamantanols and adamantanone were produced within 2 h compared to 322 μmoles in 7 h. Amberlite is not suitable compared to methylated PVP

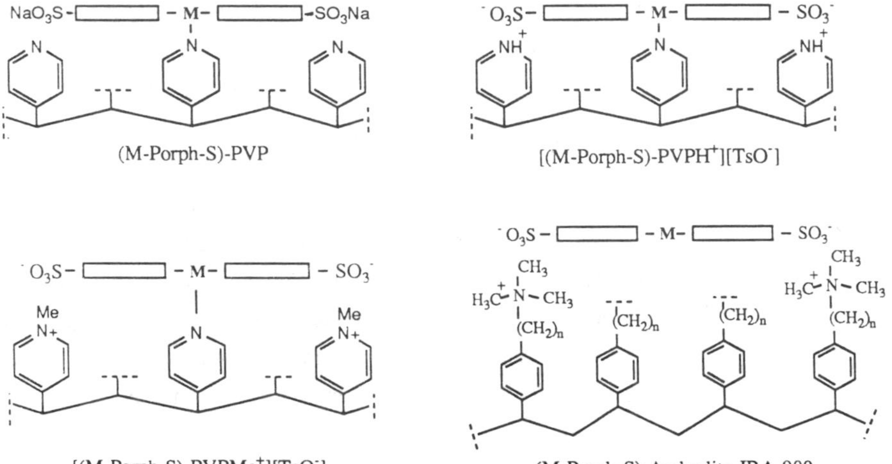

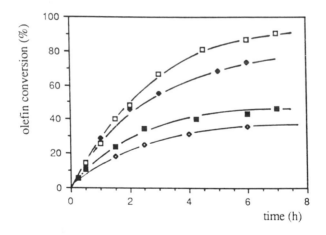

(M-Porph-S)-PVP

[(M-Porph-S)-PVPH⁺][TsO⁻]

[(M-Porph-S)-PVPMe⁺][TsO⁻]

(M-Porph-S)-Amberlite-IRA-900

Scheme II. Schematic representation of sulfonated metalloporphyrins supported on different polyvinylpyridine polymers and Amberlite-IRA-900.

Figure 1. PhIO oxidation of cyclooctene catalyzed by various sulfonated manganese porphyrin immobilized on methylated PVP, [Mn-porphyrin-S-PVPMe⁺][TsO⁻]. [MnTMPS-PVPMe⁺][TsO⁻] (◇), [MnBr₈TMPS-PVPMe⁺][TsO⁻] (◆), [MnTDCPPS-PVPMe⁺][TsO⁻] (■) and [MnCl₁₂TMPS-PVPMe⁺][TsO⁻] (□). Conditions: cyclooctene (150 µmol), PhIO (750 µmol), Mn-porphyrin-S (2 µmol) immobilized on [PVPMe⁺][TsO⁻] (200 mg of PVP treated by TsOMe) in 3 mL of dichloromethane at room temperature. (Reproduced from ref. 12a. Copyright 1992 American Chemical Society.)

(or even PVP) as support. Only 50 μmoles of adamantanols and adamantanone were formed by $MnCl_{12}TMPS$-Amberlite in 7 h compared to 322 μmoles with the same catalyst on methylated PVP.

Final Remarks. The main features of these PhIO oxygenations reactions catalyzed by manganese and iron porphyrins supported on cationic ion-exchange resins are the following:

(i) the polyvinylpyridinium resins lead to better catalysts than other simple cationic resins, because of the proximal effect due to the coordination of a pyridine unit arising from the polymer. In addition to the known favorable role of the proximal pyridine on the different steps of the oxidant activation by the metalloporphyrin (rate enhancement of both metal-oxo formation and oxygen atom transfer steps), the double interaction of sulfonated metalloporphyrin with poly(vinylpyridinium) polymers (pyridine proximal effect and sulfonato-pyridinium interactions) immobilizes the catalyst more strictly on the support than simple electrostatic interactions on classical cationic resins. In this latter case, there is probably an equilibrium between two possible types of interactions: the "coating" mode (A in Scheme III) and the "stacking" mode (B in Scheme III). In mode B, the stacking of metalloporphyrin complexes enhances the bleaching of the supported catalyst when the highly reactive metal-oxo species are formed. This phenomenon was also observed by Lindsay Smith in the case of a cationic iron porphyrin catalyst supported on anionic polymers (12b) and by our group for cationic manganese porphyrins when interacting with DNA (outside binding mode *versus* minor groove interaction) (14).

(ii) these polyvinylpyridinium supported catalysts have similar efficiency compared to the corresponding metalloporphyrin complexes in solution.

(iii) the most efficient supported metalloporphyrin catalysts for oxygenation reactions are those with manganese as central metal and with halogen atoms on β-pyrrole positions.

(iv) the best PVP-manganese catalysts can be recycled three or four times and more than 2000 cycles can be achieved in the epoxidation of cyclooctene before a significant loss of catalytic activity.

In conclusion, *the concept of the proximal effect is a key factor in metalloporphyrin-catalyzed reactions, not only for soluble complexes, but also for supported catalysts.*

Table I. Oxidation of adamantane by PhIO catalyzed by $MnCl_{12}TMPS$ immobilized on different supports [a]

Support	Products (μmol) (after 7 h)				ols/one ratio	Yield/PhIO (in %)
	Ad-1-ol	Ad-2-ol	Ad-2-one	Total		
PVP	145	82	11	238	21	33 [e]
[PVPH+][TsO-]	131	60	8	199	24	28
[PVPMe+][TsO-]	166 (159)[d]	150 (147)	6 (6)	322 (312)	53 (51)	44 (43)
[PVPMe+][TsO-][b]	120 (120)	64 (64)	6 (6)	190 (190)	31 (31)	26 (26)
Amberlite-IRA-900	30	17	3	50	16	7
Amberlite-IRA-900 [c]	28	12	3	43	13	6

[a] Conditions: adamantane (3860 μmol), PhIO (750 μmol), $MnCl_{12}TMPS$ (1 μmol) immobilized on 100 mg of PVP or 250 mg of protonated (or methylated) PVP, or on 250 mg of Amberlite-IRA-900, in 7 mL of CH_2Cl_2 under magnetic stirring, at room temperature. [b] m-Cl-PhCO$_3$H was used as oxidant (750 μmol). [c] runned in the presence of pyridine (100 μmol). [d] data after only 2 h at room temperature. [e] these data correspond to the molar ratio Ad-1-ol + Ad-2-ol + 2 x Ad-2-one / PhIO in %.

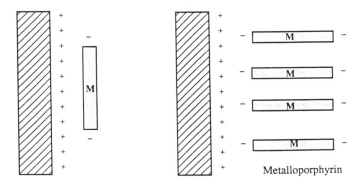

Ion-exchange resin

Coating Mode (A)   **Stacking Mode (B)**

Scheme III. Two possible mode of interactions (coating *versus* stacking) of anionic metalloporphyrins with a cationic ion-exchange resin.

Acknowledgments. This work was supported by a 'Stimulation' grant from the EEC, including a fellowship for S. C. (on leave from Padova University). Dr Anne Robert is gratefully acknowledged for fruitful discussions throughout this work.

References

(1) For review articles on metalloporphyrin-catalyzed reactions, see: (a) McMurry, T. J.; Groves, J. T. in *Cytochrome P-450: Structure, Mechanism and Biochemistry*; Ortiz de Montellano, P., Ed.; Plenum Press: New York, **1986**, Chapter I. (b) Meunier, B. *Bull. Soc. Chim. Fr.* **1986**, 578-594. (c) Montanari, F.; Banfi, S.; Quici, S. *Pure Appl. Chem.* **1989**, *61*, 1631-1636. (d) Mansuy, D. *Pure Appl. Chem.* **1990**, *62*, 741-746. (e) Meunier, B. *Chem. Rev.* in press.

(2) For some recent articles on metalloporphyrin-catalyzed oxidations, see: (a) Brown, R. B.; Hill, C. L. *J. Org. Chem.* **1988**, *53*, 5762-5768. (b) Robert, A.; Meunier, B. *New J. Chem.* **1988**, *12*, 885-896. (c) Groves, J. T.; Viski, P. *J. Am. Chem. Soc.* **1989**, *111*, 8537-8538. (d) Traylor, T. G.; Hill, K. W.; Fann, W. P.; Tsuchiya, S.; Dunlap, B. E. *ibid.* **1992**, *114*, 1308-1312. (d) Collman, J. P.; Brauman, J. I.; Hampton, P. D.; Tanaka, H.; Scott Bohle, D.; Hembre, R. T. *J. Am. Chem. Soc.* **1990**, *112*, 7980-7984. (e) Hoffmann, P.; Labat, G.; Robert, A.; Meunier, B. *Tetrahedron Lett.* **1990**, *31*, 1991-1994. (f) Murata, K.; Panicucci, R.; Gopinath, E.; Bruice, T. C. *J. Am. Chem. Soc.* **1990**, *112*, 6072-6083. (g) Ellis, P. E.; Lyons, J. E. *Coord. Chem. Rev.* **1990**, *105*, 181-193. (h) Robert, A.; Loock, B.; Momenteau, M.; Meunier, B. *Inorg. Chem.* **1991**, *30*, 706-711.

(3) (a) Leal, O.; Anderson, D. L.; Bowman, R. G.; Basolo, F.; Burwell, R. L. *J. Am. Chem. Soc.* **1975**, *97*, 5125-5129. (b) Tatsumi, T.; Nakamura, M.; Tominaga, H. *Chem. Lett.* **1989**, 419-420. (c) Rollmann, L. D. *ibid.* **1975**, *97*, 2132-2136.

(4) Tsuchida, E.; Honda, K.; Hasegawa, E. *Biochem. Biophys. Acta* **1975**, *393*, 483-495 and references therein.

(5) (a) Fukuzumi, S.; Mochizuki, S.; Tanaka, T. *Israel J. Chem.* **1987/88**, *28*, 29-36. (b) Kameyama, H.; Suzuki, H.; Amano, A. *Chem. Lett.* **1988**, 1117-1120. (c) Barloy, L.; Battioni, P.; Mansuy, D. *J. Chem. Soc., Chem. Commun.* **1990**, 1365-1367. (d) Nakamura, M.; Tatsumi, T.; Tominaga, H. *Bull. Chem. Soc. Jpn.* **1990**, *63*, 3334-3336. (e) Barloy, L.; Lallier, J. P.; Battioni, P.; Mansuy, D.; Piffard, Y.; Tournoux, M.; Valim, J. B.; Jones, W. *New J. Chem.* **1992**, *16*, 71-80. (f) Bedoui, F.; Gutiérrez Granados, S.; Devynck, J.; Bied-Charreton, C. *ibidem* **1991**, *15*, 939-941.

(6) for zeolite-encapsulated iron phthalocyanine or manganese Schiff base complexes, see: (a) Herron, N.; Stucky, G. D.; Tolman, C. A. *J. Chem. Soc., Chem Commun.* **1986**, 1521-1522. (b) Herron, N. *J. Coord. Chem.* **1988**, *19*, 25-38. (c) Bowers, C.; Dutta, P. K. *J. Catal.* **1990**, *122*, 271-279.

(7) (a) Wöhrle, D.; Gitzel, J.; Krawczyk, G.; Tsuchida, E.; Ohno, H.; Okura, I.; Nishisaka, T. *J. Macromol. Sci. Chem.* **1988**, *A25*, 1227-1254. (b) Saito, Y.; Mifume, M.; Nakayama, H.; Odo, J.; Tanaka, Y.; Chikuma, M.; Tanaka, H. *Chem. Pharm. Bull.* **1987**, *35*, 869-872.

(8) Labat, G.; Meunier, B. *J. Org. Chem.* **1989**, *54,* 5008-5011. (b) Labat, G.; Meunier, B. *New J. Chem.* **1989**, *13*, 801-804.

(9) Labat, G; Seris, J. L.; Meunier, B. *Angew. Chem., Int. Ed. Engl.* **1990**, *29*, 1471-1473.

(10) Bernadou, J.; Bonnafous, M.; Labat, G.; Loiseau, P.; Meunier, B. *Drug Metab. Disp.* **1991**, *19*, 360-365.

(11) Labat, G.; Meunier, B. *C. R. Acad. Sci. Paris* **1990**, *311 II*, 625-630.

(12) (a) Campestrini, S.; Meunier, B. *Inorg. Chem.* **1992**, *31*, 1999-2006. (b) for an other recent report on olefin epoxidations by PhIO catalyzed by iron porphyrins supported on ion-exchange resins see also: Leanord, D. R.; Lindsay Smith, J. R. *J. Chem. Soc. Perkin Trans 2*, **1990**, 1917-1923 and same authors *J. Chem. Soc. Perkin Trans 2*, **1991**, 25-30. (c) The use of a manganese porphyrin bound to colloidal anion-exchange particles in the NaOCl epoxidation of styrene appeared in the literature during preparation of the present article: Turk, H.; Ford, W. T. *J. Org. Chem.* **1991**, *56*, 1253-1260.

(13) Hoffmann, P.; Robert, A.; Meunier, B. *Bull. Soc. Chim. Fr.* **1992**, *129*, 85-97.

(14) (a) Ding, L.; Bernadou, J.; Meunier, B. *Bioconjugate Chem.* **1991**, *2*, 201-206. (b) Pitié, M.; Pratviel, G.; Bernadou, J.; Meunier, B. *Proc. Natl. Acad. Sci. USA* **1992**, *89*, 3967-3971.

RECEIVED October 30, 1992

Chapter 6

Principles and New Approaches in Selective Catalytic Homogeneous Oxidation

Craig L. Hill, Alexander M. Khenkin, Mark S. Weeks, and Yuqi Hou

Department of Chemistry, Emory University, Atlanta, GA 30322

Goals and five limitations in conjunction with the development of selective catalytic homogeneous oxidation systems are evaluated. Systems are presented that address several of the problems or goals. One involves oxidation of alkenes by hypochlorite catalyzed by oxidatively resistant d-electron-transition-metal-substituted (TMSP) complexes. A second involves oxidation of alkenes by H_2O_2 catalyzed by specific TMSP complexes, and a third addresses functionalization of redox active polyoxometalate complexes with organic groups.

The focus of this chapter is on aspects, particularly mechanistic considerations and limitations, of transition metal-catalyzed controlled delivery of oxygen to hydrocarbon substrates. The points addressed here are pertinent to heterogeneous and homogeneous systems alike but are more defensible and applicable to homogeneous systems as the detailed molecular knowledge of the latter is at a higher level. General methodologies and practical, largely nonmechanistic, considerations in transition metal-catalyzed oxidation are given in several recent books and reviews (*1-6*). The selective catalytic oxidation of organic substrates or reactants, R, by O_2 and to a lesser extent, by more expensive and less desirable oxidants, OX, equation 1, offers great opportunities for modifications or innovations with both substantive scientific and economic import (*1-6*). The general goal at present is to develop catalytic

$$R + OX \xrightarrow{\text{catalyst}} RO + RED \qquad (1)$$

processes that are selective (chemo-, regio-, and stereoselective) at high conversion of substrate, stable, minimally expensive, and environmentally friendly (green).

General Thermodynamic and Mechanistic Features of Catalytic Homogeneous Oxidations

Unlike many of the other major types of catalytic processes involving the transformation of organic materials, oxidations are usually strongly exothermic which

0097–6156/93/0523–0067$06.00/0

dictates two limitations with respect to selectivity. The first derives from the fact that the more negative the ΔG^0, the smaller the separation in the activation energies (or the differences in the free energies of activation, $\Delta\Delta G^{\ddagger}$) tends to be for the process generating the desired product (RO), $\Delta G^{\ddagger}$, versus the other products (e.g. RO'), $\Delta G^{\ddagger'}$. Increasing the temperature to capitalize on the difference in the enthalpies of activation for the derived processes, $\Delta\Delta H^{\ddagger}$, may or may not help. Although selectivity will increase with temperature if $\Delta H^{\ddagger}$ and $\Delta H^{\ddagger'}$ are substantially different, higher temperatures often lead to a loss in selectivity as side processes generating by-products with higher enthalpies of activation than $\Delta H^{\ddagger}$ become more operative. These often include destruction of the catalyst or even the components of the catalytic reactor. Second, the selectivity of oxidation processes is usually such that the activation energy ($\Delta G^{\ddagger}$ or E_a) for formation of an initial product, often the desired product, RO, is less favorable than for formation of a subsequent product(s), RO' or RO'', etc. If this situation pertains, high selectivity at high conversion of substrate can only be achieved by chemical or physical separation of RO from the reaction at a rate significantly higher than the rate at which it is formed. Physical separation of RO is not uncommon in heterogeneous catalytic oxidation systems engineered with recycling capabilities. Physical or chemical separation of a desired product from a catalytic oxidation reaction is usually more difficult and less practical in moderate-to-large scale homogeneous processes than in heterogeneous processes.

A major subset of equation 1, and the focus of much recent research for intellectual and practical reasons, is transition metal-catalyzed transfer of oxygen from oxygen donors (Table I) to organic substrates, equation 2, commonly referred to as oxo transfer oxidation or "oxygenation" in much of the recent literature. Equation 2 can proceed by one or more of several possible mechanisms. Two types of mechanism for equation 2 are predominant in the absence of autoxidation, however. The first type involves heterolytic (nonradical) activation and transfer of activated oxygen, usually from an alkylhydroperoxide, to a substrate by a metal center. The

$$R \; + \; DO \; \xrightarrow[\text{catalyst}]{L_x M^{n+}} \; RO \; + \; D \qquad (2)$$

DO = oxygen donor oxidants (see Table I);
$L_x M^{n+}$ = d-electron transition metal center.

catalyst in these cases is usually an early transition metal ion in the d^0 electronic state and the substrate is usually ligated to the transition metal during the process. This type of mechanism is characterized by the metal center functioning in an electrophilic and nonredox role. It is operable in two of the most commercially significant homogeneous catalytic oxygenation processes today, the Halcon epoxidation process (4,6) and the Sharpless chiral epoxidation technology (7-9) (R in equation 2 is alkene). The second common type of mechanism for oxygen transfer involves the formation of high-valent oxometal intermediates which then transfer oxygen to the substrate. This type of mechanism involves redox changes in the metal center, may or may not involve preequilibrium association of substrate with the metal center, and in most if not all cases, involves a solvent caged radical pair intermediate. The degree of escape of the radical from the solvent cage and/or oxidation of the intermediate caged radicals in these "oxygen atom rebound" processes (10) varies with the metal and the particular organic radical intermediate involved (11). This second type of mechanism is involved in hydrocarbon oxygenation (alkene epoxidation and alkane

C-H bond hydroxylation) catalyzed by cytochrome P-450. This class of enzymes is the dominant vehicle for oxidation and modification of hydrocarbons and other xenobiotics of low reactivity in the biosphere and remains one of the most studied of all enzymes (*12*).

Oxidants

In the development of effective catalytic oxidation systems, there is a qualitative correlation between the desirability of the net or terminal oxidant, (OX in equation 1 and DO in equation 2) and the complexity of its chemistry and the difficulty of its use. The desirability of an oxidant is inversely proportional to its cost and directly proportional to the selectivity, rate, and stability of the associated oxidation reaction. The weight % of active oxygen, ease of deployment, and environmental friendliness of the oxidant are also key issues. Pertinent data for representative oxidants are summarized in Table I (*4*). The most desirable oxidant, in principle, but the one with the most complex chemistry, is O_2. The radical chain or autoxidation chemistry inherent in O_2-based organic oxidations, whether it is mediated by redox active transition metal ions, nonmetal species, metal oxide surfaces, or other species, is fascinatingly complex and represents nearly a field unto itself (*1,13*). Although initiation, termination, hydroperoxide breakdown, concentration dependent inhibition

Table I. Oxygen Donor Oxidants, DO

DO	% Active oxygen	By-product and comments
O_2	100	None, provided reducing agent-free non-radical-chain aerobic oxygenation can be achieved -- no commercially successful system yet exists.
H_2O_2	47	H_2O. Environmentally attractive.
O_3	33	O_2. Potentially environmentally attractive.
NaClO	21.6	NaCl. Although nontoxic, inorganic salt by-products are to be avoided in general. ClO^- can produce toxic and carcinogenic chlorocarbon by-products in some cases.
t-BuOOH	17.8	t-BuOH. Commercially important in catalyzed oxygenations.
$C_5H_{11}NO_2$[a]	13.7	$C_5H_{11}NO$.
$KHSO_5$	10.5	$KHSO_4$. Water compatible but generates marginally toxic salt.
C_6H_5IO	7.3	C_6H_5I. Metal catalyzed oxidations are often quite selective but cost is prohibitive.

[a] *N*-methylmorpholine *N*-oxide (MMNO).

by initiators and other kinetic aspects of metal-mediated autoxidation processes are relatively well delineated, other aspects of these reactions are not.

After O_2, the next most desirable oxidant is H_2O_2. It is relatively inexpensive, has a high weight % of active oxygen, and, by virtue of producing H_2O in formal oxo transfer or oxygenation processes (equation 2), is more environmentally friendly than other oxidants that produce either organic compounds or inorganic salt by-products. At the same time, metal-mediated oxidations by H_2O_2 are second in complexity to those by O_2. Organic hydroperoxides such as t-butylhydroperoxide (TBHP), cumylhydroperoxide (CHP), and ethylbenzene hydroperoxide (EBHP), and various other peroxygen compounds, such as the perborates, come next in complexity. One principal difference between oxidations by H_2O_2 and these latter peroxygen oxidants derives from the difference in the richness of the chemistry between the actual metal complex and the oxidant. H_2O_2 forms more structurally characterized types of complexes with both mononuclear or polynuclear d-electron-transition metal systems than do the organo hydroperoxides. Although some N-oxides such as N-methylmorpholine N-oxide are commercially attractive oxidants in some catalyzed organic oxidations, the other oxidants below the peroxy compounds in Table I are primarily of academic interest. As some expensive and complex oxo donor compounds such as oxaziridines, some N-oxides, and to a lesser extent iodosylbenzenes, react cleanly with transition metal complex catalysts, they have been involved in elucidating many key features of both biological and abiological organic oxidations catalyzed by these transition metal centers. These oxidants will doubtless remain of value for seminal mechanistic studies of metal-catalyzed oxidations for a time.

Catalytic Aerobic Oxygenation

One major goal in selective catalytic oxidation is embodied in equations 3-6 (the ligands in these equations have been omitted for clarity). This chemistry can be conveniently termed as aerobic oxygenation as oxygen atoms are formally being transferred in the substrate oxidation process and radical chain autoxidation is absent. Although the chemistry discussed here and below is applicable, in principle, to either heterogeneous or homogeneous catalytic oxidation systems, it is likely that homogeneous ones will be developed first since more electronic, structural, and other molecular features of the catalyst itself can be adequately evaluated and systematically modified. When systems are developed in which equations 3-6 dominate the observed reaction chemistry, the substrate oxygenation process, equation 6, is selective, and the general goals of catalytic processes cited above are met (high selectivity at high conversion of substrate, stability, environmental friendliness and economic feasibility), a new era in selective oxidation will have begun. These equations facilitate selective oxidation by using both atoms of dioxygen without consuming any reducing agent. Although various systems reported thus far have facilitated some aspects of this chemistry (equations 3-6) there have been a number of demonstrable limitations with even the most promising and provocative. In the next section, we outline 5 significant classes of problems that conspire to prevent development of effective and selective catalytic oxygenation systems.

$$M^{n+} + O_2 \longrightarrow M^{(n+1)+}O_2 \tag{3}$$

$$M^{(n+1)+}O_2 + M^{n+} \longrightarrow M^{(n+1)+}OOM^{(n+1)+} \tag{4}$$

$$M^{(n+1)+}OOM^{(n+1)+} \longrightarrow 2\,M^{(n+2)+}O \tag{5}$$

$$M^{(n+2)+}O + R \longrightarrow M^{n+} + RO \tag{6}$$

Problems and Limitations

Instability of Catalyst Under Turnover Conditions. There is substantial literature on the use of transition metal complexes with organic ligands as catalysts for oxidation, including oxygenation, of organic substrates. These studies have collectively contributed greatly to our understanding of the energetic and mechanistic features of these processes. At the same time, only a handful of these systems offer the potential to be practically useful for the simple reason that organic structure, including organic ligands, is thermodynamically unstable with respect to CO_2 and H_2O. As a consequence the ligands of the catalyst, L in equation 7, are inevitably oxidized (to LO in equation 7) and/or degraded extensively at some point and the catalyst is inactivated. Alternatively, another situation can arise that can be just as deleterious: the metal center with altered ligands, $(L)_{x-1}(LO)M$, retains some catalytic activity but it is no longer selective.

$$L_xM + OX \longrightarrow LO(H) + (L)_{x-1}(LO)M + RED \tag{7}$$

Formation of Oxo-bridged Dimers and Oligomers. A facile and pervasive type of process, and one that often leads to inactivation or significant attenuation in activity of the oxygenation catalysts, particularly under mild operating conditions, involves the trapping of oxometal intermediates by some form of the catalyst to generate bridging species, often μ-oxo dimers, equation 8. In some cases this type of dehydration condensation can proceed to oligomerized or polymerized transition metal species that are marginally active. Fortunately this type of process can often be controlled by the reaction conditions (for example, it is often quite pH dependent), it is often reversible, and it can be avoided altogether by using ligands that sterically preclude formation of such units.

$$MO + M \longrightarrow M\text{-}O\text{-}M \tag{8}$$

Formation of Dead-end Intermediate Oxidation States, Particularly $M^{(n+1)+}$. For oxygenation involving oxometal species, $M^{(n+2)+}O$, the regeneration mode in the catalytic cycle is substrate oxidation via equation 6. Occasional side reactions with M-based or other redox active species can lead to the intermediate oxidation state of the catalyst, $M^{(n+1)+}$. Equations 9-12 are routes to $M^{(n+1)+}$ in the presence of O_2 as the terminal oxidant. Although the relative likelihood of equations 9-12 will depend on the particular metal complex, M, these equations are given in approximate descending probability. Since $M^{(n+1)+}$ is often incapable of reentering the catalytic cycle, equations 13-15, its generation can constitute an effective termination of the catalysis.

$$[M^{(n+1)+}O_2] \longrightarrow M^{(n+1)+} + O_2^{\cdot-} \tag{9}$$

$$MOOX \longrightarrow M^{(n+1)+} + XOO^-, \quad X = M \text{ or } H \tag{10}$$

$$M^{(n+2)+}O + RED + X^+ \longrightarrow M^{(n+1)+} + XO^- + OX \tag{11}$$

$$M^{n+} + OX \longrightarrow M^{(n+1)+} + RED \tag{12}$$

$$M^{(n+1)+} + O_2 \longrightarrow \text{No Reaction} \tag{13}$$

$$M^{(n+1)+} + OX \longrightarrow M^{(n+2)+}O \quad (\text{often very slow}) \tag{14}$$

$$M^{(n+1)+} + RED \longrightarrow M^{n+} \quad (\text{often very slow}) \tag{15}$$

Inhibition by One or More Species in the Catalytic Cycle. Inhibition of the catalyst can be manifested through binding of some reactant species to the catalytic active site, such as favorable ligand binding equilibria, equation 16, leading to coordinatively saturated species, $L_{x+1}M^{n+}$. A similar binding of some intermediates to the catalyst could also have kinetically significant consequences. The most likely form of inhibition, however, is by interaction of one of the forms of the catalyst, often the resting form, L_xM^{n+}, with one or more of the products, sometimes the principal or desired product, RO (equation 17). The resulting adduct complexes, $L_x(RO)M^{n+}$, are unreactive under the reaction conditions.

$$L_xM^{n+} + L \rightleftharpoons L_{x+1}M^{n+} \tag{16}$$

$$L_xM^{n+} + RO \rightleftharpoons L_x(RO)M^{n+} \tag{17}$$

Diffusional Problems. The likelihood of diffusional problems will be increased as the active site of the catalyst becomes more sterically hindered and physically buried. These diffusional problems can derive simply from greatly inhibited mobility of reactants and products into and out of the active site, but they can also derive exclusively from diffusional escape of the products whose sizes or other properties affecting mobility are appreciably different from those of the reactant.

Approaches/New Systems

Sterically Hindered Metalloporphyrins Capable of Direct Aerobic Oxygenation. The catalytic aerobic olefin epoxidation system of Quinn and Groves, (tetramesitylporphyrinato)Ru/O₂/olefin substrate, effects equations 3-6, that is, the direct oxygenation of substrate using O_2 as the oxidant without consumption of reducing agent (*14*). The (tetramesitylporphyrinato)Ru complex sterically precludes formation of μ-oxo dimer and exhibits the appropriate redox potentials to facilitate the processes in the equations. This system and related sterically hindered metalloporphyrin systems, however, are rapidly inactivated by inhibition and by forming a kinetic dead-end intermediate oxidation state, Ru^{III}. Alcohols such as methanol and other oxygenated organic species can bind to and inactivate or decrease the activity of the Ru center. The inherent kinetic oxidative instability of the porphyrin ligand itself has not been addressed as the system does not last long enough to assess this issue.

Polyhalogenated Metalloporphyrin Oxygenation Catalysts. Recently tetraarylporphyrins and metallotetraarylporphyrins, some of the most oxidatively resistant of organic moieties, have been polyhalogenated by several research groups, and particularly by those of Traylor and Tsuchiya, to make them more resistant to oxidation (*15-18*). Indeed, a "teflon porphyrin", one with all C-H bonds replaced by C-F bonds, has been made. Although such metalloporphyrins have been shown to be remarkably stable in strongly oxidizing media, one concern with these systems is that the more electron withdrawing the porphyrin ligand becomes, the higher the redox potential of the central transition metal ion and the less likely it will be to bind and reduce O_2 or oxygen donors which is a requisite for effective catalytic oxygenation. Many of these complexes may be thermodynamically incapable of binding O_2. Furthermore, no highly halogenated metalloporphyrins that also are sterically hindered to prevent μ-oxo dimer formation have been made and shown to interact with O_2. In any event, the cost and effort required to obtain these elegant oxidatively robust complexes is prohibitive on all but the smallest of potential applications. Despite the inherent limitations of metalloporphyrins as catalysts for sustained selective oxidation, these are fascinating materials and the approach continues to show promise. Recently some halogenated metalloporphyrins have exhibited reasonable levels of oxidative stability in some autoxidation processes despite the fact that the ligands are still thermodynamically unstable to oxidation (*18*). It should be noted that even perfluorocarbons are thermodynamically unstable to oxidation and are documented to combust in some cases (*19*).

Redox Active Zeolites. Metallophthalocyanines and other forms of d-electron redox active transition metals included within zeolites are totally inorganic systems capable of catalyzing the oxygenation of hydrocarbons including alkanes under mild conditions in solution (*13*). These systems effectively counter several of the five general limitations elaborated in the previous section. Under the modest thermal conditions of the reactions, however, diffusion of the more sterically encumbered alcohols versus the substrate alkanes in alkane oxygenation by such transition metal complex-encapsulated zeolite catalysts is a major problem (*13*). At the same time, titanium silicalite, TS-1, is a Ti-containing zeolite that has impressive catalytic properties with respect to oxidizing hydrocarbon substrates with H_2O_2. TS-1 is already in commercial operation (*20-21*). There is little additional work in this area but the potential of such systems warrants serious investigation in our view.

d-Electron-transition-metal-substituted Polyoxometalate (TMSP) Catalysts. TMSP complexes are, as the name indicates, compounds formed by substituting one or more of the d^0 early-transition-metal ions that make up the inorganic skeleton of a polyoxometalate such as $PW_{12}O_{40}^{3-}$ (*22-29*) with d^n, $n \neq 0$, transition metal ion(s). The latter is(are) chosen with electronic and structural features requisite for the catalysis of interest. Catalytic hydrocarbon oxygenation is the focus of this article. Distinct coordination environments for the d-electron-containing "active sites" are exhibited in different classes of TMSP complexes. These complexes, by virtue of their composition, are stable to oxidative degradation. Catalytic oxidation based on TMSP-type complexes can be conducted in a homogeneous mode (dissolved in solution) or in a heterogeneous mode. In the homogeneous mode, TMSP-based catalysts combine the stability advantages of inorganic metal oxide systems, which include zeolites (both are entirely inorganic and hence thermodynamically resistant to oxidative degradation under catalytic reaction conditions), with the tractability advantages of homogeneous catalysis. Since our first paper and the initial patent on this subject (*30-34*), several other groups have published a number of papers exploring hydrocarbon oxygenation with a range of oxygen donors, equation 2, catalyzed by TMSP complexes (*35-42*).

The robustness of TMSP-type catalysts allows one to address many of the limitations regarding the development of optimal catalysts for homogeneous selective oxidation, including the five general limitations articulated above: (1) inadequate stability, particularly oxidative stability, (2) μ-oxo dimer formation, (3) deadend oxidation states, (4) inhibition by H_2O and products, and (5) diffusional problems. Points 2, 3, 4, and 5, are often not significant problems in heterogeneous catalytic oxidations as operation at high temperatures and other extreme conditions can eliminate several or all these considerations. Many TMSP catalysts are likely to be stable and operational at high temperature and under extreme conditions even in the homogeneous mode thus considerably reducing the impact of these limitations. I say "likely" as few such reactions have yet to be actually examined experimentally. In addition, however, TMSP compounds are available that preclude point 2, μ-oxo dimer formation, under very mild conditions where such processes might be a serious problem. One exemplary class of TMSP complexes where μ-oxo dimer formation is sterically precluded is the d-electron-containing transition metal complexes of the polychelating HPA-23 d^0 heteropolytungstate complex ($[NaSb_9W_{21}O_{86}]^{18-}$). The latter complex has six potential binding sites *(43)*.

We report here three studies that address three separate but significant issues in the emerging area of selective catalytic oxidation by TMSP-type complexes. The first study establishes for the first time that some TMSP complexes are compatible with basic oxidants and basic conditions. The second study reports the first oxidation, in this case selective alkene epoxidation, by the economically and environmentally desirable oxidant, aqueous hydrogen peroxide, catalyzed by TMSP complexes. The third study demonstrates that redox active polyoxometalates can be derivatized with alcohols in a manner that should prove useful for fabricating future generations of more sophisticated and selective TMSP catalysts.

Results

Sustained Oxygenation of Alkenes by OCl⁻ Catalyzed by TMSP Complexes.

None of the studies on TMSP-catalyzed homogeneous oxidation reactions reported to date involve basic media. The simple reason for this is that the families of TMSP complexes most accessible and amenable to rigorous characterization, principally those derived from the common Keggin structural family of polyoxometalates, $X^{n+}M_{12}O_{40}^{(8-n)-}$, are not hydrolytically stable in basic media. Unfortunately, some terminal oxidants of interest are compatible only with basic media. One test case addressed here is hypochlorite, OCl⁻ an inexpensive although not environmentally optimal oxidant. We have found that by using two-phase conditions, keeping the TMSP complexes in the organic phase, and using complexes of reasonable stability in neutral or basic media that successful oxidation of organic substrates by OCl⁻ catalyzed by TMSP complexes can be realized. Table II summarizes the time dependence of the distribution of organic oxidation products generated by OCl⁻ oxidation of an exemplary alkene, cyclohexene, in a CH_2Cl_2/H_2O two-phase system under low turnover conditions catalyzed by two TMSP complexes, $^6Q_8[Cu(II)P_2W_{17}O_{61}]$ and $^6Q_{16}[((Ni(II))_4(P_2W_{15}O_{56})_2]$, where $^6Q = (n\text{-}C_6H_{13})_4N^+$. Although it is possible to obtain up to 40% cyclohexene conversion or under different conditions 100% selectivity to cyclohexene oxide (Khenkin, A. M.; Weeks, M. S., unpublished results), the selectivity for epoxide is generally poor principally because of the facile further reaction of initial oxidation products under these particular reaction conditions. Little effort has been made thus far, however, to optimize this chemistry. The careful control of pH, removal of products and other techniques would doubtless improve the observed selectivities. The data available at present suggest that hypochlorite exhibits less selective olefin oxidation reactions in TMSP-catalyzed reactions than some other nonbasic oxygen donor oxidants *(30-42)*.

A central issue in this exercise was to address the compatibility of TMSP catalysis with basic oxidants. A pertinent experiment involved the epoxidation of cyclohexene by hypochlorite under high turnover conditions. The spectroscopic properties of $^6Q_{16}[((Ni(II))_4(P_2W_{15}O_{56})_2]$, before and after reaction indicated minimal decomposition of this particular TMSP after 1,400 turnovers. The preliminary dynamic product data, the negligible amount of oxidation in the absence of the TMSP complex, and the stability of the catalyst under turnover conditions defines TMSP complexes to be capable, in principle, of sustained catalytic oxygenation of organic substrates by OCl⁻ and perhaps other basic oxidants. Given that several parameters could, in principle, be adjusted to further stabilize the TMSP complexes in basic conditions including the inclusion of one or more Nb or Ta atoms at key locations in the complexes, these systems are of some interest.

Table II. Oxidation of Cyclohexane by Hypochlorite Catalyzed by Two TMSP Complexes

Products	Yields[a]					
	$^6Q_8[(Cu^{II})P_2W_{17}O_{61}]$			$^6Q_{16}[(Ni^{II})_4(P_2W_{15}O_{56})_2]$		
	40 min	100 min	5 h	30 min	90 min	7 hr
(cyclohexene oxide, =O)	3.3	4.9	6.8	18.7	18.1	25.7
(2-chlorocyclohexanol, Cl/OH)	3.9	7.8	11.1	3.7	2.8	8.0
(dichlorocyclohexane, Cl/Cl)	10.7	12.7	20.2	15.0	8.3	6.4
(cyclohexanediol, OH[b]/OH)	c	c	c	c	c	c
(phenol, OH)	c	c	1.3	1.2	1.2	1.2
(cyclohexanone, =O)	1.3	3.3	2.3	0.9	1.8	0.9
(cyclohexenone, =O)	c	2.6	3.6	3.7	4.6	3.4

[a]Yields (%) based on hypochlorite after the elapsed reaction times indicated.
[b]Partitions principally into the aqueous phase. [c]Not detectable by GC ($\ll$ 1 mol %).

Another point about these hypochlorite oxidations concerns the fate of the oxidizing equivalents unaccounted for in the organic product distributions at the time of GC and GC/MS analysis. Despite the number of papers on biomimetic and other homogeneous catalyzed oxidations of hydrocarbon substrates by hypochlorite, the product balance with respect to this oxidant is rarely satisfactory. More telling, however, is that there may be no report addressing the fate of the equivalents of active oxygen that are not accounted for in the organic products. In the case of the oxidations studied here, perchlorate is generated in sufficient quantities to account for most of the missing active oxygen, a result implicating that the TMSP, and by inference, a host of conventional transition metal complexes with oxidizable organic ligands, may be capable of catalyzing disproportionation of hypochlorite.

The use of benzene as a solvent eliminates the chlorinated organic products in Table II. This point and the products in the table are consistent with radical abstraction (hydrogen atom transfer) from the substrate, cyclohexene, by the high valent (and likely oxometal) intermediate form of the TMSP complexes, followed by chlorine abstraction from the solvent by intermediate organic radicals. Separation and analysis of the two phases after the reaction reveals that the polyoxometalate is intact, but resides in the aqueous phase. This probably results from one or more of four time-dependent phenomena: (1) the increasing concentration of polar organic products (largely diol) in the aqueous phase (shifting the partitioning coefficient of the TMSP anions further toward the aqueous phase), (2) the generation of protons from allylic hydroxylation and possibly other oxidations resulting in replacement of H^+ for Q^+ as the counterions of the TMSP complexes (the free acid forms of polyoxometalates are much more soluble in water than the Q forms), (3) the generation of chloride, both from the oxidation reactions and from hypochlorite disproportionation (this could favor a shifting to the right of the following metathetic equilibrium: Q^+TMSP^- (CH_2Cl_2 phase) + Na^+(aq phase) + Cl^-(aqueous phase) $\rightleftharpoons$ Q^+Cl^-(CH_2Cl_2 phase) + Na^+(aq phase) + $TMSP^-$(aqueous phase), and (4) the partial oxidative destruction of the hydrophobic CH_2Cl_2-solubilizing Q counter ions. The latter is unlikely as there is no evidence for Q destruction. The oxygenation and other oxidation chemistry going on in the aqueous phase at later reaction times has not been investigated but would be of much interest given the long term desirability of water-compatible catalysts and catalytic processes. The efficiency and selectivity in cyclohexene oxidation is dependent on the nature of the transition metal, the polyoxometalate and the solvent.

Alkene Epoxidation by H_2O_2 Catalyzed by TMSP Complexes. The principal chemistry on epoxidation of alkenes by H_2O_2 involving polyoxometalates is that of Venturello (WO_4^{2-}/PO_4^{3-}/H_2O_2/chlorocarbon solvent + PTC) (44-48) and Ishii ($PW_{12}O_{40}^{3-}$/H_2O_2/chlorocarbon solvent + PTC) (49-51), where PTC = phase transfer catalyst. Although definitive mechanistic work is lacking on these systems, it is clear that they are related and involve common epoxidizing intermediates. A dominant feature of this chemistry is that breakdown of the Keggin heteropolytungstate, $PW_{12}O_{40}^{3-}$, is facile in the presence of the commercially available aqueous H_2O_2. For this reason, the conventional wisdom has been that polyoxometalates, in general, would be too susceptible to hydrolytic and peroxolytic degradation to be of interest as catalysts for H_2O_2 reactions other than to function as the precursor of the active epoxidizing species in the Venturello/Ishii chemistry. This study shows this is unlikely to be the case. Table III summarizes preliminary results on H_2O_2 oxidation of cyclohexene in CH_3CN catalyzed by three TMSP complexes with reasonable hydrolytic stability. The reactions catalyzed by the first two complexes are quite attractive and rival those seen in the toxic chlorocarbon containing Venturello/Ishii systems. Most interestingly, examination of the $[(Fe^{II})_4(B-PW_9O_{34})_2]^{10-}$ catalyst by ^{31}P NMR and other spectral methods before, during, and after 2087 turnovers indicates part of the catalyst had been degraded after this considerable chemistry. These data coupled with evaluation of initial reaction rates of cyclohexene epoxidation by the TMSP complexes in Table III and authentic species from the Venturello system argue strongly that different epoxidizing species are involved in these different systems. Given all the desirable attributes of H_2O_2, this chemistry warrants rigorous investigation.

Modification of Catalytically Active Polyoxometalates with Organic Groups. The covalent attachment of organic groups to polyoxometalates would introduce organic structure and hence potential oxidative instability. At the same time, however, such groups could be used not only to change solubilities, partition coefficients and other bulk properties of relevance in catalysis but also to render catalytically active polyoxometalates capable of molecular recognition and highly selective catalysis. One example of the latter is chiral groups to make derivatized polyoxometalate catalysts for sustained catalyzed asymmetric oxygenation and other oxidation reactions. Our initial experiments have sought to assess the feasibility of attaching polyfunctional organic groups to the surfaces of polyoxometalates. Several sites of attachment between organic groups and the polyoxometalate would enhance not only the stability but also molecular recognition capabilities of the adduct, the organic derivatized complex.

Table III.   Oxidation of Cyclohexene by H_2O_2 Catalyzed by TMSP Complexes

TMSP[a]	$[H_2O_2]_0/$ $[H_2O_2]_f$ (M)	Time, h	Products, (Yield %)[b]	S^c %	T^d
$^6Q_{10}[(Ni^{II})_4(B\text{-}PW_9O_{34})_2]^e$	1.0/0.3	16	oxide (30.4) enol (0.2) enone (0.2)	99	216
$^4Q_{10}[(Fe^{II})_4(B\text{-}PW_9O_{34})_2]$	0.2/0.02	3	oxide (30.7) enol (0.8) enone (2.0)	95	15
$^4Q_4[(Fe^{II})PW_{11}O_{39}]$	0.2/0.16	27	oxide (3.7) enone (45)	15	2.5

[a]$^6Q = (n\text{-}C_6H_{13})_4N^+$; $^4Q = (n\text{-}C_4H_9)_4N^+$. [b]Yield on ΔH_2O_2; [c]Selectivity of cyclohexene oxide formation = mol of cyclohexene oxide/mol of all products. [d]Turnover number = mol of all products/mol catalyst. [e][TMSP] = 1 mM.

The condensation of 1,1,1-tris(hydroxymethyl)ethane with $[(n\text{-}C_4H_9)_4N]_5H_4[P_2V_3W_{15}O_{62}]$ proceeds to give the complex with the tris unit covalently attached to the three oxygens that bridge the three V atoms residing on one end of the polyoxometalate (the "cap" position of the Wells-Dawson structural family of polyoxometalates), $[(n\text{-}C_4H_9)_4N]_5H[CH_3C(CH_2)_3P_2V_3W_{15}O_{62}]$. Preliminary experiments have indicated that this C_{3v} complex with a tri-linked "hat" is quite stable to hydrolysis and capable of catalyzing the oxidation of tetrahydrothiophene (THT) to the corresponding sulfoxide, (THTO) by t-butylhydroperoxide (TBHP) for 30 turnovers without apparent hydrolytic or oxidative degradation of the capping organic group (Hou, Y.; Hill, C. L. unpublished work).

Some TMSP systems that preclude μ-oxo dimer formation and exhibit the right proton-linked redox potentials should facilitate selective catalytic aerobic oxidation such as that defined in equations 3-6. Selective catalytic aerobic oxidation has yet to be achieved, however. The few TMSP systems examined thus far in conjunction with O_2-based oxidations function as redox initiators of autoxidation.

Experimental Section

Materials and Methods. The lacunary heteropolytungstates $Na_{12}P_2W_{15}O_{56} \cdot 18H_2O$ (52) and $K_{10}P_2W_{17}O_{61} \cdot 22H_2O$ (52); and their transition metal substituted polyoxotungstate derivatives $Na_{16}[(Ni^{II})_4(P_2W_{15}O_{56})_2] \cdot 22H_2O$ (52), $K_8[(Cu^{II})P_2W_{17}O_{61}]$ (53), $K_{10}[(Fe^{II})_4(B\text{-}PW_9O_{34})_2]$ (54-55), $K_{10}[(Ni^{II})_4(B\text{-}PW_9O_{34})_2]$ (54-55), and $Na_5[(Fe^{II})PW_{11}O_{39}]$ (54-55), were prepared and purified by literature methods. Purity of the complexes in all cases was judged to be satisfactory by ^{31}P NMR, FTIR, and elemental analysis. Stock solutions of the tetra-n-alkylammonium salts, either tetra-n-hexyl (^{6}Q) or tetra-n-butyl (^{4}Q) ammonium salts were prepared by direct extraction or by precipitation from water (56,57). All catalysts for the hypochlorite reactions were used as volumetric stock solutions in dichloromethane. Both the sodium hypochlorite (5.25%, Clorox) and the 30% aqueous H_2O_2 (Fisher) were commercial samples and titrated iodometrically using a standard literature procedure (58). The cyclohexene was reagent grade from Fluka and had a purity greater than 99% by gas chromatography (GC).

The distributions of organic oxidation products were unambiguously identified through the use of GC (Hewlett-Packard 5890 instrument; 5% phenyl methyl silicone fused-silica capillary column; nitrogen carrier gas, temperature programming; FID detection) and by gas chromatography-mass spectrometry (GC-MS, Hewlett-Packard 5971 A MSD instrument). The electronic absorption spectra were recorded on a Hewlett-Packard (H/P) Model 8451A diode-array UV-visible spectrometer. ^{31}P broad-band proton-decoupled NMR spectra were measured on an IBM WP-200-SY spectrometer, operating at a frequency of 81 MHz and reported relative to 0.1% trimethyl phosphate in CD_3CN as the external reference.

Catalytic OCl⁻ oxidations. In a typical reaction with TMSP (either $^{6}Q_{16}[(Ni^{II})_4(P_2W_{15}O_{56})_2]$ or $^{6}Q_8[(Cu^{II})P_2W_{17}O_{61}]$), aqueous hypochlorite and alkene, 4 mL of dichloromethane, and 1.7×10^{-5} mol of the phase transfer catalyst tetra-n-hexylammonium chloride (^{6}QCl) were placed in a 25-mL round-bottom schlenk flask. The solution was thoroughly degassed and then left under Ar. The reaction was initiated by addition of 1.70 mmol of cyclohexene, 0.01 mmol of the TMSP, and 0.17 mmol of hypochlorite in 1.0 mL of water (pH 10.8). Two homogeneous phases were exhibited throughout the duration of each of these reactions. The products in the organic layers were analyzed GC and GC/MS as indicated above while the reactions were stirred at room temperature for 7 h.

Catalytic H_2O_2 Oxidations. The reactions were carried out in 10-mL tubes equipped with serum cap and a stirring bar. The catalyst, hydrogen peroxide, and substrate were dissolved in of CH_3CN. Trimethylacetonitrile was added to the reaction as an internal standard. All reactions were done under argon, each was purged by three freeze-thaw cycles and GC analysis was performed on aliquots withdrawn directly from the reaction mixture. Typically, alkene (1 mmol) was added to the solution of the TMSP (0.004 mmol) in 1 mL of CH_3CN then 25 μL of 30% aqueous H_2O_2 was added and reaction mixture was stirred at 20 °C.

Synthesis of $[(n\text{-}C_4H_9)_4N]_5H[CH_3C(CH_2)_3P_2V_3W_{15}O_{62}]$. This compound was made by condensing a slight molar excess of 1,1,1-tris(hydroxymethyl)ethane with $[(n\text{-}C_4H_9)_4N]_5H_4[P_2V_3W_{15}O_{62}]$. Elemental analysis, ^{1}H, ^{31}P, ^{51}V, and ^{183}W NMR established the identity and purity of the compound. The hydrolytic stability of the complex was assessed using the ^{1}H and ^{31}P NMR. The stability of the complex during oxidation of tetrahydrothiophene (THT) to the corresponding sulfoxide, (THTO) by t-butylhydroperoxide (TBHP) was assessed using ^{51}V and ^{31}P NMR.

Acknowledgments. Support for the research was provided by the National Science Foundation (oxometal complexes and hydrocarbon chemistry), the U.S. Army Research Office (decontamination catalysis and derivatized polyoxometalates).

Literature Cited.

(1) Sheldon, R. A.; Kochi, J. K. *Metal-Catalyzed Oxidations of Organic Compounds,* Academic Press, New York, 1981, Chapter 3.

(2) *Organic Syntheses by Oxidation with Metal Compounds,* Mijs, W. J.; de Jonge, C. R. H. I., Eds., Plenum: New York, 1986.

(3) Parshall, G. W. *Homogeneous Catalysis. The Applications and Chemistry of Catalysis by Soluble Transition Metal Complexes* Wiley-Interscience: New York, 1980.

(4) Sheldon, R. A. *ChemTech* **1991**, 566.

(5) Sheldon, R. A. In *Dioxygen Activation and Homogeneous Catalytic Oxidation* Simandi, L. I.; Ed.; Elsevier: Amsterdam, 1991, pp 573-594.

(6) Jorgensen, K. A. *Chem. Rev.* **1989**, *89*, 431, and references cited in each.

(7) Katsuki, T.; Sharpless, K. B. *J. Am. Chem. Soc.* **1980**, *102*, 5976.

(8) Woodard, S. S.; Finn, M. G.; Sharpless, K. B. *J. Am. Chem. Soc.* **1991**, *113*, 106.

(9) Finn, M. G.; Sharpless, K. B. *J. Am. Chem. Soc.* **1991**, *113*, 113.

(10) Groves, J. T. *J. Chem. Ed.* **1985**, *62*, 928.

(11) Hill, C. L. In *Activation and Functionalization of Alkanes*, Hill, C. L. Ed. Wiley: New York, 1989, Chapter VIII.

(12) Ortiz de Montellano, P. R., Ed.; *Cytochrome P-450*, Plenum: New York, New York, 1986.

(13) Tolman, C. A.; Druliner, J. D.; Nappa, M. J.; Herron, N. "In *Activation and Functionalization of Alkanes*, Hill, C. L. Ed. Wiley: New York, 1989, Chapter X and references cited.

(14) Groves, J. T.; Quinn, R. *J. Am. Chem. Soc.* **1985**, *107*, 5790.

(15) Traylor, P. S., Dolphin, D., Traylor, T. G. *J. Chem. Soc., Chem. Commun.,* **1984**, 279.

(16) Traylor, T. G., Tsuchiya, S. *Inorg. Chem.* **1987**, *26*, 1338.

(17) Traylor, T. G. *Pure and Appl. Chem.* **1991**, *64*, 265.

(18) Ellis, P. E.; Lyons, J. E. *Coord. Chem. Rev.* **1990**, *105*, 181.

(19) Christe, K. O. *Chem. & Eng. News* **1991**, [October 7th issue], 2.

(20) Notari, B. *Stud. Surf. Sci. Catal.* **1988**, *37*, 413.

(21) Romano, U.; Esposito, A.; Maspero, F.; Neri, C.; Clerici, M. *Stud. Surf. Sci. Catal.* **1990**, *55*, 33.

(22) Pope, M. T. *Heteropoly and Isopoly Oxometalates*; Springer Verlag: New York, 1983.

(23) Day, V. W.; Klemperer, W. G. *Science* **1985**, *228*, 533.

(24) Pope, M. T.; Müller, A. *Angew. Chem. Intern. Ed. Engl.* **1991**, *30*, 34.

(25) Jeannin, Y.; Fournier, M. *Pure Appl. Chem.* **1987**, *59*, 1529.

(26) Misono, M. *Catal. Rev.-Sci. Eng* **1987**, *29*, 269.

(27) Kozhevnikov, I. V.; Matveev, K. I. *Russ Chem. Rev. (Engl Transl.),* **1982**, *51*, 1075.

(28) Kozhevnikov, I. V.; Matveev, K. I. *Appl. Catal.,* **1983**, *5*, 135.

(29) Hill. C. L. In *Metal Catalysis in Hydrogen Peroxide Oxidations*, Strukul, G., Ed. Reidel, 1992, Chapter 8, in press.

(30) Hill, C. L.; Brown, Jr., R. B. *J. Am. Chem. Soc.* **1986**, *108,* 536.

(31) Faraj, M.; Lin, C.-H.; Hill, C. L. *New J. Chem.* **1988**, *12*, 745.

(32) Hill, C. L.; Brown, Jr., R. B.; Renneke, R. F. *Prepr. Am. Chem. Soc. Div. Pet. Chem.* **1987**, *32*, [No. 1] 205.

(33) Hill, C. L.; Renneke, R. F.; Faraj, M. K. Brown, Jr., R. B. *The Role of Oxygen in Chemistry and Biochemistry (Stud. Org. Chem. / Amsterdam)*, Ando, W., Moro-oka, Y., Eds, Elsevier: New York, 1988, p 185.

(34) Hill, C. L. U.S. patent 4,864,041 (1989).
(35) Neumann, R.; Abu-Gnim, C. *J. Chem. Soc., Chem. Commun.* **1989**, 1324.
(36) Neumann, R.; Abu-Gnim, C. *J. Am. Chem. Soc.,* **1990**, 112, 6025.
(37) Katsoulis, D. E.; Pope, M. T. *J. Chem. Soc., Chem. Commun.* **1986**, 1186.
(38) Katsoulis, D.; Pope, M. T. *J. Chem. Soc., Dalton Trans.* **1989**, 1483.
(39) Lyon, D. K.; Miller, W. K.; Novet, T.; Domaille, P. J.; Evitt, E.; Johnson, D. C.; Finke, R. G. *J. Am. Chem. Soc.* **1991**, *113*, 7209.
(40) Mansuy, D.; Bartoli, J.-F.; Battioni, P.; Lyon, D. K.; Finke, R. G. *Ibid.* **1991**, *113*, 7222.
(41) Lyons, J. E.; Ellis, P. E., Durante, V. A. *Stud. Surf. Sci. Catal.* **1991**, *67*, 99.
(42) Rong, C.; Pope, M. T. *J. Am. Chem. Soc.* **1992**, *114*, 2932.
(43) Michelon, M.; Hervé, G.; Leyrie, M. *J. Inorg. Nucl. Chem.* **1980**, *42*, 1583.
(44) Venturello, C.; Alneri, E.; Ricci, M. *J. Org. Chem.* **1983**, *48*, 3831.
(45) Venturello, C.; D'Aloiso, R.; Bart, J. C.; Ricci, M. *J. Mol. Cat.* **1985**, *32*, 107.
(46) Venturello, C.; Ricci, M. *J. Org. Chem.* **1986**, *51*, 1599.
(47) Venturello, C.; D'Aloiso, R. *J. Org. Chem.* **1988**, *53*, 1553.
(48) Venturello, C.; Gambaro, M. *Synthesis,* **1989**, *4*, 295, and references cited.
(49) Ishii, Y.; Yamawaki, K.; Yoshida, T.; Ura, T.; Ogawa, M. *J. Org. Chem.* **1987**, *52*, 1686.
(50) Ishii, Y.; Yamawaki, K.; Ura, T.; Yamada, H.; Yoshida, T.; Ogawa, M. *J. Org. Chem.* **1988**, *53*, 3587.
(51) Ishii, Y.; Ogawa, M. in *Reviews on Heteroatom Chemistry, Vol. 3* , Oae, S., Ed.; MYU: Tokyo; 1990, pp 121-145 and references cited.
(52) Finke, R. G.; Droege, M. W.; Domaille, P. J. *Inorg. Chem.* **1987**, *26*, 3886.
(53) Malik, S. A.; Weakley, T. J. R. *J. Chem. Soc. (A),* **1968**, 2647.
(54) Finke, R.G.; Droege, M.W.; Domaille, P.J. *Inorg. Chem.* **1987**, *26*, 3886.
(55) Tourné, C.M.; Tourné, G.F. *Bull. Soc. Chim. Fr.* **1969**, 1124.
(56) Corigliano, F.; Pasquali, S. D. *Inorg. Chim. Acta* **1975**, *12*, 99.
(57) Katsoulis, D. E.; Pope, M. T. *J. Am. Chem. Soc.* **1984**, *106*, 2737.
(58) Lee, W. A.; Bruice, T. C. *Inorg. Chem.* **1986**, *25*, 131.

RECEIVED October 30, 1992

Chapter 7

Nature of the Co–Mn–Br Catalyst in the Methylaromatic Compounds Process
Kinetic and Thermodynamic Studies

W. Partenheimer[1] and R. K. Gipe[2]

[1]Amoco Chemical Corporation, P.O. Box 3011, Naperville, IL 60566–7011
[2]Consultant, 72 Evans Road, Brookline, MA 02146

We have used the reaction of m-chloroperbenzoic acid with Co/Mn/Br as a model system to attempt to understand the nature of this important autoxidation catalyst. Using stopped-flow and UV-VIS kinetic techniques, we have determined the step-wise order in which the catalyst components react with each other. The cobalt(II) is initially oxidized to Co(III) by the peracid, the cobalt(III) then oxidizes the manganese to Mn(III), which then oxidizes the bromide. The order of these redox reactions is the opposite to that expected from thermodynamics. Suggestions will be made of the relationship of this model to the known characteristics of autoxidation processes.

A highly efficient, general method to produce aromatic acids is via the liquid phase reaction of methylaromatic compounds with dioxygen:

$$C_6H_{(6-n)}(CH_3)_n + O_2 \xrightarrow[\substack{HOAc \\ 25\text{-}250°C \\ 1\text{-}15 \text{ atms}}]{Co(OAc)_2, Mn(OAc)_2, HBr} C_6H_{(6-n)}(COOH)_n + n\,H_2O \quad (1)$$

Hundreds of different carboxylic acids have been produced via this method *(1)*. The industrial process, dubbed the Amoco MC process, produces billions of pounds of terephthalic acid, isophthalic acid and trimellitic acid annually. Industrial processes using just cobalt as the catalyst rather than Co/Mn/Br have also been developed *(2-3)*. The characteristics of the reaction suggest that it is, at least partially, a free radical chain mechanism involving peroxy

0097–6156/93/0523–0081$06.00/0

and alkoxy radicals, peroxides and peracids *(2-5)*. The easily observable intermediates are alcohols, acetates of the alcohol (since the solvent is acetic acid) and aldehydes.

We are greatly indebted to the pioneering work of Jones who characterized much of the chemistry and reported the initial kinetic data*(6-9)*. He has shown that reaction of m-chloroperbenzoic acid (MCPBA) with Co(II) acetate initially generates an active form of cobalt, labeled Co(III)a, which slowly re-arranges to a less active form Co(III)s. The difference in rates of these forms with bromide and manganese(II) are 30,000 and 6,000 respectively. There have been numerous studies of solid and solution forms of Co(III) acetate which suggest various different polynuclear compounds form *(10)*.

Peracids form as transient species from the oxidation of benzaldehyde during autoxidation. For convenience we have chosen m-chloroperbenzoic acid (MCPBA) as our oxidant since this would be similar to the peracid formed from the very important intermediate 4-carboxybenzaldehyde formed during the oxidation of p-xylene *(2)*. MCPBA would be formed in very low concentrations during oxidation hence we normally study the reaction of MCPBA with an excess of catalyst components i.e. MCPBA < < Co,Mn,Br (pseudo first order conditions). The sequence of reactions that occurs when MCPBA is reacted with Co(II), Mn(II), and HBr has been previously discussed by Jones *(9)* in the presence of 5% water in acetic acid. We have repeated much of this work in 10% H_2O/HOAc solutions and in general agree with his findings when one accounts for differences in temperatures, concentrations, and water concentrations.

The Sequence of Reactions that Occurs when MCPBA is Added to a Mixture of Co(II), Mn(II), and Bromide in 10% Water Acetic Acid.

All of the rate data is given on Table I. Under these conditions, MCPBA reacts with an excess of Co(II) acetate, Mn(II) acetate, and bromide compounds to form Co(III)a, Co(III)s, manganese(III) acetate and tribromide. These species have characteristic spectra given on Figure 1 which make kinetic studies easy to perform. MCPBA was reacted INDIVIDUALLY with Co(II), KBr, and Mn(II) to give Co(III), KBr₃, and Mn(III), see examples 1-4. MCPBA reacts two orders of magnitude faster with Co(II) than with either Mn(II) or bromide and seven orders of magnitude faster than its own thermal decomposition. Thus MCPBA will react predominantly with Co(II) in a mixture of Co(II), Mn(II), and KBr. What happens to Co(III)a when it is generated? It can rearrange to the stable form, Co(III)s, react with Mn(II), or react with Br-. The fastest reaction is the oxidation of Mn(II) to give Mn(III), see examples 5-7. This reaction is 11 times faster than the reaction with Br- and 940 times as fast as the rearrangement to Co(III)s. The major pathway is therefore MCPBA preferentially reacting with Co(II) to generate Co(III)a which then reacts with Mn(II) to give Mn(III). Mn(III) is subsequently

Table I. Rate Constants for Selected Reactions of MCPBA, Co(II), Co(III), Mn(II), Mn(III), and Bromide in 10% Water/Acetic Acid [a]

Reactants	Products	k,s-1 [b]	Temp,C	Comments
1. MCPBA + Co(II)	MCBA + Co(III)	66(2)	30	
2. MCPBA + KBr	MCBA + KBr$_3$	0.08	23	
3. MCPBA + Mn(II)	MCBA + Mn(III)	0.017	23	autocatalytic rxn [c]
4. MCPBA [d]	MCBA	2x10^{-6}	25	0.1% H$_2$O
5. Co(III)a	Co(III)s	0.0070	25	
		(0.002)		
6. Co(III)a + Mn(II)	Co(II) + Mn(III)	6.6(.1)	23	
7. Co(III)a + NaBr	Co(II) + NaBr$_3$ [e]	0.59(.01)	23	
8. MCPBA + (Co+Mn) [f]	MCBA + Co(III)a	43(.2)	25	1st of 2 rxns
9. Co(III)a + Mn(II)	Co(II) + Mn(III)	7.1(.1)	25	2nd of 2 rxns
10.MCPBA + (Co+Mn+NaBr) [g]	MCBA + Co(III)a	36(1)	25	1st of 2 rxns
11.Co(III)a + Mn(II)	Co(II) + Mn(III)	7.3(.1)	25	2nd of 3 rxns
12.Mn(III) + HBr	Mn(II) + HBr$_3$	0.0066	25	3rd of 3 rxns
		(0.0004)		
13.MCPBA + Co(II)	MCBA + Co(III)a	307(12)	60	
14.MCPBA + (Co+Mn)	MCBA + Co(III)a	232(18)	60	1st of 2 rxns
15.Co(III)a +Mn(II)	Co(II) + Mn(III)	123(16)	60	2nd of 2 rxns
16.MCPBA + (Co+Mn+NaBr)	MCBA + Co(III)a	330(25)	60	1st of 3 rxns
17.Co(III)a + Mn(II)	Co(II) + Mn(III)	82(2)	60	2nd of 3 rxns
18.Mn(III) + NaBr	Mn(II) + NaBr$_3$	0.11(.01)	60	3rd of 3 rxns
19.Co(III)ox [h] + (Co+Mn)	Co(II) + Mn(III)	1.19	60	
20.Co(III)a	Co(III)s [i]	0.24(.04)	60	
21.Co(III)s + p-xylene	Co(II)	0.00012 [j]	80	

[a] [MCPBA]o = 0.0005 M and all others 0.0100M unless otherwise stated. All data have been measured in our labs using traditional and stopped flow apparatus unless otherwise stated. Initial compounds are Cobalt(II) and manganese (II) acetate tetrahydrates.

[b] Standard deviation based on at least three independent measurements. Standard deviation in parenthesis ().

[c] Autocatalytic reaction, rate refers to fast part of S curve.

[d] Rate of thermal decomposition reported in ref (11) for perbenzoic acid. A number of other peracids give approximately the same rates.

[e] [NaBr]o = 0.01M, [Co(II)]o = 0.005M, [MCPBA]o = 0.0003M.

[f] Refers to MCPBA being added to a mixture of Co(II) + Mn(II) acetates.

[g] Refers to MCPBA being added to a mixture of Co(II) + Mn(II) acetates and NaBr.

[h] A sample of Co(III) prepared via ozone. [Co(II)]o = 0.01M, [Co(III)s]o = 0.001M, [Mn(II)]o = 0.01 M.

[i] [Co(II)]o = 0.01M, [Co(III)]o = 0.001.

[j] [Co(III)]o = 0.001M, [p-xylene]o = 1.0 M, under nitrogen.

reduced by bromide, example 8. This sequence is shown in Figure 2. To confirm this sequence we have added MCPBA to mixtures of Co(II)/Mn(II) (examples 8,9;14,15) and to Co(II)/Mn(II)/NaBr mixtures (examples 10-12;16-18) and observe the two and three consecutive reactions respectively. We have repeated the latter two experiments at 35, 45, and 60°C in which we obtain the same type of absorbance changes and similar relative rates. The data at 60°C are given on Table I.

The Importance of the Conversion of Co(III)a to Co(III)s

As pointed out in the introduction, Co(III)s is much less reactive than Co(III)a by several orders of magnitude. CLEARLY, WE DO NOT WANT Co(III)s TO FORM DURING CATALYZED REACTIONS BECAUSE OF ITS GREATLY REDUCED REACTIVITY. Table I illustrates that the rate of re-arrangement of Co(III)a to Co(III)s is much slower than the reaction of Co(III)a with Mn(II). Hence Co(III)s does not play an important role in Co/Mn/Br mixtures. This is confirmed in example 19 where a sample of Co(III) acetate, prepared by ozonolysis and having a uv-vis spectrum nearly identical to Co(III)s, is reacted with Mn(II). The rate of reaction is 80 times slower at 60°C than with Co(III)a. Jones concludes (6) that in processes using only cobalt as the catalyst (no manganese or bromide) the rate of re-arrangement of Co(III)a to Co(III)s is now faster than the rate of Co(III) with p-xylene. The p-xylene is forced to react with the less active form of cobalt(III). This is illustrated on Figure 2.

The Thermodynamics of the Sequence of Reactions when MCPBA is Reacted with a Co(II)/Mn(II)/Br- Mixture.

The free energies of the appropriate reactions are given on Table II. As can be seen, the most easily oxidizable substance is bromide ion followed by Mn(II) and finally Co(II). We are assuming MCPBA to be the strongest oxidant since it completely reacts with Co(II) to give Co(III). When MCPBA reacts with a mixture of Co/Mn/Br however, it does not react with the most easily oxidizable substance - the bromide ion- but rather the hardest to oxidize substance-Co(II). Similarly Co(III) does not choose the most spontaneous choice available to it (the bromide ion) but rather reacts with Mn(II).

The peroxides and peracids formed in autocatalytic systems are highly energetic molecules. We now see that the Co/Mn/Br catalyst serves to rapidly relax this energy in increasingly lower steps winding up with a highly selective bromide(O) radical (probably as a complex with the metal). The bromide(O) transient species quickly reacts with methylaromatic compounds to form $PhCH_2^{\bullet}$ radicals and hence continues to propagate the chain sequence.

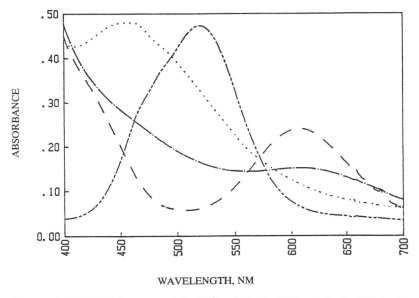

WAVELENGTH, NM

Figure 1. UV-VIS Spectra of Co(III)a (---), Co(III)s (-.-), Co(II) (--..), and Mn(III)(...). Co(III)a = Co(III)s = 0.00100M, Co(II) = 0.0200M, Mn(III) = 0.00200M

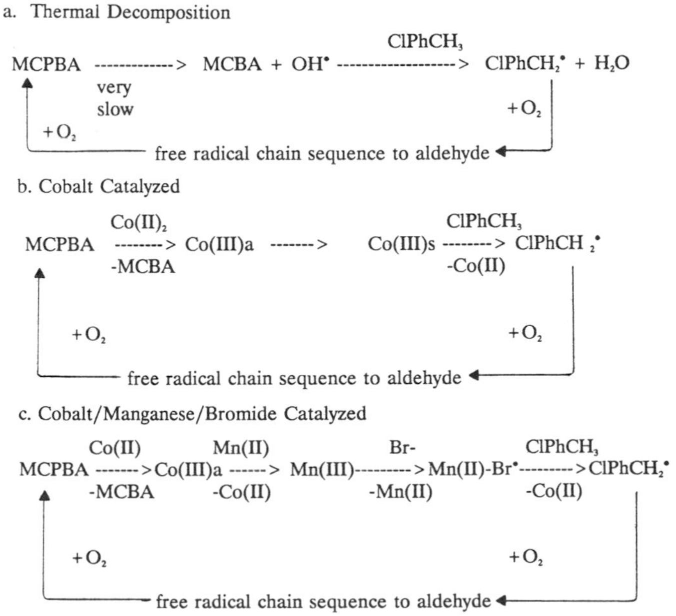

a. Thermal Decomposition

$$
\begin{array}{c}
\hspace{6em} \text{ClPhCH}_3 \\
\text{MCPBA} \ \text{------------>} \ \text{MCBA} + \text{OH}^\bullet \ \text{------------------>} \ \text{ClPhCH}_2^\bullet + \text{H}_2\text{O} \\
\end{array}
$$

very
slow $+\text{O}_2$
$+\text{O}_2$

free radical chain sequence to aldehyde ◄

b. Cobalt Catalyzed

Co(II)₂   ClPhCH₃
MCPBA   --------> Co(III)a ------->   Co(III)s --------> ClPhCH₂•
        -MCBA   -Co(II)

$+\text{O}_2$ $+\text{O}_2$

free radical chain sequence to aldehyde ◄

c. Cobalt/Manganese/Bromide Catalyzed

Co(II)   Mn(II)   Br-   ClPhCH₃
MCPBA -------->Co(III)a ------> Mn(III)---------->Mn(II)-Br•--------->ClPhCH₂•
       -MCBA   -Co(II)   -Mn(II)   -Co(II)

$+\text{O}_2$ $+\text{O}_2$

free radical chain sequence to aldehyde ◄

Figure 2. Schematic Diagram Illustrating the Difference Between Thermal, Cobalt Catalyzed, and Co/Mn/Br Catalyzed Oxidation of M-Chlorotoluene

Table II. Selected Reduction Potentials Corresponding to Reactions of MCPBA with a Co(II)/Mn(II)/Bromide Mixture

Reaction	Reduction Potentials, Kcal/mol	
	100% H_2O [a]	10%H_2O/HOAc
MCPBA----> m-chlorobenzoic Acid	< -42	< -28
Co(III) -----> Co(II)	-42	-28 [b]
Mn(III)-------> Mn(II)	-35	-18 [c]
Br- ------------> Br_2	-25	---

[a] Weast, R. C., Editor, *"Handbook of Chemistry and Physics,"* The Chemical   Rubber Company, 67th Edition.

[b] Roelofs, M. G., Wasserman, E., and Jensen, J. H., *J. Am. Chem. Soc.,* **1987** pp 109, 4207.

[c] Raju, U. G. K., Venkat, V., Sethuram, R. B., and Navaneeth Rao, T., *J. Electroanal. Chem.* **1982** pp 133, 317-322.

The Enhancement in Activity and Selectivity in Cobalt Catalyzed Oxidation of MCPBA.

Table III illustrates that cobalt behaves as an extraordinary catalyst in its reaction with MCPBA increasing the rate by a factor of 400,000 and reducing the activation from 27 to 9.5 kcal/mol. However, cobalt also greatly enhances the selectivity in the system (Table III). The yield to the desired acid increases from 89% to 100% with the expected decrease in the by-products. The thermal decomposition of MCPBA, equation 2, releases the hydroxyl radical which can easily attack the acetic acid forming carbon dioxide and methyl acetate.

$$ClPh(=O)OOH \quad ------> \quad ClPh(=O)O^{\bullet} + OH^{\bullet} \qquad (2)$$

The formation of chlorobenzene and carbon dioxide is probably coming from the decarboxylation of the ClPhCOO$^{\bullet}$ radical *(11)*. This is avoided when MCPBA is reacted with Co(II) presumably because the cobalt(II) exists as a dimer which can absorb both electron equivalents of the MCPBA and avoid a radical reaction *(7)*:

$$ClPh(=O)OOH + Co(II)_2 \quad ------> \quad ClPh(=O)OH + Co(III)_2 \qquad (3)$$

Table III. Comparison of the Reaction of M-Chloroperbenzoic Acid with and without Cobalt at 60°C

	Thermal Decomposition	With Cobalt(II) Acetate
Half-life, sec	900 [a]	0.0023
Ea,Kcal/mol	27 [a]	9.5 [c]
m-chlorobenzoic acid, mol% yld.	89 [a]	100 [a]
Chlorobenzene, mol% yld.	4.8	<0.7
Carbon dioxide, mmol [b]	0.70	0.25
Methylacetate, mmol [b]	0.034	None detected

[a] As reported in *(11)* in 0.1% water/acetic acid.

[b] Solutions were 0.0328 M in MCPBA and .0668 M in Co(II) at 72°C in 10% water/acetic acid.

[c] As measured by us in 10% water/acetic acid. Jones *(8)* reports the same value in 5% water/acetic acid.

Summary

In a free-radical chain mechanism we want to 1) produce a given product selectively, 2) simultaneously produce radical species which will further propagate the chain. Consider the autoxidation of m-chlorotoluene to m-chlorobenzoic acid in the three ways given on Figure 2. For the sake of argument, we initially start with MCPBA. We will also assume the free radical chain mechanism sequence does not contain a rate determining step.

The thermal decomposition of MCPBA is slow and unselective. When cobalt catalyzed, the initial reaction is very fast and selective but the reaction is hindered by the re-arrangement of Co(III)a to Co(III)s and by the slow reaction with m-chlorotoluene. These reactions are also characterized by a high steady state concentration of Co(III). High concentrations of Co(III) are not desirable because Co(III) is known to react with the acetic acid solvent and also decarboxylate aromatic acids *(2)*.

The Co/Mn/Br now eliminates the bottleneck caused by the presence of Co(III)s. The steady state concentration of Co(III) is also much lower caused by its rapid reduction by Mn(II). This reduces carboxylic acid decomposition. We have measured the rate of Mn(III) oxidaion of bromide in the presence and absence of p-xylene and do not find any difference in rate. Hence the system also eliminates the slow Co(III) + chlorotoluene reaction. This sequence of reactions is overall faster and more selective than either the thermal or cobalt catalyzed oxidation of m-chlorotoluene.

Literature Cited

(1) The author has a file containing about 270 different reactants that have been oxygenated in this way.

(2) Blackburn, D. W, *"Catalysis of Organic Reactions,"* Marcel Dekker, Inc., **1990**, Chapter 20.

(3) Weissermel, K. and H.-J. Arpe, H.-J., *"Industrial Organic Chemistry,"* Verlag Chemie, Weinheim, New York, N.Y. **1978**, pp 342.

(4) Sheldon, R. A. and Kochi, J. K. *"Metal-Catalyzed Oxidations of Organic Compounds,"* Academic Press, New York, N.Y. **1981**.

(5) Emanuel, N. M., Devisov, E. T. and Maizus, Z. K., *"Liquid-Phase Oxidation of Hydrocarbons,"* translated from the Russian, Plenum Press, New York, NY **1967**. Press, pp 196.

(6) Jones, G. H. *J. Chem. Research* (S), **1981** pp 228-229.

(7) Jones, G. H. *J. Chem. Soc.*, Chem. Commun., **1979** pp 536.

(8) Jones, G. H. *J. Chem. Research* (M), **1981** pp 2801.

(9) Jones, G. H. *J. Chem. Research* (5), **1982** pp 207.

(10) Blake, A. B., Chipperfield, J. R., Lau, S. and Webster, D. E., *J Chem Soc* Dalton Trans; pp 3719 **1990**.

(11) Hendriks, C. F., van Beek, H. C. A., and Heertjes, P. M., *Ind. Eng. Chem. Prod. Res. Dev.*, **1979** Vol. 18; No. 1, pp 38.

RECEIVED October 30, 1992

REACTIVITY OF SINGLE CRYSTALS AND WELL-DEFINED CRYSTAL FACES

Chapter 8

Reactivity of Oxygen Adatoms on the Au(111) Surface

Mark A. Lazaga[1], David T. Wickham[2], Deborah H. Parker[3], George N. Kastanas[1], and Bruce E. Koel[1]

[1]Department of Chemistry, University of Southern California, Los Angeles, CA 90089—0482
[2]TDA Research, 12421 West 49th Avenue, Wheat Ridge, CO 80033
[3]Chemistry Division, Argonne National Laboratory, Argonne, IL 60439

The adsorption and reaction of CO, CO_2, NO_2, H_2O, CH_3OH, and C_2H_4 were studied on clean and oxygen-precovered Au(111) surfaces. High coverages of oxygen adatoms, Θ_O, up to one monolayer were formed under clean, UHV conditions by exposure of Au(111) to ozone. Neither CO nor CO_2 adsorbed on clean or oxygen-precovered Au(111) surfaces, but the CO oxidation reaction occurs readily at low temperatures. The Langmuir-Hinshelwood (LH) mechanism was found to be in operation over the entire range of oxygen coverage; we observed negative apparent activation energies ($E_{app} = -2.5$ kcal/mol) with no strong dependence on oxygen coverage. The activation of NO_2, H_2O, CH_3OH, and C_2H_4 by oxygen adatoms on Au(111) is also discussed and compared to Au(110). We find that H_2O and C_2H_4 are unreactive at low temperatures, and CH_3OH has an intermediate reactivity, being less reactive than CO.

Copper and silver are important industrially as oxidation catalysts. Higher selectivity in oxidation catalysis might be achieved by using a less reactive metal catalyst, such as Au, if reactive oxygen can still be made available. Schwank (1) has reviewed many of the applications of elemental gold in catalysis, and striking activities are found for some gold-catalyzed oxygen transfer reactions. For example, gold catalyzes the reduction of nitric oxide and the carbonyl group in acetone, and the oxidative dehydrogenation of alcohols to aldehydes. Furthermore, gold films are being used commercially as gas sensors (2), atmospheric NO_y sensors have been developed that utilize CO oxidation in Au catalytic tubes (3) and there is renewed interest in supported Au catalysts which carry out H_2 and CO oxidation at room temperature (4). New catalyst development can be aided by a fundamental understanding of the interactions of oxygen with gold surfaces and the role of oxygen in surface reactions of molecules on Au.

Very few studies of chemical reactions on Au surfaces have been carried out under clean, ultrahigh vacuum (UHV) conditions utilizing surface analysis to insure that no impurities are present at the surface. This is largely due to the low reactivity of gold. Formic acid, HCOOH, has been studied several times on gold single crystal surfaces, showing molecular adsorption on clean Au(110) (5, 6) and Au(111) (6). Madix and coworkers have made pioneering studies of the adsorption and reaction of CO and CO_2 on clean and oxidized Au(110) (7), and the activation of HCOOH, H_2CO, CH_3OH, C_2H_2, H_2O, and C_2H_4 by surface oxygen on Au(110) (5, 8). In general, these molecules show no reactivity on the clean surface of gold, but can be oxidized when coadsorbed with oxygen adatoms (8). The reactivity trends are similar to those seen on Ag surfaces (9). A correlation of the reactivity of oxygen toward these molecules on Cu, Ag, and Au has been established by Madix and coworkers based on their gas phase acidities and the interactions described in terms of Brönsted acid-base reactions (8, 10, 11). While the framework for understanding reactivity is certainly in place, no reactivity measurements are available for surface oxygen coverages on Au(110) exceeding 0.25 monolayers, which could certainly occur under a variety of catalytic conditions, or for Au(111) surfaces, which are the most stable Au crystal faces and thus should comprise most of the catalyst metal surface area. This situation has prompted us to explore the interactions of several molecules with Au(111) and oxygen adatoms on Au(111) single crystal surfaces. Specifically, we want to explore the relative reactivity of Au(111) versus Au(110) and the reactivity of high coverages (up to one monolayer) of adsorbed oxygen.

In this paper, we give a brief overview of experiments that probe the reactivity of CO, CO_2, NO_2, H_2O, CH_3OH, and C_2H_4 on clean and oxygen-covered Au(111) surfaces. Where applicable, we interpret our results in the context of Brönsted acid-base reactions. These studies are made possible by using ozone to cleanly form oxygen adatoms on Au(111) under UHV conditions. We note that an interesting, low temperature reaction of H_2O+ NO_2 also forms surface oxygen on Au(111).

Experimental methods

The experiments were performed in stainless steel UHV chambers which were equipped with the instrumentation necessary to perform Auger Electron Spectroscopy (AES), X-ray Photoelectron Spectroscopy (XPS), UV Photoelectron Spectroscopy (UPS), Low Energy Electron Diffraction (LEED), work function measurements ($\Delta\phi$), High Resolution Electron Energy Loss Spectroscopy (HREELS), and Temperature Programmed Desorption (TPD). The Au(111) crystal was heated resistively and cooled by direct contact of the crystal mounting block with a liquid nitrogen reservoir. The temperature of the Au(111) crystal was monitored directly by means of a chromel-alumel thermocouple pressed firmly into a small hole drilled into the side of the crystal.

The Au(111) crystal was cleaned using the procedure of ref. (12) by heating in 2×10^{-8} torr of NO_2 at 800 K to remove carbon from the surface. Cycles of cleaning in NO_2 followed by argon ion sputtering at 800 K were repeated until no "oxide" was formed by NO_2 exposure and no carbon was detected. The crystal was then annealed to 800 K and gave the expected LEED pattern for the reconstructed surface.

The ozone used for dosing oxygen was prepared in our laboratory using a commercial ozone generator and concentrated on a silica gel trap (13). The generator produces a mixture of about 5% ozone and 95% unreacted O_2 with an O_2 flow of 5 cm^3/min. The ozone is concentrated by passing this mixture for several hours through a gas absorption bottle containing dried silica gel suspended in a dewar filled with an ethanol-dry ice bath at -80 °C. The ozone may be kept indefinitely at -80 °C, but there is a chance of explosion if it is allowed to warm up and come in contact with the ethanol in the bath. Since a great deal of oxygen is present as an impurity, the ozone was purified by cooling the bath to -110 °C and pumping out the O_2 with a mechanical pump. The bath was then allowed to warm up to -80 °C. The stainless steel gas dosing lines on the UHV chamber were passivated extensively by ozone exposure and the lines were recharged with fresh ozone frequently during experiments. A glass microcapillary array was used for dosing the ozone directly to the sample. All dosing of ozone was performed with the Au crystal at 300 K.

Gas exposures are reported in units of Langmuir (1 Langmuir = 1 L = 1×10^{-6} torr·sec), uncorrected for ion gauge sensitivity and doser enhancement. Coverages, Θ, are reported relative to the unreconstructed Au(111) surface atom density ($\Theta = 1$ corresponds to 1.39×10^{15} atoms/cm^2).

Results and Discussion

Adsorption of Oxygen on Au(111) by Ozone Exposure. The clean Au(111) surface is inactive for any O_2 adsorption at temperatures down to 100 K under UHV conditions. Also, it is impossible to dissociatively adsorb O_2 on Au surfaces under UHV conditions due to a kinetic barrier for O_2 dissociation. However, we have discovered that oxygen adatom coverages up to approximately one monolayer, $\Theta_O = 1$, can be formed cleanly and routinely in UHV by exposure of ozone, $O_{3(g)}$, on Au(111) at 300 K (13). We studied the oxygen layer prepared in this manner by TPD, AES, XPS, work function measurements, LEED, and HREELS. HREELS establishes that no adsorbed O_2 is present. LEED observations show no ordered overlayer for any coverage studied, but indicate one dimensional disordering of the adlayer at low coverages and disordering of the Au substrate atoms at higher coverages. Work function versus coverage data confirms occupation of adatoms sites and electron transfer from the substrate into the adlayer, but the magnitude of the work function change is strongly dependent on the oxygen coverage. AES results show a linear relationship of the O(KVV) signal with oxygen coverage as determined by TPD.

Figure 1 shows XPS spectra for atomic oxygen adsorbed on a Au(111) surface. The binding energy of the O(1s) peak is observed between 529.8 eV and 530.2 eV, with the O(1s) peak shifting to higher binding energies with increasing oxygen coverage.

O_2 TPD spectra from the atomic oxygen adlayer shows an O_2 desorption peak maximum that shifts from 525 K to higher temperatures for increasing oxygen coverages for $0 \leq \Theta_O \leq 0.1$ and then remains constant at 550 K for higher coverages. Analysis of the TPD data indicates that the desorption of O_2 from Au(111) can be described well by first-order kinetics. The activation energy for desorption exhibits a sharp increase with increasing coverage from 16 kcal/mol at $\Theta_O = 0$ to 28 kcal/mol at $\Theta_O = 0.15$ and then exhibits a slow increase to 29 kcal/mol near saturation. This results in an upper limit for the Au-O bond dissociation energy of D(Au-O) ≤ 74 kcal/mol at $\Theta_O = 1$. We can not determine this bond energy more accurately at this time since the value of the activation energy for dissociative O_2 adsorption is not known. Sault et al. (*14*) observed similar O_2 desorption kinetics from the oxygen adlayer on the Au(110) surface, but they observed a peak maximum about 40 K higher than we observe on Au(111).

Coadsorption and low temperature reaction of H_2O and NO_2 on Au(111).

Large coverages of atomic oxygen on Au(111) can also be prepared by a NO_2 + H_2O surface reaction that occurs much below room temperature. Preliminary experiments have been performed in which H_2O was adsorbed on NO_2 precovered Au(111) at 95 K. Figure 2 shows a few of the TPD results for increasing precoverages of NO_2. The bottom curves are following water adsorption on clean Au(111). A desorption state characteristic of physisorbed water is seen in the left panel, and, of course, no NO_2 desorption is observed in the right panel. Precoverage of the surface with NO_2 induces two new water desorption peaks at 180 and 215 K, and leads to the <u>formation of oxygen adatoms at the surface</u>. Thus, we observe a large O_2 desorption in TPD at 550 K (identical to O_2 TPD spectra of ref. (*13*))which increases up to about $\Theta_O = 0.5$ with increasing NO_2 precoverage. This is a significant result since both H_2O and NO_2 are completely reversibly adsorbed, i.e., form no surface oxygen, when adsorbed separately. The water TPD state at 180 K probably has its origin in water stabilized by hydrogen bonding with oxygen adatoms as we have seen before. We postulate that the peak at 215 K is due to disproportionation of surface hydroxyl species to form H_2O. NO_2 is very reactive, as it looks like surface oxygen from the reaction continues to accumulate even after NO_2 multilayers are formed. It is premature at this point to speculate on the mechanism of this intriguing reaction. We plan HREELS studies of this system in the future.

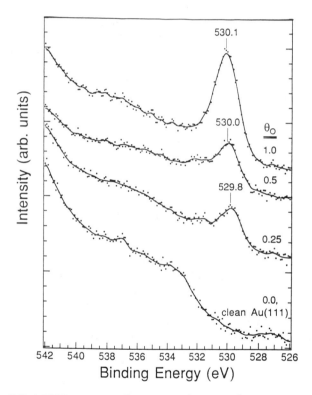

Figure 1. O(1s) XPS spectra of oxygen adatoms after ozone exposure on clean Au(111) at 300 K.

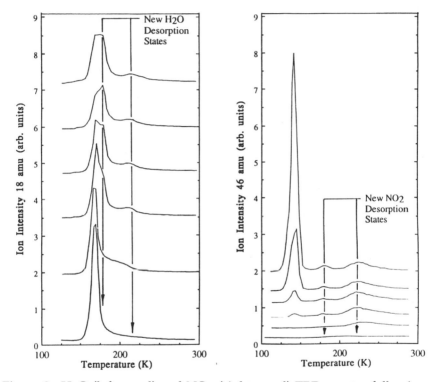

Figure 2. H_2O (left panel) and NO_2 (right panel) TPD spectra following the adsorption of H_2O on Au(111) at 110 K with increasing NO_2 precoverages.

Finally, we note that H_2O, because of its low gas phase acidity, does not react with oxygen adatoms on the Au(111) surface (as discussed in a later section). If we can find conditions or reagents (such as NO_2) that will activate water, i.e., abstract hydrogen to form surface hydroxyl species, then we can expect to see reactions with a large variety of molecules with higher gas phase acidities such as methanol, etc.

Adsorption and Reaction of CO and CO_2. The clean and oxygen-precovered Au(111) surface does not chemisorb CO at temperatures down to 100 K. This was established by CO uptake experiments utilizing XPS and TPD. C(1s) and O(1s) XPS spectra taken after a 10 L CO exposure to Au(111) with $\Theta_O = 0.0$, 0.5 and 1.0 at 100 K showed no evidence for a new carbon-containing surface species, i.e., no adsorbed CO, CO_2, or carbonate species is stable on the Au(111) surface under these conditions. The area of the O(1s) peak after CO exposure was slightly smaller for both oxygen precoverages, but this is attributed to background CO oxidation and the formation of CO_2 which desorbs from the surface at 100 K (vide infra).

The clean surface of a Au(111) crystal at 100 K and a surface covered with $\Theta_O = 0.12$, 0.25, 0.75, and 1.0 was dosed with 10 and 50 L CO_2. XPS studies of the C(1s) and O(1s) regions did not reveal any significant new peaks after CO_2 exposure. Appreciable CO_2 chemisorption does not occur on the clean or oxygen-dosed Au(111) surface, nor does a stable surface carbonate form under these conditions.

CO oxidation is one of the simplest catalytic reactions, and thus it is of interest from a fundamental as well as technological viewpoint. The reactions of surface oxygen with CO, or other organic molecules, can often be explored easily by titration transient techniques that involve introducing gas phase reactants and monitoring the products and the surface oxygen coverage as a function of time. When CO was exposed to oxygen on the Au(111) surface, CO_2 was formed and desorbed from the surface in a clean-off reaction. AES was used to monitor the surface oxygen coverage and thus measure the reaction rate as a function of oxygen coverage. We have studied the transient reaction kinetics of CO oxidation with adsorbed oxygen adatoms to form CO_2 over Au(111) single crystals at temperatures from 250 - 375 K and CO pressures between 2×10^{-8} and 1×10^{-7} Torr (15). Kinetics were obtained for high precoverages of atomic oxygen (up to $\Theta_O = 1$) by using O_3 decomposition . This is the first study of any reaction on Au(111) under clean, UHV conditions and such high oxygen coverages.

The CO oxidation reaction occurs rapidly at room temperature and below. As an example, on a Au(111) surface at 250 K with Θ_O near unity exposed to a constant CO pressure of $P_{CO} = 2 \times 10^{-8}$ Torr, the reaction rate expressed as turn-over frequency (TOF; [molecules CO_2 (Au atom·s)$^{-1}$]), is approximately 2.5×10^{-3} immediately after the reaction has been initiated and then declines at a relatively constant rate reaching a value of 3×10^{-4} after the reaction has proceeded approximately 800 seconds. Reaction order studies of the CO oxidation rate were carried out as a function of oxygen coverage at CO pressures of 2, 5, and 10 x 10^{-8} Torr and as a function of CO pressure at oxygen coverages of Θ_O = 0.3, 0.5, 0.7 at three reaction temperatures of 250, 300, and 375 K. The data for a reaction temperature of 300 K is shown in Figure 3a. The CO oxidation rate on oxygen covered Au(111) was close to first order in CO pressure and first order in Θ_O over the entire range of Θ_O.

Analysis of the temperature dependence of this reaction by Arrhenius plots at each oxygen coverage, shown in Figure 3b, yielded reaction apparent activation energies (E_{app}) of -2.5 ± 0.7 kcal/mole for all conditions studied, more or less independent of Θ_O or CO pressure. The negative value for E_{app} results from the reaction rate increase as the temperature decreases over the range of 250 - 375 K. This result is in contrast to most studies of CO oxidation, where the rate increases with increasing temperature. The implications of this result is that a mechanism in which a single elementary reaction step is rate-limiting is ruled out since a single elementary reaction step can not have a negative activation energy. Importantly, this very clearly and unambiguously rules out an Eley-Rideal mechanism, even though this system might be considered as a great candidate since CO does not chemisorb on the clean or oxygen-covered Au surface and the reaction rate is first order in Θ_O up to one monolayer of surface oxygen. In summary, the Langmuir-Hinshelwood (LH) mechanism was found to be in operation for CO oxidation over the Au(111) surface over the entire range of oxygen coverage. This is an interesting result since the preferred CO oxidation reaction mechanism occurs via an adsorbed CO species, even though CO is not appreciably activated by adsorption on Au(111), i.e., CO is only physisorbed with heat of adsorption less that 5 kcal/mol, and the activation energy for the CO + O reaction must be very small (E_{LH} = E_{app} + E^{CO}_{des} = -2.5 + 5 = 2.5 kcal/mol).

Surface science studies of CO oxidation on Au(110) single crystals have been made previously in which a Pt filament was used to adsorb oxygen adatoms (Θ_O ≤ 0.25) on the Au(110) surface and a CO titration was performed subsequently (7). CO did react to form CO_2 with E_{app} = 2 ± 1 kcal/mole. Since CO was not observed to adsorb on Au(110) at 125 K, it is only physisorbed (as on Au(111)) and we can estimate that E_{LH} = 7 kcal/mol on Au(110). The difference from Au(111) is probably due to a weaker Au-O bond on Au(111) which leads to a lower barrier for reaction. No surface carbonate was formed from CO_2 + O_a on Au(110) either (7). This is in contrast to the behavior on Ag. Exposing oxygen covered Ag(110) (16) or Ag(111) (17) to CO_2 produces carbonate species which are stable to 485 K on the surface.

Adsorption and reaction of NO_2. In contrast to the lack of reactivity of CO_2 on the clean Au(111) surface, NO_2 is molecularly chemisorbed via its two oxygen atoms on clean Au(111) at temperatures of 175 K and below to form a O,O'-nitrito surface chelate with C_{2v} symmetry (12). The NO_2 sticking

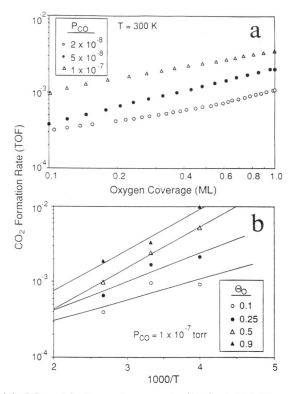

Figure 3. (a) CO oxidation rate over Au(111) at 300 K versus oxygen coverage at CO pressures of 2, 5, and 10 x 10^{-8} Torr. (b) Arrhenius plot of the CO oxidation rate over Au(111) at a constant CO pressure of 1 x 10^{-7} Torr and constant oxygen coverages of 0.1, 0.25, 0.5, and 0.9 monolayer.

coefficient on Au(111) in this temperature range is unity, and the chemisorbed saturation coverage is estimated to be $\Theta_{NO2} = 0.4$. NO_2 is reversibly adsorbed, i.e. no decomposition occurs when the surface is heated, and NO_2 desorbs in TPD with first-order kinetics and with a peak temperature of 220 K. Since adsorption is not an activated process, the activation energy for desorption of 14 kcal/mol also represents the adsorption energy. NO_2 exposures that are larger than required to saturate the chemisorbed state on Au(111) at 100 K populate an N_2O_4 physisorbed state. Also, when the chemisorbed NO_2 surface chelate on Au(111) at 100 K is exposed to NO, N_2O_3 is formed on the surface. N_2O_3 is adsorbed in an upright configuration with a vibrational spectrum very close to the gas-phase N_2O_3 species. Chemisorbed N_2O_3 decomposes upon heating to 175 K to yield $NO_{(g)}$ and the unperturbed NO_2 surface chelate.

TPD and HREELS were used to probe the interaction of oxygen adatoms and NO_2 on Au(111) at 100 K (*18*). The NO_2 sticking coefficient retains a constant value of unity at 100 K regardless of the oxygen atom precoverage, Θ_O, but oxygen adatoms inhibit the formation of the O,O'-nitrito surface chelate. Oxygen atoms reduce the chemisorbed NO_2 coverage to zero when $\Theta_O = 1$, but does not alter the desorption energy of the coadsorbed surface chelate that forms at lower Θ_O. Alternative bonding geometries (linkage isomers) on this chemically modified surface, such as N-bonded nitro or O-bonded nitrito, were not observed.

Figure 4 shows that surface oxygen adatoms cause a new NO_2 desorption peak in TPD at 175 K. This state appears to follow first-order desorption kinetics with a peak significantly higher in temperature than that shown in the bottom curve of Figure 4 for the decomposition of N_2O_4 at 150 K on clean Au(111). The species responsible for this peak is formed with unit probability at temperatures below 175 K, forming a saturation coverage of $\Theta_{NO2} = 0.6$ when the surface is presaturated with oxygen atoms at $\Theta_O = 1$. Several of these observations concerning the adsorption and desorption kinetics implicate the formation of at least a transient NO_3 species, but the unambiguous spectroscopic assignment of a stable NO_3 species could not be made. The clear identification of the chemical nature of adsorbed NO_2 in the first monolayer on the oxygen-precovered Au(111) surface at 100 K is not possible at this time, since important differences exist between the observed vibrational spectrum and that expected either for NO_3 or physisorbed N_2O_4 (*18*). However, if an NO_3 species is formed, it is much less stable than on Ag(110) where surface bound NO_3 does not decompose until 310 K (*19*). This is also consistent with the aforementioned lower stability of CO_3 on Au(111) compared to Ag.

Consideration of the adsorption and reaction of NO_2 on Au is very important in its own right, since chemisorbed NO_2 can potentially serve as a labile source of oxygen for carrying out catalytic oxidation reactions. The ON-O bond strength in the gas phase is 73 kcal/mol and this should be reduced by another 5 kcal/mol for the chemisorbed species due to the stabilization of the NO product. Thus, the bonds that must be broken (ON-O or Au-O bond) in order to form oxidation products may be comparable in strength and have similar reactivities, depending on the activation energy barriers that exist.

Adsorption of H_2O and the $O_{(a)} + H_2O$ interaction. H_2O is a product of many oxidation reactions and its interaction with these surfaces is important to understand. We have studied water adsorption on clean and oxygen precovered Au(111) using principally TPD and XPS, and these results are summarized below.

H_2O Adsorption on clean Au(111). Water is very weakly adsorbed on clean Au(111), desorbing in a peak near 160 K in TPD studies at low H_2O exposures. Intermediate exposures cause two states, at 160 and 164 K, and at much higher exposures water desorption exhibits zero order kinetics indicative of the formation of solid water multilayers. We assign desorption spectra from exposures prior to 1 L to "monolayer" desorption, since the low temperature, zero order state increases preferentially at larger exposures. Interestingly, no clearly distinct monolayer desorption peak is observed. A rough estimate of 10-11 kcal/mol for the activation energy for monolayer desorption can be made using Redhead analysis with first order desorption kinetics. This also represents the adsorption energy for water on Au(111). No O_2 desorption was observed in these experiments, indicating reversible, molecular water adsorption.

An O(1s) XPS spectrum of water exposed to clean Au(111) is shown as the bottom curve in Figure 5. Water adsorbed on clean Au(111) is characterized by an O(1s) peak with a binding energy of 533.2 eV. This result is consistent with O(1s) binding energies of monolayer coverages of molecularly adsorbed water on other metal surfaces, e.g. 533.3 eV BE on Cu(111)(20) and 532.2 eV BE on Pt(111) (21). The O(1s) peak for multilayer water (not shown) is at 534.3 eV BE. Since H_2O is so weakly adsorbed, the 1.1 eV shift is due primarily to final state screening effects which are much greater for the adsorbed monolayer.

H_2O TPD on Au(110) has been studied by Outka and Madix (8). They observed monolayer and multilayer desorption states at 190 and 185 K, respectively. The differences in these temperatures from those reported here on Au(111) are most likely due to a small temperature measurement error in ref.(8). They assigned rather unusual zero-order desorption kinetics to the monolayer H_2O peak. While kinetic orders lower than first order for monolayer desorption can be due to hydrogen bonding and clustering on the surface, the leading edges of the TPD peaks do not appear to align correctly and it is possible that first order desorption occurs on both Au(111) and Au(110). H_2O adsorption on Au(110) is also completely reversible, as on all of the Group IB metals (8).

H_2O adsorption on oxygen-precovered Au(111). Water was exposed to 0.25, 0.50, and 1.0 monolayer atomic oxygen on Au(111), and two sets of the subsequent TPD spectra are shown in Figure 6. For these particular experiments three masses were monitored, H_2 (m/e = 2), H_2O (m/e = 18)

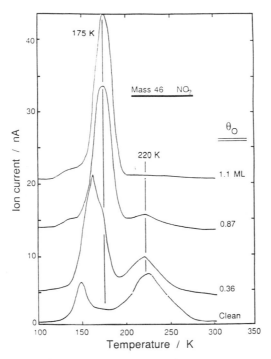

Figure 4. NO_2 TPD spectra after 3.2 L NO_2 exposure on oxygen-precovered Au(111) at 100 K.

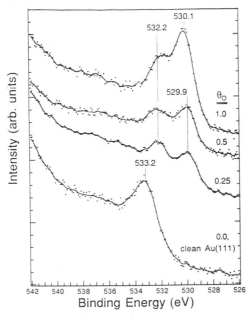

Figure 5. O(1s) XPS spectra following 0.12 L H_2O exposure on Au(111) at 110 K with several oxygen precoverages.

and O_2 (m/e =32). Hydrogen was not observed to desorb at temperatures up to 700 K. Oxygen desorbed with the same quantity and at the same temperature, 529-542 K, as from a clean gold surface indicating that surface oxygen was either unreactive on the surface or that water reacting on the surface to form hydroxyl groups reversibly desorbed at lower temperatures without increasing the oxygen adatom coverage.

For water exposures on surfaces with Θ_O = 0.25 and 0.50, two desorption peaks were observed near 183 K and 167 K. The low temperature state displays zero order desorption kinetics and is ascribed to physisorbed water multilayers. The high temperature peak can be attributed to oxygen-stabilized water or recombination and disproportionation of surface hydroxyl groups. The high temperature state is populated prior to the low temperature state. If hydroxyl formation occurred it is reasonable to expect that there would be an increase in hydroxyl formation, i.e., an increase in the area of the high temperature state, with an increase in oxygen coverage. This increase is not observed. Studies on Ru(001) show that the ratio of water bound to adsorbed oxygen on a 0.25 and 0.50 monolayer oxygen covered surface is 3.5:1 and 1:1, respectively (22). Similar results are observed on Ag(110) (16). Thus, the desorption peak at 182 K is tentatively attributed to oxygen-stabilized, nondissociatively adsorbed water. The 30 K increase in peak desorption temperature with respect to water desorption from the clean surface is due to the stabilization of water on the surface by hydrogen bonding with adsorbed oxygen or increased interaction with the metal surface due to the oxygen-induced formation of $Au^{\partial+}$ sites. The stabilization is not very large and the activation energy for desorption can be estimated using Redhead analysis to be 10.8 kcal/mol.

On Au(111) with Θ_O =1.0, a single H_2O desorption peak is observed at 168 K, as shown in the right panel of Figure 6. This peak is due to physisorbed water on the oxygen monolayer. This coverage of oxygen blocks the high temperature desorption state, which indicates that the direct adsorption of water to the Au(111) surface is blocked.

In order to distinguish between the possible explanations for the 183 K peak, spectroscopic studies using XPS were carried out. For H_2O exposures on the Au(111) surface with preadsorbed oxygen, a new O(1s) peak is observed at 532.2 eV BE as shown in Figure 5. The 1.0 eV shift could possibly be attributed to formation of a hydroxyl species, changes in hydrogen bonding interactions, or changes in the surface electronic state. The dissociation of water on oxygen precovered metal surfaces to form hydroxyl has been observed in many studies. The O(1s) peak is at 531.5 eV on Cu(111) (20) and at 530.5 eV on Pt(111) (21), corresponding to 2.0 eV and 1.7 eV shifts, respectively, when compared with water adsorbed on the clean surface. These shifts are roughly 1 eV greater, or twice as large, as the shift we observe in Figure 5. Also, if water dissociated on the oxygen covered Au(111) surface, the intensity of the preadsorbed oxygen O(1s) peak would diminish from the conversion of oxygen adatoms to hydroxyl species by H atom abstraction from H_2O. This is not observed in Figure 5.

The ratio of water to oxygen is sufficient to consume the entire amount of preadsorbed oxygen by forming hydroxyl species. Hydrogen bonding between adsorbed water molecules has been observed to cause 0.2 eV shifts (23). The increased electron density on the oxygen adatoms may cause adsorbed water to have a stronger interaction in hydrogen bonding with oxygen adatoms compared to coadsorbed water and account for the 0.9 eV shift.

Our TPD and XPS results show that water is nondissociatively adsorbed on clean Au(111) at 100 K. This is in good agreement with previous studies on other Group IB surfaces, Cu(111) (20), Ag(110) (24) and Au(110) (8). Monte Carlo simulations predict that water can adsorb in clusters on metal surfaces (25). Electron energy loss spectra for water adsorbed on Ag(110) indicate that extensive clustering of water does occur at 100 K (24). Electron stimulated desorption ion angular distribution (ESDIAD) experiments on Ru(001) show that at low coverages, less than 0.2 monolayers, interactions between neighboring molecules are weak, but lateral interactions increase with increasing coverage (26). TPD of water on clean Pt(111) suggests the formation of islands with sufficient exposures (21). Thus, the complex desorption kinetics are often affected by strong lateral interactions and possibly island formation due to hydrogen bonding.

Our XPS results for water adsorbed on oxygen precovered Au(111) show that water is molecularly adsorbed on the surface at 100 K. While we can not at this time rule our transient hydroxyl formation at higher temperatures during TPD, oxygen adatoms at all coverages on Au(111) are less reactive toward water than on the other Group IB metal surfaces (20, 24). On Ag and Cu, oxygen adatoms abstract H from water at temperatures below 150 K to form adsorbed hydroxyl species. At higher temperatures hydroxyl species recombine to desorb as molecular water.

Adsorption and reaction of CH_3OH. CH_3OH was dosed on the surface of a Au(111) crystal at 100 K with Θ_O = 0, 0.1, 0.25, 0.5, 0.75, and 1. Several of the CH_3OH TPD spectra from these experiments are shown in Figure 7. On the clean Au(111) surface, CH_3OH is only weakly and reversibly chemisorbed, desorbing molecularly at low coverages in a peak at 184 K in TPD. Formation of physisorbed CH_3OH multilayers at higher exposures of 1 L or larger leads to a desorption peak at 168 K with an onset near 140 K. Surface oxygen shifts the most tightly bound CH_3OH to a peak at 220 K, thus increasing the adsorption energy of CH_3OH, and also alters the desorption spectra of the weakly bonded states in the temperature range of 170 - 200 K. On the surface with Θ_O = 1, the CH_3OH monolayer desorbs at 190 K.

Coadsorbed oxygen adatoms activate CH_3OH dissociation to produce CO_2 and H_2O, as shown in Figure 8 for 0.2 L CH_3OH exposures. For Θ_O = 0.1 and 0.25, CH_3OH desorption occurs in two peaks at 201 and 224 K. These peaks arise from an increase in the binding energy of methanol on the surface and/or CH_3OH recombination. The surface with Θ_O = 1 is still reactive, desorbing the majority of the CO_2 product at 120 K. This shift of product formation to lower temperature indicates that high oxygen

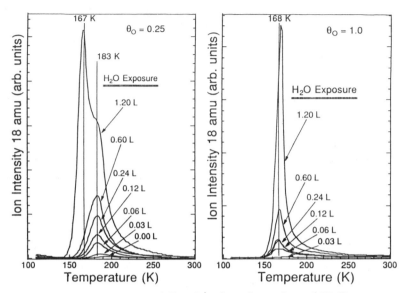

Figure 6. H$_2$O TPD spectra following the adsorption of H$_2$O on oxygen-precovered Au(111) at 110 K with Θ_O = 0.25 (left panel) and Θ_O = 1.0 (right panel).

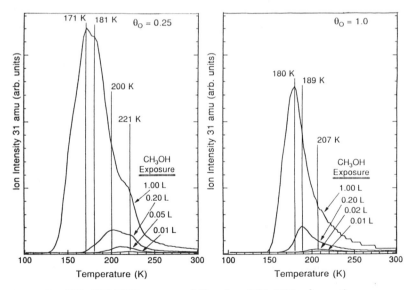

Figure 7. CH3OH TPD spectra following CH3OH adsorption on oxygen-precovered Au(111) at 95 K with Θ_O = 0.25 (left panel) and Θ_O = 1.0 (right panel).

coverages may lower the activation energy for methanol oxidation on Au catalysts.

Figure 9 shows C(1s) and O(1s) XPS spectra taken during annealing studies following 0.2 L CH_3OH exposure on Au(111) at 95 K with $\Theta_O = 0.25$. A CH_3OH exposure of 0.2 L on clean Au(111) at 95 K (not shown) gave a C(1s) peak at 286.3 eV BE, which disappeared completely by 180 K. On Au(111) with $\Theta_O = 0.25$, a peak at 285.9 eV BE is seen at 95 K and some C-containing species with a C(1s) peak at 285.3 is formed at 185 K. The O(1s) peak for CH_3OH adsorbed on Au(111) at 95 K with $\Theta_O = 0.25$ is at 532.5 eV BE. This peak is largely removed by heating to 185 K, but a broad peak remains centered near 531 eV BE that could be due to hydroxyls or some other partially oxidized intermediate. Vibrational spectroscopy will have to be used to make this identification and to further elaborate on the decomposition mechanism.

The XPS studies of the C(1s) and O(1s) regions at 100 K are not conclusive in determining whether reaction occurs immediately upon adsorption at 95 K to form hydroxyl or other oxidized products. However, during TPD, oxygen adatoms activate CH_3OH decomposition to yield CO_2 and H_2O desorption. This chemistry on Au(111) appears to be very similar to that on Au(110) (*8*).

Adsorption and reaction of C_2H_4. Ethylene reversibly adsorbed on clean Au(111) at 95 K. Desorption of ethylene in TPD from the clean Au(111) surface was observed from approximately 100 - 250 K, with a large peak at 104 K. The reason for the broad desorption peak is uncertain. Outka and Madix (*8*) have shown that a broad range of desorption is typical of C_2 hydrocarbons on Au(110). They propose that these molecules are weakly bound and therefore occupy a variety of binding sites with different activation energies, consequently they desorb at different temperatures. We observed no H_2 evolution from the surface up to temperatures of 700 K and AES showed no residual carbon on the surface after heating. Thus, C_2H_4 is reversibly, and most likely molecularly, adsorbed on clean Au(111).

XPS studies of ethylene exposed to clean Au(111) at 95 K were also carried out. The C(1s) peak occurred at 284.7 eV BE. Warm-up experiments show that the carbon peak was still present at 165 K, confirming that the long tail observed in the TPD spectra was due to adsorbed ethylene on the Au(111) surface. Only trace amounts of carbon were observed at 200 K, and none was observed at 300 K. These results are consistent with the thermal desorption results. Also, LEED showed no new ordered structure was formed due to adsorbed ethylene.

Ethylene was exposed on oxygen covered Au(111) at 95 K with $\Theta_O = 0.1, 0.5$ and 1.0. Subsequent TPD spectra showed only ethylene desorption in a peak shifted to a slightly higher temperature of 111 K, due to a small stabilization by coadsorbed oxygen, again with a long tail extending to higher temperatures. Masses appropriate for H_2, C, H_2O, C_2H_4O, CO, and CO_2 were also monitored, but no desorption was observed upon heating the sample to 700 K. AES showed that carbon did not exist on the surface

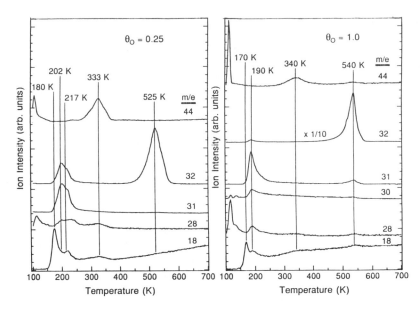

Figure 8. TPD spectra following the adsorption of 0.2 L CH3OH on oxygen-precovered Au(111) at 95 K with Θ_O = 0.25 (left panel) and Θ_O = 1.0 (right panel).

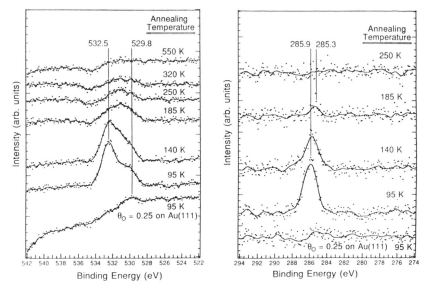

Figure 9. C(1s) (left panel) and O(1s) (right panel) XPS spectra taken during annealing studies following 0.2 L CH3OH exposure on Au(111) at 95 K with Θ_O = 0.25.

following TPD and oxygen desorption occurred at the same temperature as without ethylene adsorption. The amount of oxygen desorbed after ethylene exposure was 10, 94 and 96 percent of the amount adsorbed prior to ethylene exposure for 0.1, 0.5 and 1.0 monolayer coverages, respectively. Since the oxygen loss can be attributed to reactions with background CO to form CO_2, there is no clear evidence that ethylene reacts on the oxygen precovered surface.

In similar studies, ethylene (4.0 L) was also exposed to a Au(111) surface at 300 and 400 K with $\Theta_O = 0.25$. Subsequently, O_2 desorbed with greater than 70 percent of the amount adsorbed prior to ethylene exposure. Again, we find no evidence for reaction of ethylene with adsorbed oxygen on the Au(111) surface at these pressures.

These results are in good agreement with similar experiments carried out on Au(110) (8). The lack of reactivity on the clean gold surface is not surprising as gold is known to have a low activity for hydrocarbon dissociation. Ethylene was unreactive with the oxygen covered Au(111) surface for temperatures up to 400 K. These results agree with the gas phase acidity - reactivity correlations proposed previously (8, 10, 11). That is, ethylene with a high gas phase acidity enthalpy (low acidity) is predicted to be less reactive on gold than molecules which have lower gas phase acidity enthalpies (greater acidity). For water and ethylene the gas-phase acidity enthalpies are 391 and 416 kcal/mol, respectively. Since water was observed to be unreactive on the oxygen covered Au(111) surface, it was predicted that ethylene was also unreactive.

Summary

Ozone can be used to produce oxygen adatoms on Au surfaces under UHV conditions. This allows for new surface science studies utilizing surface oxygen on Au model catalyst surfaces. In studies of up to one monolayer of oxygen adatoms on Au(111), no strong evidence for special reactivity of these high coverages of oxygen was observed. It is not enough to simply have a collision of an incident reactant with the oxygen adatoms; Au surface sites are required for reaction. CO and CO_2 do not adsorb or form a stable surface carbonate species on the clean or oxygen covered Au(111) surface at 100 K. CO oxidation, however, occurs readily below room temperature. The CO oxidation rate on Au(111) was nearly first order in CO pressure and first order in oxygen coverage over the monolayer range. A Langmuir-Hinshelwood mechanism was found to be in operation over the entire range of oxygen coverage; we observed negative apparent activation energies ($E_{app} = -2.5$ kcal/mol) with no strong dependence on oxygen coverage. NO_2 chemisorbs on Au(111), but a surface nitrate species is not very stable, if it forms at all. Carbonate and nitrate species are much less stable on Au than on Ag. Water and ethylene are weakly adsorbed on the clean and oxygen covered surface at 100 K, but do not react with coadsorbed oxygen adatoms on Au(111). Methanol is weakly and reversibly adsorbed on clean Au(111), and methanol combustion to form CO_2 and

H_2O was observed on oxygen covered Au(111). The reactivity of these molecules with oxygen coadsorbed on Au(111) is shown to be less than on copper and silver surfaces, but similar to that on Au(110). The observed reactivity on oxygen covered Au(111) is consistent with the gas phase acidities of the reactant molecules. Finally, an interesting low temperature reaction between NO_2 and H_2O was observed, suggesting that surface bound NO_2 is a strong H-abstraction agent (more reactive than oxygen adatoms) and thus has interesting potential as an oxidizing reagent for use with Au catalysts.

Acknowledgments

Acknowledgment is made to the Donors of The Petroleum Research Fund, administered by the American Chemical Society, for partial support of this work. GNK gratefully acknowledges the Arizona State Research Institute Postdoctoral Fellowship Program for partial financial support .

Literature Cited

1. Schwank, J. *Gold Bulletin* **1983**, *16*, 103.
2. McNerney, J.J.; Buseck, P.R.; Hanson, R.C. *Science* **1972**, *178*, 611.
3. Fahey, D. W.; Eubank, S. S.; Hübler, G.; Fehsenfeld, F. C. *J. Atmos. Chem.* **1985**, *3*, 435.
4. Haruta, M.; Yamada, N.; Kobayashi, T.; Iijima, S. *J. Catal.* **1989**, *115*, 301.
5. Outka, D.A.; Madix, R.J. *Surf. Sci.* **1987**, *179*, 361.
6. Chtaib, M.; Thiry, P.A.; Pireaux, J.J.; Delrue, J.P.; Caudano, R. *Surf. Sci.* **1985**, *162*, 245.
7. Outka, D.A.; Madix, R.J. *Surf. Sci.* **1987**, *179*, 351.
8. Outka, D.A.; Madix, R.J. *J. Am. Chem. Soc.* **1987**, *109*, 1708.
9. Barteau, M.A.; Madix, R.J. In *The Chemical Physics of Solid Surfaces and Heterogeneous Catalysis;* King, D.A.; Woodruff, D.P., Eds.; Elsevier, New York, 1982, *Vol. 4;* p. 95.
10. Barteau, M.A.; Madix, R.J. *Surf. Sci.* **1982**, *120*, 262.
11. Jorgensen, S.W.; Madix, R.J. *Surf. Sci.* **1983**, *130*, L291.
12. Bartram, M.E.; Koel, B.E. *Surf. Sci.* **1989**, *213*, 137.
13. Parker, D.H.; Koel, B.E. *Surf. Sci.*, submitted.
14. Sault, A.G.; Madix, R.J.; Campbell, C.T. *Surf. Sci.* **1986**, *169*, 347.
15. Wickham, D.T.; Parker, D.H.; Kastanas, G.N.; Koel, B.E. *Surface Sci.*, submitted.
16. Bowker, M.; Barteau, M.A.; Madix, R.J. *Surf. Sci.* **1980**, *92*, 528.
17. Felter, T.E.; Weinberg, W.H.; Latushkina, G.Y.; Boronin, A.I.; Zhdan, P.A.; Boreskov, G.K.; Hrbek, J. *Surf. Sci.* **1982**, *118*, 369.
18. Bartram, M.E.; Parker, D.H.; Koel, B.E. *J. Am. Chem. Soc.*, submitted.
19. Outka, D.A.; Madix, R.J.; Fisher, G.B.; Dimaggio, C. *Surf. Sci.* **1987**, *179*, 1.
20. Au, C.; Breza, J.; Roberts, M.W. *Chem. Phys. Lett.* **1979**, *66*, 340.
21. Fisher, G.B.; Gland, J.L. *Surf. Sci.* **1980**, *94*, 446.

22. Kretzschmer, K.; Sass,J.K.; Bradshaw, A.M.; Holloway, S. *Surf. Sci.* **1982,** *115,* 183.
23. Sexton, B.A. *Surf. Sci.* **1980,** *94,* 435.
24. Stuve, E.M; Madix, R.J.; Sexton, B.A. *Surf. Sci.* **1981,** *111,* 11.
25. Clementi, J. *J.Phys.Chem.* **1985,** *89,* 4426.
26. Madey, T.E.; Yates, J.T.,Jr. *Chem. Phys. Lett.* **1977,** *51,* 77.

RECEIVED November 6, 1992

Chapter 9

Control of the Methanol Reaction Pathway by Oxygen Adsorbed on Mo(112)

Tetsuya Aruga, Ken-ichi Fukui, and Yasuhiro Iwasawa

Department of Chemistry, Faculty of Science, The University of Tokyo, Hongo, Bunkyo-ku, Tokyo 113, Japan

The effect of oxygen adatoms on the reaction path of methanol on the Mo(112) surface has been examined in relation to the genesis of solid catalysis as well as the creation of new active surfaces. It has been found that the formation of a p(1x2)-O layer results in a new methanol dehydrogenation path, which differs from the oxidative dehydrogenation usually observed on molybdenum oxides. The CO adsorption experiment indicated that half the first-layer Mo atoms on the Mo(112)-p(1x2) surface are completely blocked while the rest are almost free from the electronic effect of oxygen modifiers, suggesting that the new dehydrogenation path is due to the selective blocking of the second-layer Mo atoms, leaving one-dimensional rows of bare Mo atoms.

It has been one of the long-sought goals of the surface chemistry to optimize the structure and electronic properties of catalyst surfaces for particular catalytic reactions. To this end, considerable efforts have been devoted to achieve a full understanding of microscopic principles of the catalysis. Practically, a complete set of techniques for surface modification should be established to modify the catalyst surfaces and control the reaction paths. In order to establish reliable means to modify the electronic properties and steric confinement of the surface, we have examined the modification of the Mo(112) surface (Figure 1) by atomic oxygen. Molybdenum, both in metallic and oxide forms, is used in many industrial catalysts. This is partly because molybdenum exhibits a wide range of chemical reactivity according to its various oxidation states. MoO_3 and iron/molybdenum oxides are used as industrial catalysts for methanol oxidation to form formaldehyde selectively. The iron/molybdenum oxide catalyst consists of $Fe_2(MoO_4)_3$ and MoO_3, and shows kinetics and selectivity similar to that of MoO_3 (1), suggesting that Mo-O sites play a dominant role in the methanol oxidation. MoO_3 has a layered structure along the (010) plane. The (010) surface is not chemically active because there are no dangling bonds and unsaturated Mo atoms. Actually, Sleight et al. (2) studied methanol adsorption on the (010) surface of a MoO_3 single crystal and found that no methanol chemisorbs on the $MoO_3(010)$ surface. On the other hand, methanol is decomposed completely to CO and H_2 on metallic Mo (3,4), suggesting oxidative dehydrogenation of methanol occurs on partially oxidized Mo sites or defect sites. It would be interesting if the active site for the selective dehydrogenation of alcohols can be prepared on well-defined crystal surfaces.

0097–6156/93/0523–0110$06.00/0

Molybdenum has a body-centered cubic lattice, and its (112)-(1x1) surface is composed of densely packed ($d_{\text{Mo-Mo}}$ = 2.73 Å) atomic rows separated by 4.45 Å from each other. Oxygen atoms are expected to occupy trough sites and hence the first-layer atoms are accessible for gas-phase molecules, allowing directly probing the electronic effect of oxygen adatoms on the first-layer Mo atoms by the adsorption of simple molecules. The oxygen adsorption on Mo(112) results in the successive formation of ordered structures as a function of oxygen coverage. Upon annealing the Mo(112) surface with a very high coverage of oxygen atoms, MoO_3 grows epitaxially (5). The modification of Mo(112) by oxygen adatoms will change the methanol chemistry drastically, which provides a model for the industrial catalysts for methanol dehydrogenation. In the present chapter, we will survey the experimental findings on the modification of Mo(112) by submonolayer-coverage oxygen adatoms and its effects on the reaction of methanol. The results presented here indicate that the selective blocking of the second-layer Mo atoms results in a novel dehydrogenation reaction of methanol on this surface. This dehydrogenation reaction ($CH_3OH \rightarrow H_2CO + H_2$) differs from the oxidative dehydrogenation reaction ($CH_3OH + O \rightarrow H_2CO + H_2O$) observed for molybdenum oxides. We also discuss the reaction scheme for methanol dehydrogenation on oxygen-modified Mo(112) surfaces.

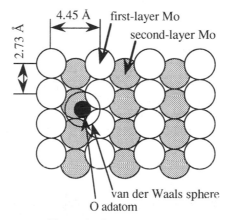

Figure 1. Oxygen on Mo(112)

Preparation of Ordered Oxygen Overlayers on Mo(112)

The clean Mo(112) surface exhibits a sharp p(1x1) pattern in low-energy electron diffraction (LEED), indicating that the surface preserves the bulk structure as shown in Figure 1. The exposure of the clean surface to oxygen at room temperature, followed by annealing to 600 K, results in a series of ordered structures as observed by LEED (6). The LEED patterns observed with increasing oxygen coverage include p(2x1), p(1x2), three patterns (A-C) before p(2x1), and two patterns (D,E) between p(2x1) and p(1x2). The oxygen coverage was monitored by Auger electron spectroscopy (AES) and was calibrated to that for the p(2x1)-O surface, 0.5 ML. The Mo(112)-p(1x2)-O structure was found to be completed at θ_O=1.0 ML. When the surface was further exposed to oxygen, the oxygen coverage gradually increased as monitored by AES. The p(1x2) pattern remained unchanged, while the spots were diffused slightly. Upon annealing to around 700 K, the p(1x2) surface with excess oxygen was changed to p(1xn) ($n>3$) and sometimes the surface was faceted, while the p(1x2)-O surface of θ_0=1.0 remained unchanged even after annealing to 1000 K. The oxygen modified surfaces of $\theta_0 \leq 1.0$ were stable up to 1100 K.

There are several possible adsorption sites for oxygen atoms on Mo(112). If one assumes that oxygen atoms tend to occupy high coordination sites, the most plausible adsorption site can be assumed to be either a quasi-threefold site composed of two first-layer and one second-layer Mo atoms or a quasi-threefold site composed of one first-layer and two second-layer Mo atoms. Whereas the adsorption site of oxygen on Mo(112) has not yet been determined, Rabalais et al. (7) carried out ion scattering and recoiling experiments on the W(112)-p(2x1)-O and p(1x2)-O surfaces. They concluded that the quasi-threefold site composed of two first-layer and one second-layer atoms is most probable on both surfaces. We therefore suppose that the oxygen atoms on Mo(112) occupy the quasi-threefold hollow sites composed of two first-layer atoms and one second-layer atom as shown in Figure 1.

Based on the observed LEED patterns, models for the ordered structures have been proposed (6). At an initial stage of the oxygen adsorption, oxygen atoms form atomic rows along the [111] direction. The spacing betwen adjacent oxygen atoms along the [111] row is twice as long as that in the close-packed Mo rows. The spacing and phase between adjacent rows are random. With increasing coverage, the [111] oxygen rows tend to occupy every second trough, while their phase is random. At oxygen coverage of 0.5, every Mo trough is occupied by a [111] oxygen row. This corresponds to the Mo(112)-p(2x1)-O structure. The CO adsorption experiment (6) showed that O atoms occupy either of two equiprobable quasi-threefold sites in a Mo trough as shown in Figure 2. When oxygen coverage exceeds 0.5, domains of c(4x2)-O and then p(1x2)-O structures start to grow. At θ_0=1.0, the p(1x2) LEED spots are most clearly observed. If all oxygen atoms occupy equivalent sites, the structure should be p(1x1). One possible model for the p(1x2)-O structure is a pairing row model as shown in Figure 2, in which every second Mo row is coordinated by oxygen atoms on both sides, while the other Mo rows have no oxygen atoms directly coordinated. Although there is a possibility of the reconstruction of the substrate, the results of the CO adsorption experiments, as summarized below, provide support for these structure models.

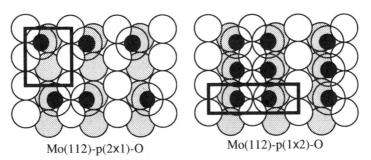

Mo(112)-p(2x1)-O   Mo(112)-p(1x2)-O

Figure 2. Models of oxygen-modified Mo(112) surfaces.

The structures of oxygen adlayers on Mo(112) should be compared to those for oxygen adsorption on the W(112) surface which has a structure almost identical to that of Mo(112). The adsorption and ordering of oxygen on the W(112) surface has been studied by several investigators (8-10). There are several differences in the oxygen structure between Mo(112) and W(112) surfaces. When W(112) is exposed to oxygen at room temperature, a series of LEED patterns are observed. At first, elliptic subspots elongated to k direction are observed at (h+1/2,k), which are changed to sharp spots at θ_0=0.5, indicating the completion of the p(2x1)-O structure. The elliptic spots are observed in a wide coverage range below 0.5 on W(112), while similar spots on Mo(112) are observed only when θ_O approaches 0.5. At 1 ML, oxygen atoms form a p(1x2) structure on Mo(112), while a p(1x1) LEED pattern is observed on W(112) (8-10). On W(112), p(1x2) is observed at θ_O=1.5. Two models have been proposed for the p(1x2)-O structure on W(112). One of them assumes that 1-ML oxygen atoms occupy every trough sites and 0.5-ML oxygen atoms are adsorbed on every second row of the topmost W layer (11). The other model assumes that every second trough is filled with two oxygen rows in both sides while the other trough is filled with a single oxygen row (10). Anyway, these structures are completely different from the Mo(112)-p(1x2)-O structure, while the p(2x1)-O structures on both surfaces are almost the same.

CO Adsorption on Oxygen-Modified Surfaces

The adsorption and dissociation of carbon monoxide has been widely studied on surfaces of metals and metal oxides. The mechanism of bonding between CO and substrates has been thoroughly investigated. The CO adsorption can therefore be used to characterize novel oxygen-modified Mo(112) surfaces. The results of temperature-programmed desorption (TPD) of CO from the oxygen-precovered Mo(112) surfaces are reproduced in Figure 3. The desorption of CO_2 was not observed at any conditions. The CO desorption peaks above 800 K are due to the recombinative desorption of dissociatively adsorbed CO (6). These peaks decreased in intensity with increasing oxygen precoverage and was almost completely suppressed at θ_O=0.5, indicating that the dissociation of CO is inhibited by the presence of preadsorbed oxygen atoms. CO molecules adsorbed associatively on the clean Mo(112) surface exhibit a single desorption peak (α_1) at 310 K. With increasing θ_O, the α_1-peak is shifted slightly to lower temperature and a new peak (α_2) appears at 220 K. The α_2-CO peak grows with increasing θ_O, while the α_1 peak is suppressed concomitantly. The α_1:α_2 peak area ratio reached 1:2 at θ_O=0.5 (p(2x1) surface). This result can be ascribed to the weakening of the Mo-CO bonds due to the coadsorpiton of electronegative oxygen atoms. The further increase of θ_O, however, results in the suppression of the α_2 peak and the regrowth of the α_1 peak. At θ_O=1.0, where the p(1x2) structure is completed, the α_2 peak disappears completely and only the α_1 peak is observed at 300 K. The behavior of CO desorption peaks from oxygen-modified Mo(112) beyond θ_O=0.5 is quite unnatural, since electronegative oxygen atoms are usually believed to weaken the adsorption bonding of coadsorbed CO molecules on many metal surfaces. For instance, the desorption temperature of molecular CO on Mo(100) was reported (12) to decrease monotonically by oxygen modification, which was explained in terms of the withdrawal of electrons from Mo by oxygen, resulting in the reduction in the back-donation capability of the surface. In principle, the effect of oxygen on coadsorbed molecules can be described as the synergistic effect of electrostatic interaction and indirect charge transfer via the substrates (13).

In order to elucidate the results of the CO TPD experiment, the detailed structure of the oxygen-modified Mo(112) surfaces and the adsorption sites of CO on these surfaces have been considered. Zaera et al. (14) investigated the CO adsorption on the Mo(110) surface by high-resolution electron-energy-loss spectroscopy (HREELS) and found v(M-CO) at 2100 cm-1, which can be assigned to CO adsorbed at atop sites. Francy et al. (15) also found a 2100 cm^{-1} loss for CO on W(100) and assigned it to atop CO. Recently, He et al. (16) indicated by infrared reflection-absorption spectroscopy that at low exposures CO is likely bound to the substrate with the C-O axis tilted with respect to the surface normal. They, however, have also shown that CO molecules adsorbed on O-modified Mo(110) exhibi v_{C-O} at 2062 and 1983 cm^{-1}. characteristic to CO adsorbed on atop sites. Thus it is supposed that CO adsorbs on top of the first layer Mo atoms.

As already noted, oxygen atoms on Mo(112) are supposed to occupy the quasi-threefold hollow sites composed of two first-layer Mo atoms and one second-layer atom. As shown in Figure 1, there are two equivalent quasi-threefold sites in a p(1x1) unit mesh. (The Mo-O distance is assumed to be 2.1 Å in Figures 1 and 2, since M-O distance for a variety of metal surfaces (17) lies in 2.0-2.2 Å.) If all CO molecules occupy the sites in the same side, only one CO species would be formed and it is difficult to explain the two CO desorption peaks from the p(2x1)-O surface. We, therefore, suppose that each oxygen atom occupies randomly either of the two equiprobable sites as shown in Figure 2. A kinematical calculation of the diffraction intensity showed that this structure exhibits a sharp p(2x1) diffraction pattern. On this p(2x1)-O structure, three types of toplayer Mo atoms with different coordination environments are formed: those coordinated by two oxygen atoms (denoted as Mo_{2C}), those coordinated by one oxygen atom (Mo_{1C}) and those not coordinated (Mo_{NC}). The ratio of the three species is 1:2:1 when oxygen atoms occupy either of two equivalent sites completely randomly. The van der Waals spheres (4) of O atoms are shown by dotted lines in Figure 2, which suggests that CO cannot be adsorbed on Mo_{2C} but on Mo_{1C} and Mo_{NC}. The α_1- and α_2-CO species are then assigned to CO adsorbed on Mo_{NC} and Mo_{1C}, respectively. The α_1 peak observed for the p(1x2)-O surface is assigned to CO adsorbed on Mo_{NC}. The fact that the α_2-CO is not observed on the p(1x2)-O surface can be explained by the structure model shown in Figure 2; there coexist Mo_{NC} and Mo_{4C}. The amount of CO species on the Mo(112)-p(2x1)-O and p(1x2)-O surfaces can also be *quantitatively* explained by the models in Figure 2. The absolute coverage for the α_1- and α_2-CO species in the present model can be estimated by assuming that the neighboring Mo sites cannot be occupied at the same time. This leads to CO coverages of 0.125 for the α_1-CO species and 0.25 for α_2-CO on the p(2x1)-O surface, and 0.25 for the α_1-CO species on the p(1x2)-O surface. These are in agreement with the result of TPD.

The desorption energy of the α_1- and α_2-CO species was estimated to be 72 and 56 kJ mol^{-1}, respectively. The difference, 16 kJ mol^{-1}, can be ascribed to the electronic modification effect due to one oxygen atom, since the steric blocking is considered to be very small for both Mo_{NC} and Mo_{1C}. This electronic effect is restricted to Mo atoms directly coordinated by O(a) and can be explained in terms of the electrostatically-enhanced through-metal charge transfer (13).

It should also be pointed out that the dissociative adsorption of CO is also influenced by the oxygen modification. The amount of dissociated CO decreased linearly with increasing oxygen coverage. The CO dissociation was suppressed almost completely at θ_0=0.5, where the Mo(112)-p(2x1)-O structure is completed. The identical result has been obtained for the dissociative adsorption of H$_2$. These results imply that the ensembles required for the dissociation of CO and H$_2$ are blocked completely by the p(2x1)-O layer.

Methanol Chemistry on Clean Mo(112)

The adsorption and reaction of methanol on metal surfaces has been widely studied (18-34). Methanol has C-O, C-H, and O-H bonds, serving as one of the simplest systems for the selective activation of chemical bonds. The methoxyl (CH$_3$O(a)) species has been considered as an intermediate of the methanol decomposition. On many transition metal surfaces, adsorbed methanol molecules are usually decomposed to H$_2$ and CO, although Ag and Cu are used as catalysts for the conversion of methanol to formaldehyde. The adsorption and reaction of alcohol molecules on Mo surfaces has been studied on the (100) (4) and (110) (35) surfaces. Alcohol molecules are decomposed effectively also on these surfaces.

The results of temperature-programmed reaction (TPR) of methanol on the clean Mo(112) surface (6) indicate that methoxyl species is also formed on Mo(112) and is decomposed to give rise to C(a), O(a), and H$_2$(g). Figure 4 shows the TPR results for the clean Mo(112) surface exposed to 4-L methanol (CH_3OD) at 130 K. The peaks below 200 K are due to methanol adsorbed on the sample holder. The small peak of 30 amu around 200 K is ascribed to CH_2O^+ from methanol desorbed intact. The desorption peak of 16 amu around 200 K is due to methane, which is produced in the titanium sublimation pump during methanol exposure, and adsorbed on the sample holder. The main desorption products are H_2 around 400 K and CO above 800 K. The H_2 desorption peak is much larger than that observed upon the H_2 adsorption up to saturation on the clean surface. The H_2 desorption peaks seem to consist of three components: two relatively small components at 350-400 K and around 500 K and a sharp peak at 410 K. For the HD desorption trace, only the peak at 350-400 K is seen. These results suggests that the methanol decomposition reaction on the clean Mo(112) surface proceeds as follows.

$$CH_3OH(a) \xrightarrow{<300 \text{ K}} CH_3O(a) + H(a) \tag{1}$$

$$H(a) \xrightarrow{350\text{-}400 \text{ K}} 1/2 \, H_2(g) \tag{2}$$

$$CH_3O(a) \xrightarrow{410 \text{ K}} 3/2 \, H_2(g) + C(a) + O(a) \tag{3}$$

$$CH_3O(a) \xrightarrow{500 \text{ K}} 1/2 \, H_2(g) + 1/2 \, CH_4(g) + 1/2 \, C(a) + O(a) \tag{4}$$

$$C(a) + O(a) \xrightarrow{>800 \text{ K}} CO(g) \, . \tag{5}$$

CH_3OD (CH_3OH) molecules are first dissociated to $CH_3O(a)$ species and D(a) (H(a)) at low temperatures. Most of D(a) (H(a)) species are recombinatively desorbed at 350-400 K with H(a) adsorbed from the residual gas. The methoxyl decomposition occurs at 410 K, giving rise to a sharp reaction-limited desorption peak of H_2. The methoxyl decomposition also takes place around 500 K to form $CH_4(g)$ and $H_2(g)$. The CO desorption peaks above 800 K coincide with those for dissociatively adsorbed CO after the CO exposure. Methoxy species has been commonly assumed as an intermediate of the methanol decomposition on various metal surfaces. Ko et al. (*3,36*) suggested that methoxyl species is formed during the reaction of methanol on Mo(100) and W(100). Serafin et al. (*37*) detected methoxyl species on oxygen modified Mo(110) by using XPS. The TPR results for the Mo(112) surface suggest that CO(g) dose not desorb but C(a) and O(a) are formed during the methoxyl decomposition. This is in contrast to the cases of noble matels (*33*) where the simultaneous desorption of CO and H_2 is observed during the methoxyl decomposition. The formation of C(a) and O(a) species, rather than the evolution of CO(g), on Mo(112) implies that the C-O bond breaking competes well with the C-H breaking during the methoxyl decomposition on Mo(112), which may be related to the large bonding energy for Mo-C and Mo-O bonds.

Methanol Chemistry on Oxygen-Modified Mo(112)

The surface modification by O, S, and C atoms and its effects on the methanol chemistry has been investigated on several metal surfaces. It has been reported that these modifier atoms stabilize methoxyl species on Fe(100) (*18,19*), Ni(110) (*28*), Mo(100) (*4*) and W(112) (*38*). The stabilization of methoxyl species has been ascribed to the blocking of the sites necessary for the methoxyl decomposition.

The TPR experiment for the Mo(112)-p(2x1)-O surface gave rise to a result essentially identical to that for the clean Mo(112) surface, while the desorption features of hydrogen were somewhat changed. The result indicated that adsorbed methanol molecules are first dissociated to form methoxyl species, which then decomposes to $H_2(g)$, $CH_4(g)$, C(a) and O(a). This suggests that oxygen atoms in p(2x1) structure exert little influence on the reaction path of the methoxyl decomposition and that the methoxyl decomposition can be catalyzed by small Mo ensembles as compared to that required for the dissociative adsorption of CO and H_2. The main decomposition temperature of methoxyl species on the Mo(112)-p(2x1)-O surface was 440 K. This is higher by 30 K than that for the clean Mo(112) surface, indicating that the surface methoxyl species is stabilized by the p(2x1)-O layer. The stabilization can be ascribed either to the thermodynamical effect due to the modification of the Mo electronic states by oxygen or to the kinetical effects due to the steric confinement of the decomposition sites and/or the restraint of the surface migration by the ordered oxygen layer.

To the contrary, an amazingly different reaction path is opened on the p(1x2)-O surface (θ_0=1.0). Figure 5 shows the TPR results for CH_3OD on the Mo(112)-p(1x2)-O surface. The desorption peaks are completely different from those from clean Mo(112) and Mo(112)-p(2x1)-O surfaces. The desorption peak of dissociated CO above 800 K is considerably reduced, indicating that the complete decomposition of methoxyl is suppressed on this surface. At 560 K, the simultaneous desorption of formaldehyde (H_2CO, shown by the 30-amu curve), H_2 (2 amu), CO (28 amu), and CH_4 (16 amu) is observed. The relative yields of the carbon-containing species produced from 0.14 ML of methanol are shown in Figure 6. No desorption of water was observed at temperatures from 130 to 1100 K. This suggests that the formation of formaldehyde on this surface is *not* due to the oxidative dehydrogenation usually observed on molybdenum oxides (*1,39*). The oxygen modification of Mo(112) thus resulted in the new dehydrogenation mode of methanol to produce H_2CO and H_2. We have examined the methanol decomposition on the Mo(112)-p(1x2)-^{18}O surface. It was found that the desorbed reaction products do not contain any ^{18}O-labeled species. This result suggests that the oxygen adatoms do not directly participate in the dehydrogenation reaction but play a subsidiary role as a modifier. For the molybdenum oxide catalysts, the direct participation of substrate oxygen is postulated (*1,39*). The difference may be ascribed to the fact that the oxygen atoms on the Mo(112)-p(1x2)-O surface occupy highly-stable multi-coordination sites. It would be interesting to extend the study to higher oxygen coverages.

The desorbed species at 560 K does not contain deuterated molecules when CH_3OD is used, indicating that methoxyl groups are also formed on p(1x2)-O at low temperatures and are decomposed at 560 K. The main reaction path on the p(1x2)-O surface can then be summarized as

$$CH_3OH(a) \quad \xrightarrow{<300\ K} \quad CH_3O(a) \ + \ H(a) \tag{6}$$

$$H(a) \quad \xrightarrow{350\text{-}400\ K} \quad 1/2\ H_2(g) \tag{7}$$

$$CH_3O(a) \quad \xrightarrow{560\ K} \quad H_2CO(g) \ + \ 1/2\ H_2(g). \tag{8}$$

The isotope scrambling experiment has also been carried out to gain an insight into the reaction mechanism. After a mixture of $CH_3O(a)$ and $CD_3O(a)$ species of approximately 1:1 molar ratio was formed on the Mo(112)-p(1x2)-O surface, the isotope distribution of the desorption products were measured by QMS. Figure 7 shows the relative amounts of isotopes in each product desorbed at 560 K. The results show that H and D atoms are mixed almost statistically for hydrogen (H_2:HD:D_2 = 1:2:1), while for formaldehyde incomplete mixing (H_2CO:HDCO:D_2CO $\approx$ 1:1:1) is observed. For methane, the relative yield of CH_2D_2 is significantly small as compared with CH_4, CH_3D, CD_3H, and CD_4.

The statistical distribution in desorbed hydrogen molecules implies that H(a), which is produced by the C-H bond scission of methoxyl species, migrates quickly on

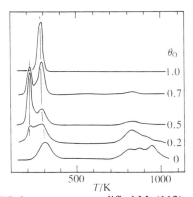

Figure 3. TPD of CO from oxygen-modified Mo(112). Adapted from Ref. 6.

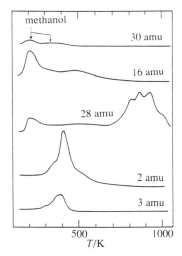

Figure 4. TPR results for CH$_3$OD on clean Mo(112). The small peaks of 30 amu around 200 K are due to CH$_2$O$^+$ from methanol desorbed intact. Adapted from Ref. 6.

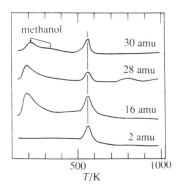

Figure 5. TPR results for CH$_3$OD on Mo(112)-p(1x2)-O. The small peaks of 30 amu around 200 K are due to CH$_2$O$^+$ from methanol desorbed intact. Adapted from Ref. 6.

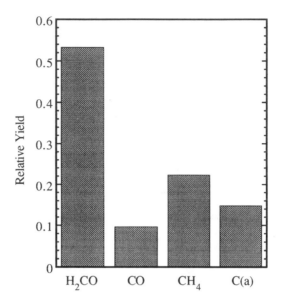

Figure 6. The Relative yields for C-containing species from 0.14 ML of methanol on Mo(112)- p(1x2)-O.

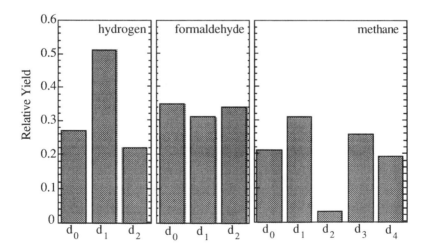

Figure 7. Isotope distribution in the products from a 1:1 mixture of $CH_3O(a)$ and $CD_3O(a)$.

the surface and initiates other elementary reactions, leading to the simultaneous desorption of all the products. The isotope distribution for methane indicates that methane is formed by the reaction of H(a) with methoxyl species, or by the reaction of H(a) with CH_3(a) produced from methoxyl species. The fact that few CH_2D_2 is contained in the desorbed methane suggests that the hydrogen exchange between H(a) and CH_3O(a) or CH_3(a) is negligible.

The formation of formaldehyde (reaction 8) can be considered to be composed of two elementary steps,

$$CH_3O(a) \longrightarrow CH_2O(a) + H(a) \tag{9}$$

$$CH_2O(a) \longrightarrow H_2CO(g). \tag{10}$$

The formation of a considerable amount of HDCO(g) suggests that the H-D exchange reaction via CHO(a) species,

$$CH_2O(a) \rightleftharpoons CHO(a) + H(a), \tag{11}$$

competes with the desorption of formaldehyde (step 10). As shown in Figure 6, the selectivety to formaldehyde is higher than to CO, suggesting that the reverse reaction of step 11 proceeds faster than the further dehydrogenation of CHO(a) species,

$$CHO(a) \longrightarrow CO(a) + H(a) \tag{12}$$

It should be noted that step 12 proceeds much faster than the reverse reaction of step 11 on many transition metal surfaces (*33*), producing CO(g) and H_2(g) simultaneously.

The Role of Oxygen in the Methanol Dehydrogenation

On the clean and oxygen-modified surfaces, adsorbed methanol first dissociates into CH_3O(a) and H(a) and then undergoes the following reaction steps. On the clean Mo(112) surface, the methoxyl decomposition to form H_2(g), C(a) and O(a) takes place at 410 K. On the Mo(112)-p(2x1)-O surface, almost the same decomposition reaction takes place at 440 K. On the Mo(112)-p(1x2)-O surface, methoxyl species undergoes a completely different reaction path at 560 K. The large upward shift of the decomposition temperature by 150 K as compared with the clean Mo(112) surface suggests that the methoxyl adsorbates are considerably stabilized on the Mo(112)-p(1x2)-O surface. The nature of the stabilization is apparently different from that for the Mo(112)-p(2x1)-O surface because the stabilization on the Mo(112)-p(1x2)-O layer generated the new dehydrogenation reaction path.

The peak position of the simultaneous desorption peaks of H_2CO, H_2, CO, and CH_4 was the same for CH_3OH and CH_3OD, while a simultaneous upward shift by 10 K was observed when CD_3OD was used, indicating the kinetic isotope effect due to the C-H (C-D) bond scission. This suggests that the surface reactions around 560 K are initiated by the C-H bond breaking. It is interesting to recall that the extensive C-O bond scission was observed during the methoxyl decomposition on the low-θ_O surfaces. By contrast, the CO desorption peak above 800 K was much smaller for the Mo(112)-p(1x2)-O surface, indicating that the amount of the C(a) and O(a) species formed during the methoxyl decomposition was small compared with the cases of the low-θ_O surfaces. This also suggests that the C-O bond scission is restrained on the Mo(112)-p(1x2)-O surface. As already stated, the electronic structure of the toplayer Mo_{NC} atoms does not differ between the Mo(112)-p(2x1)-O and p(1x2)-O surfaces. The suppression of the C-O bond scission during the methoxyl decomposition on the

Mo(112)-p(1x2)-O surface can then be ascribed to the steric blocking by oxygen atoms of the second-layer Mo atoms. We propose that the second-layer Mo atoms play an essential role in the C-O bond cleavage of methoxyl species and hence only the C-H bond scission can be catalyzed on the Mo(112)-p(1x2)-O surface, where only the first-layer Mo atoms are available. As already pointed out (40), metal atoms of high coordination sites exhibits high electronic fluctuation (i.e., fluctuation in charge, spin, etc.) and hence play a dominant role in many catalytic reactions. On the Mo(112)-p(1x2)-O surface, the selective blocking of the second layer Mo atoms resulted in the inhibition of the C-O bond cleavage to lead to the stabilization of methoxyl species. Thus the simple dehydrogenation of methane can proceed on the isolated one-dimensional Mo structure characteristic to the Mo(112) surface.

Conclusion

The results presented here indicate that a new methanol dehydrogenation reaction path is opened when the Mo(112) surface is modified by a p(1x2) oxygen layer. The result of the CO adsorption experiment suggests that main electronic effect of oxygen modifier is restricted to the metal atoms directly bonded with the oxygen atoms. This leads to a concept of the selective blocking of the surface atoms to create new active structures, which can provide a powerful mean to control catalytic reaction paths.

Literature Cited

(1) Farneth, W. E.; Ohuchi, F.; Staley, R. H.; Chowdhry, U.; Sleight, A. W. *J. Phys. Chem.* **1985**, *89*, 2493.

(2) Chowdhry, U.; Ferretti, A.; Firmant, L. E.; Machiels, C. J.; Ohuchi, F.; Sleight, A.W.; Staley, R. H. *Appl. Surf. Sci.* **1984**, *19*, 360.

(3) Ko, E. I.; Madix, R. J. *Surf. Sci.* **1981**, *112*, 373.

(4) Miles, S. L.; Bernasek, S. L.; Gland, J. L. *J. Phys. Chem.* **1983**, *87*, 1626.

(5) Zhang, C.; Van Hove, M. A.; Somorjai, G. A. *Surf. Sci.* **1985**, *149*, 326.

(6) Aruga, T.; Fukui, K.; Iwasawa, Y. *J. Am. Chem. Soc.* **1992**, *114*, 491; Fukui, K.; Aruga, T.; Iwasawa, Y. to be published.

(7) Bu, H.; Grizzi, O.;Shi, M.;Rabalais, J. W. *Phys. Rev. B* **1989**, *40*, 10147.

(8) Chang, C. C.; Germer, L. H. *Surf. Sci.* **1967**, *8*, 115.

(9) Tracy, J. C.; Blakeley, J. M. *Surf. Sci.* **1969**, *15*, 257.

(10) Benziger, J. B.; Preston, R. E. *Surf. Sci.* **1985**, *151*, 183.

(11) Wang, G. -C.; Pimbley, J. M.; Lu, T. -M. *Phys. Rev. B* **1985**, *31*, 1950.

(12) Ko, E. I.; Madix, R. J. *Surf. Sci.* **1981**, *109*, 221.

(13) Aruga, T.; Sasaki, T.; Iwasawa, Y. In *Ordering at Surfaces and Interfaces;* Yoshimori, A.; Shinjo, T.; Watanabe, H. Eds.; Springer-Verlag: Berlin, Heidelberg, 1992; pp 237-244.

(14) Zaera, F.; Kollin, E.; Gland, J. L. *Chem. Phys. Lett.* **1985**, *121*, 454.

(15) Franchy, R.; Ibach, H. *Surf. Sci.* **1985**, *155*, 15.

(16) He, J.-W.; Kuhn, W. K.; Goodman, D. W. *Surf. Sci.* **1992**, *262*, 351.

(17) MacLaren, J. M.; Pendry, J. B.; Rous, P. J.; Saldin, D. K.; Somorjai, G. A.; Van Hove, M. A.; Wedensky, D. D. *Surface Crystallographic Information Service - A Handbook of Surface Structures;* D. Reidel, Dordrecht, 1987.

(18) Lu, J. -P.; Albert, M.; Bernasek, S. L.; Dwyer, D. S. *Surf. Sci.* **1989**, *218*, 1.

(19) Lu, J. -P.; Albert, M.; Bernasek, S. L.; Dwyer, D. S. *Surf. Sci.* **1990**, *239*, 49.

(20) Wang, J.; Masel, R. I. *J. Catal.* **1990**, *126*, 519.
(21) Wang, J.; Masel, R. I. *J. Vac. Sci. Technol.* **1991**, *A9*, 1879.
(22) Wang, J.; Masel, R. I. *Surf. Sci.* **1991**, *243*, 199.
(23) Wachs, I. E.; Madix, R. J. *J. Catal.* **1978**, *53*, 208.
(24) Bowker, M.; Madix, R. J. *Surf. Sci.* **1980**, *95*, 190.
(25) Sexton, B. A.; Hughes, A. E.; Avery, N. R. *Appl. Surf. Sci.* **1985**, *22/23*, 404.
(26) Sexton, B. A.; Hughes, A. E.; Avery, N. R. *Surf. Sci.* **1985**, *155*, 366.
(27) Wachs, I. E.; Madix, R. J. *Surf. Sci.* **1978**, *76*, 531.
(28) Bare, S. R.; Stroscio, J. A.; Ho, W. *Surf. Sci.* **1985**, *155*, L281.
(29) Huntley, D. R. *J. Phys. Chem.* **1989**, *93*, 6156.
(30) Sexton, B. A. *Surf. Sci.* **1981**, *102*, 271.
(31) Russel, Jr., J. N.; Gates, S. M.; Yates, Jr., J. T. *Surf. Sci.* **1985**, *163*, 516.
(32) Solymosi, F.; Tarnoczi, T. I.; Berko, A *J. Phys. Chem.* **1984**, *88*, 6170.
(33) Parmeter, J. E.; Jiang, X.; Goodman, D. W. *Surf. Sci.* **1990**, *240*, 85.
(34) Hrbek, J.; De Paola, R.; Hoffmann, F. M. *Surf. Sci.* **1986**, *166*, 361.
(35) Shiller, P.; Anderson, A. B. *J. Phys. Chem.* **1991**, *95*, 1396.
(36) Ko, E. I.; Benziger, J. B.; Madix, R. J. *J. Catal.* **1980**, *62*, 264.
(37) Serafin, J. G.; Friend, C. M. *J. Am. Chem. Soc.* **1989**, *111*, 8967.
(38) Benziger, J. B.; Preston, R. E. *J. Phys. Chem.* **1985**, *89*, 5002.
(39) Iwasawa, Y. *Adv. Catal.***1987**, *35*, 87.
(40) Falicov, L. M.; Somorjai, G. A. *Proc. Natl. Acad. Sci. USA* **1985**, *82*, 2207.

RECEIVED November 6, 1992

Chapter 10

Structure Sensitivity in Selective Oxidation of Propene over Cu$_2$O Surfaces

Kirk H. Schulz[1] and David F. Cox

Department of Chemical Engineering, Virginia Polytechnic Institute and State University, Blacksburg, VA 24061

Studies of the interaction of propene (CH$_2$=CHCH$_3$) with Cu$_2$O(111) and (100) single crystal surfaces have demonstrated that each step in the allylic oxidation to acrolein (CH$_2$=CHCHO) exhibits structure-sensitivity. Oxygen vacancies (i.e., surface defects) on the nonpolar, Cu$_2$O(111) surface have been found to be energetically-favorable sites for the dissociation of propene to allyl when compared to (100) and non-defective (111) surfaces. The oxygen insertion reaction requires the presence of coordinately-unsaturated surface lattice oxygen. The final reaction step is hydrogen elimination from the carbon α to oxygen in the σ-bonded allylic intermediate (identified as a surface allyloxy, CH$_2$=CHCH$_2$O-). The activation energy for this final reaction varies by over 7 kcal/mol depending on which Cu$_2$O surface is investigated.

The partial oxidation of propene (CH$_2$=CHCH$_3$) to acrolein (CH$_2$=CHCHO) is a useful model for the class of allylic oxidation reactions of olefins. Several mixed oxides catalyze this reaction, but Cu$_2$O is the only reported single-component oxide catalyst to exhibit significant activity and selectivity (1-3). Because of its single-component nature, single crystal Cu$_2$O was chosen for investigating the structure sensitivity and site requirements for the propene selective oxidation reaction.

The basic steps in the reaction pathway of propene oxidation to acrolein have been widely studied. In the rate determining step over Cu$_2$O and bismuth-molybdate catalysts, a methyl hydrogen is abstracted from propene to produce a symmetric, π-allyl intermediate (4,5). The order in which the second and third steps occur is not well understood, but involves a second hydrogen abstraction and lattice oxygen insertion into the symmetric π-allyl to form an oxygen-containing σ-allyl species. Grasselli and

[1]Current address: Department of Chemical Engineering, University of North Dakota, Box 8101, University Station, Grand Forks, ND 58202

coworkers (3) advocate the formation of a σ-allyl on bismuth-molybdate catalysts where oxygen insertion occurs prior to the final hydrogen abstraction (i.e., an allyloxy intermediate, $CH_2 = CHCH_2O$-). However, several groups advocate the formation of a σ-allyl species on bismuth-molybdate (1) and Cu_2O (6) catalysts where lattice oxygen insertion occurs after the final hydrogen abstraction.

Surfaces Studied

The use of Cu_2O single crystal surfaces as model catalysts allows for the testing of site requirements for propene adsorption and oxidation. The two low-index surfaces investigated differ both in the availability of surface lattice oxygen and the Cu^+ coordination numbers. The ideal, stoichiometric, nonpolar, $Cu_2O(111)$ surface exposes singly- and doubly-coordinate Cu^+ cations (bulk Cu^+ coordination $= 2$) in the second atomic layer in the ratio of 1 to 4, respectively. However, the top atomic layer is composed exclusively of threefold-coordinate oxygen anions (bulk lattice O^{2-} coordination $= 4$). The ideal, stoichiometric $Cu_2O(111)$ surface is illustrated in Figure 1a. An oxygen-deficient $Cu_2O(111)$ surface may also be prepared by exposure to reducing gases. Such treatments lead to a ($\sqrt{3}x\sqrt{3}$)R30° LEED periodicity due to an ordered one-third of a layer of oxygen vacancies (7). Each oxygen vacancy gives rise to a threefold site of singly-coordinate Cu^+ cations. The $Cu_2O(111)$-($\sqrt{3}x\sqrt{3}$)R30° surface is illustrated in Figure 1b.

The $Cu_2O(100)$ surface used in this study is a polar, Cu^+-terminated, reconstructed surface which displayed a ($3\sqrt{2}x\sqrt{2}$)R45° LEED periodicity with many missing spots (7). Although there is no definitive model of the structure of the reconstructed surface, the periodicity of the reconstruction suggests a relaxation of top atomic layer Cu^+ cations, possibly associated with a weak Cu^+-Cu^+ bonding interaction (7). In contrast to the (111) surface, the ideal, Cu^+-terminated (100) surface exposes no lattice oxygen in the top atomic layer. All top layer Cu^+ cations are singly coordinated. The ideal (i.e., unreconstructed) Cu^+ terminated surface is illustrated in Figure 2a. An oxygen terminated surface can also be prepared by 10^9 L (1 L = 10^{-6} Torr· sec) exposures to O_2 at room temperature. This preparation lifts the reconstruction on the (100) surface to form a (1x1), oxygen-terminated surface which contains doubly-coordinate oxygen in the outer atomic layer. This O-terminated (100) surface is shown in Figure 2b. A more complete description of the characterization of the $Cu_2O(100)$ and (111) surfaces has been reported previously (7).

Propene Adsorption

Adsorption at Low Pressure (P < 10^{-6} Torr). The adsorption of propene has been studied with thermal desorption spectroscopy (TDS) on all of the different forms of the (100) and (111) surfaces and under several different conditions of exposure. For exposures at low pressure (P < 10^{-6} Torr), no selective oxidation is observed. For small exposures (< 5 L) at low-temperature (100K-120K), four propene desorption states are observed from the $Cu_2O(111)$ surface compared to two desorption states from the $Cu_2O(100)$-Cu surface. These TDS results are shown in Figure 3, and give a clear indication of a structure-sensitive interaction of propene with Cu_2O.

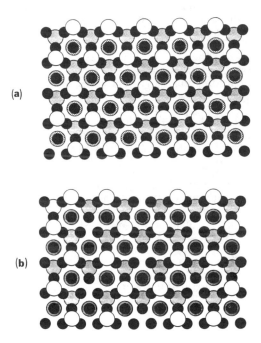

Figure 1. Ball model illustrations of (a) the ideal, stoichiometric Cu$_2$O(111) surface and (b) the oxygen-deficient Cu$_2$O(111)-($\sqrt{3}$x$\sqrt{3}$)R30° surface. The small filled circles represent Cu$^+$ cations, and the larger open circles represent O^{2-} anions. For clarity, only the top four atomic layers are shown.

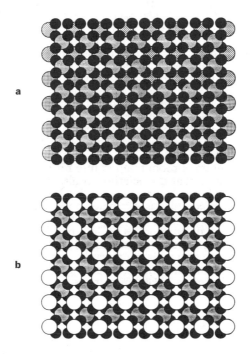

Figure 2. Ball model illustrations of (a) the ideal (i.e., unreconstructed) Cu$^+$-terminated (100) surface and (b) the O-terminated (100) surface.

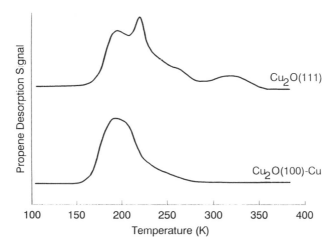

Figure 3. Propene thermal desorption traces from the (111) and Cu$^+$-terminated (100) surface following 0.3 L doses in UHV at 110 K.

For the low adsorption temperatures, propene dissociation is observed only on the $Cu_2O(111)$ surface. Dissociation and non-selective oxidation to CO occurs at low temperatures, and gives rise to desorption-limited CO as a product. Dissociation to a surface allyl (C_3H_5) has been confirmed also by adsorption of propene on a deuterium-predosed surface to yield singly-labeled propene (C_3H_5D). A comparison of the desorption behavior between $Cu_2O(111)-(1x1)$ and $Cu_2O(111)-(\sqrt{3}x\sqrt{3})R30°$ surfaces demonstrates that the dissociation of propene to allyl and/or the recombination of allyl with surface hydrogen occurs at oxygen vacancies (i.e., defects) on the (111) surface associated with three-fold cation sites. The highest temperature desorption feature at 325 K shown for the (111) surface in Figure 3 is characteristic of this process.

Propene adsorption at room temperature for large exposures $(3x10^3$ L) at low pressure shows that dissociation to allyl occurs on all surfaces investigated at higher temperature, but the probability for dissociative adsorption at room temperature is estimated to be less than 10^{-5}. Even though dissociation to allyl is observed, no selective oxidation products are formed during thermal desorption. The low pressure results allow for bounds to be put on the energetics of propene dissociation to allyl assuming that the dissociative adsorption is exothermic. The projected limits are given in Table I, and indicate that the activation energy for propene dissociation to allyl is less than 22 kcal/mol on the surfaces studied (8).

Table I. Energetics for Propene Dissociation to Allyl on Cu_2O

$Cu_2O(100)$ 16.9 kcal/mol $< E_a <$ 21.1 kcal/mol

$Cu_2O(111)$ $E_a <$ 21.8 kcal/mol

The lack of selective oxidation products even when propene is dissociated to allyl suggests that the "activation" of propene for selective oxidation involves not just dissociation, but also the separate process involving oxygen insertion. Using the different single crystal surfaces, different preparations (7,8) and different adsorption temperatures allow one to test the interaction of propene with different forms of surface oxygen (i.e., molecular, doubly-coordinate and triply coordinate). Since no selective oxidation products are observed, lower limits on the energies for "activation" by surface oxygen can be estimated from the low pressure thermal desorption results. In all cases, no selective oxidation is observed, and the maximum activation energies which can be accessed experimentally are limited by the desorption of reactants. The results are summarized in Table II. The results suggest that the activation energy for oxygen insertion into an allyl is greater than that for dissociation to allyl. All the results discussed above for low-pressure propene adsorption have been reported in detail elsewhere (8).

Table II.   Energetics of Propene Activation for Oxygen Insertion
($P < 10^{-6}$ Torr)

Surface	form of oxygen	reaction	limiting process	lower limit on E_a
$Cu_2O(100)$-Cu	molecular O_2	no reaction	propene desorbs	$E_a > 15$ kcal/mol
$Cu_2O(100)$-O	atomic, doubly coordinate	no reaction	allyl-hydrogen recombination	$E_a > 21$ kcal/mol
$Cu_2O(111)$	atomic, triply coordinate	CO formation, no selective oxidation	allyl-hydrogen recombination	$E_a > 24$ kcal/mol

Adsorption at Atmospheric Pressure. The lack of selective oxidation products following adsorption at low pressure indicates the existence of a "pressure gap" for the selective oxidation process. To overcome the low pressure limitations, propene adsorption at 1 atm. was investigated (9). Different single crystal Cu_2O surfaces were exposed to flowing propene at room temperature and 1 atm. for short periods (about 2 seconds), then returned to ultrahigh vacuum for thermal desorption studies. These conditions of exposure are reducing because hydrogen is released to the surface as a result of propene dissociation. Propane is the primary C_3 product evolved. Non-selective oxidation products (CO, CO_2 and H_2O) are also observed. However, following adsorption at atmospheric pressure, selective oxidation products including acrolein and allyl alcohol are observed also in TDS. Figure 4 shows the TDS traces for propene and the selective oxidation products from the O-terminated $Cu_2O(100)$ surface following propene adsorption at atmospheric pressure. The selectivity to acrolein is near 2% for the O-terminated (100) surface and the (111) surface, but less than 0.1% (and in many cases undetectable) for the Cu^+-terminated (100) surface. The variation in acrolein yields from the different surfaces indicate that the primary site requirement for oxygen insertion and selective oxidation is the presence of coordinately-unsaturated surface oxygen anions like those present on the (111) and O-terminated (100) surfaces. The coordination number of these oxygen anions appears to make little difference in the selective oxidation process, with two-coordinate and three-coordinate anions both capable of producing acrolein. The lack of coordinately-unsaturated anions at the Cu^+-terminated (100) surface accounts for the minimal selective oxidation products formed.

Identification of the σ-Allyl Intermediate

X-ray photoelectron spectroscopy (XPS) of the surface following propene adsorption at 1 atm. indicates that the oxygen insertion reaction occurs at room temperature (9). A high binding energy contribution to the C 1s XPS signal is observed at 286.5 eV on the (111) and O-terminated (100) surfaces, but not on the Cu^+-terminated (100). The binding energy is similar to that observed for alkoxides on Cu_2O surfaces (10), and is attributed to the oxygenated carbon of the oxygen-containing σ-allyl postulated to be the final surface intermediate in the selective oxidation to acrolein (1,3,6). The similarity of the σ-allyl binding energy in XPS to that of surface alkoxides suggests that the final surface intermediate is an allyloxy ($CH_2 = CHCH_2O$-).

To verify this pathway, the product distribution and kinetic parameters observed in TDS for acrolein formation from the σ-allyl formed by 1-atm. propene exposures were compared to those for the reaction of acrolein and allyl alcohol under ultrahigh vacuum conditions. The dissociative adsorption of allyl alcohol ($CH_2 = CHCH_2OH$) to allyloxy ($CH_2 = CHCH_2O$-) allows one to investigate the anticipated final intermediate if oxygen insertion precedes the final hydrogen abstraction. Adsorbed acrolein has the same composition as the anticipated σ-allyl formed if the final hydrogen abstraction precedes oxygen insertion. The thermal desorption data for adsorbed acrolein and for acrolein produced from allyl alcohol and propene oxidation on the $Cu_2O(100)$ surface is shown in Figure 5. For this data, the acrolein signal from adsorbed acrolein and from allyl alcohol are from UHV experiments, while the signal due to propene oxidation is the result of a 1 atm. exposure. For the two oxygenate exposures in UHV, the Cu^+ terminated surface is used since the adsorbate itself provides the oxygen required for the associated oxygen-containing surface species. For propene oxidation, the O-terminated (100) surface is used since oxygen must be supplied by the lattice. Note that the acrolein product from propene and allyl alcohol is clearly reaction-limited by comparison to the desorption trace for adsorbed acrolein. Using the Redhead equation (11) assuming a normal preexponential of 10^{13} sec^{-1}, the first-order activation energy for acrolein production in TDS from the propene-derived σ-allyl (470 K) and from the allyl alcohol-derived allyloxy (525 K) using a 2 K/sec temperature ramp is essentially the same: 31.5 ± 1.8 kcal/mol.

In comparing the product distributions, it is noted that allyl alcohol is observed as a reaction product from propene TDS, as shown in Figure 4. The surface chemistry of acrolein has been checked to ensure that allyloxy (and therefore allyl alcohol and reaction-limited acrolein) is not due to simply to the subsequent hydrogenation of acrolein formed by an alternate pathway in the propene TDS run. The reactivity of acrolein and hydrogen with Cu_2O *does not* produce allyl alcohol (12). Hence, the propene oxidation pathway clearly proceeds through an allyloxy surface intermediate. The final step in the reaction sequence is a unimolecular hydride elimination from allyloxy ($CH_2 = CH-CH_2O$-) at the carbon α to the oxygen heteroatom. Hence, oxygen insertion precedes the final hydrogen abstraction in the allylic oxidation of propene over Cu_2O. Details of the results for atmospheric pressure exposures are given elsewhere (9).

Allyloxy Decomposition

The similarity between the measured activation energies for the reaction-limited production of acrolein over $Cu_2O(100)$ from allyl alcohol in UHV or propene following a 1 atm. exposure gives a clear indication that these reactions involve the same surface intermediate, an allyloxy. This similarity also suggests that the surface intermediates formed by these two routes behave in a chemically similar fashion. For the (100) surface, the Cu^+-alkoxide surface complex is similar regardless of whether oxygen from

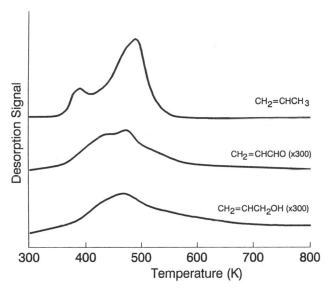

Figure 4. Desorption of C_3 compounds from the O-terminated $Cu_2O(100)$ surface following propene exposure at 1 atmosphere.

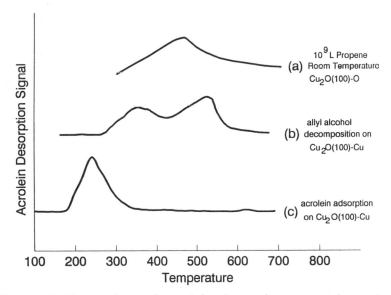

Figure 5. Comparison of acrolein desorption traces from the $Cu_2O(100)$ surface following (a) exposure to propene at 1 atm., (b) allyl alcohol adsorption in UHV and (c) acrolein adsorption in UHV.

the O-terminated (100) surface is incorporated into propene or allyl alcohol dissociates to the alkoxide on the Cu^+-terminated (100) surface.

On the (111) surface, however, the activation energy for allyloxy decomposition is significantly lower than that for the (100) surface. This structure-sensitive variation in the activation energy for allyoxy decomposition is illustrated in Figure 6 which shows the reaction-limited acrolein product traces for UHV exposures of allyl alcohol over the stoichiometric (111) and Cu^+-terminated (100) surfaces. The highest temperature features from each trace (400 K on the (111) surface and 525 K on the (100) surface) are known to be due to the decomposition of allyloxy to acrolein (10). For both surfaces, these high temperature desorption features are observed in conjunction with allyl alcohol desorption via recombination of allyloxy and surface hydrogen, behavior characteristic of an alkoxide decomposition (10). On the (111) surface, the activation energy is more than 7 kcal/mol lower than observed on the (100) surface. A similar difference in the activation energy for methoxy decomposition in the oxidation of methanol has also been observed over these two surfaces (13). Hence, as with the structure sensitivity displayed in both the adsorption of propene and the oxygen insertion step, the final elementary step in the reaction pathway to acrolein also displays structure sensitivity.

Catalytic Consequences of the Structure Sensitivity

The dissociation of propene to allyl at low temperatures on defect sites on the (111) surface suggests that point defects such as oxygen vacancies may provide lower energy sites for the initial dissociation reaction. However, since both surfaces investigated are capable of dissociating propene to allyl at room temperature and 1 atmosphere (relatively mild conditions), it is unlikely that such energetically favorable point defect sites would significantly effect the overall rate under catalytic reaction conditions. However, the effect of such sites on the selectivity of the reaction could be considerable. Note that even for adsorption at low temperatures on the stoichiometric $Cu_2O(111)$-(1x1) surface, propene dissociation results in non-selective oxidation to CO. Surface allyl is verified only in the presence of point defects. While these low pressure and low temperature results for the (111) surface clearly cannot be assumed to be characteristic of real powder catalysts under reaction conditions, there are interesting parallels to another study in the literature. From studies of propene oxidation on polycrystalline Cu_2O, Wood et al. found that a copper rich (i.e., oxygen deficient) surface favors selective oxidation to acrolein, while an oxygen rich surface favors nonselective oxidation (14). This compositional observation suggests that the potential importance of point defects like oxygen vacancies in controlling the selectivity of Cu_2O catalysts cannot not be overlooked.

The site requirements for oxygen insertion, coordinately-unsaturated surface lattice oxygen, would not be expected to significantly limit the activity of Cu_2O catalysts under reaction conditions. Under conditions for selective oxidation, the feed of oxidant with propene would be expected to sustain a steady state concentration of such species even on polar surfaces like the (100). Hence while it is possible to sustain Cu^+ terminated surfaces like $Cu_2O(100)$-Cu in our TDS experiments, the possibility of sustaining such surfaces under reaction conditions is small. Also, since exchange of oxygen between the bulk and the surface occurs at elevated temperatures (9,10,12), the availability of surface oxygen for selective

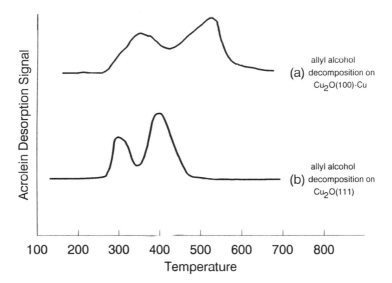

Figure 6. Comparison of acrolein desorption traces from allyl alcohol decomposition over (a) the $Cu_2O(100)$-Cu surface and (b) the stoichiometric (111) surface.

oxidation is not expected to be compromised under reaction conditions. Since the coordination number of the surface oxygen does not determine its ability to produce acrolein, the different reactivity observed in TDS for the Cu^+ and O terminated (100) surfaces is probably best thought of as compositional sensitivity rather than structure sensitivity, to the extent that they can be distinguished. As with the point defects discussed above, however, different selectivities associated with Cu rich or oxygen rich surfaces may effect the catalyst performance.

The differences observed in the activation energy for allyoxy decomposition to acrolein over different crystallographic surfaces is also expected to have a negligible effect on the activity of the catalyst. It has been well established that under catalytic reaction conditions the rate-limiting step in propene oxidation over Cu_2O involves the abstraction of a methyl hydrogen in the initial dissociation step (4,5). Since the decomposition of allyloxy is not rate-limiting, the 7-8 kcal/mol difference in the activation energies for this reaction over different crystallographic planes should have no significant effect on the overall kinetics under catalytic reaction conditions. As for possible effects on the selectivity, differences in the selectivity for the allyloxy decomposition reaction over different surfaces has not yet been fully investigated.

Acknowledgments

We gratefully acknowledge the National Science Foundation for support of this work through CBT-870876. We also thank Professor L. Tapiero for providing the single crystal used in this study.

Literature Cited

1. Keulks, G.W.; Krenzke, L.D. and Noterman, T.M. *Advan. Catal.* **1978**, *27*, 183.
2. Hucknall, D.J., *Selective Oxidation of Hydrocarbons;* Academic Press: New York, 1974.
3. Grasselli, R.K. and Burrington, J.D., *Advan. Catal.* **1981**, *30*, 133.
4. Adams, C.R. and Jennings, T.J., *J. Catal.* **1964**, *3*, 549.
5. Voge, H.H.; Wagner, C.D. and Stevenson, D.P., *J. Catal.* **1963**, *2*, 58.
6. Adams, C.R. and Jennings, T.J., *J. Catal.* **1963**, *2*, 63.
7. Schulz, K.H. and Cox, D.F., *Phys. Rev. B.* **1991**, *43*, 3061.
8. Schulz, K.H. and Cox, D.F., *Surf. Sci.* **1992**, *262*, 318.
9. Schulz, K.H. and Cox, D.F., *J. Catal.*, Submitted.
10. Schulz, K.H. and Cox, D.F., *J. Phys. Chem.*, Submitted.
11. Redhead, P.A., *Vacuum* **1962**, *12*, 203.
12. Schulz, K.H. and Cox, D.F., *J. Phys. Chem.*, Submitted.
13. Cox, D.F. and Schulz, K.H., *J. Vac. Sci. Technol. A* **1990**, *8*, 2599.
14. Wood, B.J.; Wise H. and Yolles, R.S., *J. Catal.* **1969**, *15*, 355.

RECEIVED October 30, 1992

Chapter 11

Structure and Reactivity of Alkali-Promoted NiO Model Catalysts

Surface Vibrational Spectroscopic Characterization

J. G. Chen, M. D. Weisel, F. M. Hoffmann, and R. B. Hall

Corporate Research Laboratories, Exxon Research and Engineering Company, Annandale, NJ 08801

As previously reported by Otsuka et al. [1] and Ungar et al. [2], alkali salts-promoted NiO catalysts exhibited very high selectivity in the oxidative coupling of methane to C_2 hydrocarbon species. In an effort to understand the fundamental aspects of alkali-promoted activation of methane on NiO, we have carried out a detailed characterization of a K/NiO/Ni(100) model catalyst by using a variety of surface spectroscopies. Our results provide some detailed understanding concerning the formation, morphology, thermal stability and surface reactivity of NiO on Ni(100). In addition, our results demonstrate that potassium interacts strongly with oxygen on Ni(100), as indicated by the HREELS detection of the formation of K-O bonds, by the TPD observation of an oxygen-induced, high-temperature potassium desorption peak, and by the observation of a potassium-accelerated oxidation of Ni(100). FTIR measurements using CO as a probing molecule also indicate the presence of a new type of surface binding sites due to the interaction of potassium with NiO/Ni(100), which could most likely be attributed to the formation of Ni^{3+} sites. This argument is further supported by the FTIR results obtained on lithium-promoted NiO/Ni(100). Furthermore, surface reactivities of clean and K-doped NiO/Ni(100) towards H_2 and CH_4 have also been investigated. These results indicate that, although the presence of potassium reduces the reactivity towards H_2, it enhances the reactivity of NiO with CH_4.

The activation of methane on alkali-promoted metal oxides has been the subject of a number of recent investigations [1-7]. As reported by Otsuka et al. [1] and by Ungar et al. [2], powder NiO catalysts react with CH_4 to form only the deep oxidation products of CO_2 and H_2O. However, by adding alkali salts in the catalyst formulation, deep oxidation can be suppressed, while oxidative coupling, i.e., the formation of C_2H_6 (and C_2H_4), is enhanced. The formation of C_2 hydrocarbons from CH_4 is obviously a catalytic reaction of great industrial importance, since potentially it provides a useful method for converting abundant, remote gas reserves to useful petrochemicals and fuels [8, 9].

0097–6156/93/0523–0133$06.50/0

In an effort to understand the details concerning the alkali-promoted selectivity in the oxidative coupling of methane, we have carried out a detailed characterization of a model K/NiO/Ni(100) catalyst under well-controlled, ultrahigh vacuum conditions. Our systematic approach involved the following procedures: (1) detailed investigation of the formation and stability of NiO on a clean Ni(100) surface; (2) spectroscopic characterization of K-doped NiO by *in-situ* deposition of potassium onto well-characterized NiO/Ni(100) substrates; and (3) determination of the reactivities of NiO/Ni(100) and K/NiO/Ni(100) towards H_2 and CH_4. Such an approach allowed us to employ various surface sensitive spectroscopies including high-resolution electron energy loss spectroscopy (HREELS), Fourier-transform infrared reflection-absorption spectroscopy (FTIR), low energy electron diffraction (LEED), fluorescence-yield near-edge X-ray absorption spectroscopy (FYNES), X-ray photoelectron spectroscopy (XPS), Auger electron spectroscopy (AES), and temperature programed desorption (TPD). These measurements enabled us to gain some detailed understandings of the structure, stability and reactivities of the K-doped NiO/Ni(100) model catalysts.

Several recent studies [10-13] have applied surface science techniques to investigate the properties of model alkali-promoted oxide catalysts. For example, Badyal et al. [10] have applied a combination of XPS, UPS and AES to investigate the interaction of lithium on a NiO film produced on a Ni(111) surface. Their XPS and UPS results suggest that the doping of lithium produces some novel oxygen species on surfaces [10]. The XPS investigation by Peng et al. of Li/MgO catalysts allowed these authors to conclude that the Li^+O^- centers are the active sites for the activation of methane [11]. Our earlier HREELS, LEED, XPS, AES and TPD investigation of the interaction of potassium with NiO/Ni(100) also demonstrated that surface science techniques could be utilized to derive the nature of the potassium-NiO interactions [13].

In this paper, we will use the model K/NiO/Ni(100) system as an example to demonstrate that a detailed, complementary characterization of the model catalyst could best be achieved by using a combination of a variety of surface techniques: The methods of HREELS, LEED, XPS and AES could be applied to obtain properties on and near the surface regions; the technique of FYNES, being a photon-in/photon-out method [14], could be utilized to investigate the bulk properties up to 2000 Å into the surface; the method of FTIR using CO as a probing molecule is, on the other hand, sensitive only to the properties of the top-most surface layer. The results to be presented in this paper will be mainly those obtained by using the two vibrational spectroscopies (HREELS and FTIR). Results from other surface techniques will also be discussed or presented when they provide additional information to the vibrational data. As will be demonstrated in this paper, HREELS is a very powerful technique for this type of studies, since it allows one to readily detect vibrational modes due to K-O, K-Ni and Ni-O oscillators, which generally appear in the low-frequency range (50-600 cm-1). In addition, the excellent spectroscopic resolution of the FTIR method enables one to obtain distinct $v(CO)$ features which could be related to the presence of surface Ni sites with different structures and with different oxidation states.

We should also point out that, although all alkali salts promoted NiO catalysts demonstrated a general enhancement in the selectivity, lithium salt doped NiO catalysts gave rise to the best results for the oxidative coupling of methane [1,2]. It is therefore more relevant to characterize the promoting effect of lithium on the model NiO/Ni(100) catalysts. The main reason that the K/NiO/Ni(100) system was chosen over Li/NiO/Ni(100) in the current study is due to the different thermal stabilities of these two model systems. We have found that a large fraction of the deposited lithium overlayer disappeared via both diffusion into the bulk Ni(100) and desorption in the form of Li_2O at about 800 K. These properties made it rather difficult for an accurate investigation of the reaction of Li-promoted NiO/Ni(100) with methane, which generally occurs at temperatures of 800 K or above. On the other hand, the K/NiO/Ni(100) model catalyst is thermally stable at temperatures up to 1000 K, and is therefore better suited for the investigations reported here. We strongly believe that the promoting effect of potassium observed for the K/NiO/Ni(100) model system could be generalized to the modification effects of other alkali metals on the NiO catalysts.

Experimental

The experiments reported in the present paper were carried out in four separate ultra-high vacuum (UHV) chambers. The HREELS experiments were performed in a multi-level chamber equipped with facilities for HREELS, LEED and TPD studies [15]. The HREELS spectra reported here were collected at an incident beam energy of 3.0 eV with a resolution (FWHM) of the elastic peak in the range of 35 - 55 cm^{-1}. The FTIR measurements were carried out in a UHV chamber equipped with FT-infrared reflection-absorption spectrometer, LEED, AES and TPD facilities [16]. The FTIR spectra were obtained with a Perkin Elmer 1800 FTIR in single reflection with a wide-band MCT detector by adding 400 scans at 4 cm^{-1} resolution. The XPS data were collected in a third UHV chamber equipped with XPS, UPS, LEED, and TPD facilities. The FYNES measurements were obtained at the U1 beam line of the National Synchrotron Light Source, Brookhaven National Laboratory [14]. The oxygen K-edge fluorescence yield was measured by using a differentially-pumped, UHV-compatible proportional counter as described in detail previously [14].

The atomically clean, well-ordered Ni(100) surface was obtained by using a standard cleaning procedure as described in detail elsewhere [17]. The deposition of potassium was achieved by evaporation from a SAES getter source [18]. Reagent grade (99.999% purity) gases of O_2, H_2 and CH_4 were introduced into the UHV chambers through leak-valves.

Results and Discussion

The Formation, Morphology and Stability of NiO on Ni(100). The interaction of oxygen with a Ni(100) surface has been investigated by many authors using a variety of surface techniques [19-23]. We have also investigated the formation and thermal stability of NiO on Ni(100) using a combination of HREELS, LEED, FYNES, XPS and AES [17, 24]. In brief, the interaction of oxygen with Ni(100) at 300 K occurs in two stages: a dissociative chemisorption stage at oxygen coverages less than or equal to 0.5 ML (monolayer), followed by the formation of NiO at higher oxygen coverages. There is a rapid uptake of oxygen up to the point at which a p(2x2)-O ordered overlayer is observed in LEED. This corresponds to 0.25 monolayer of O atoms which occupy the 4-fold hollow site on the Ni(100) surface. The reactive sticking probability declines rapidly in reaching this coverage, and addition of sufficient oxygen to form the c(2x2)-O ordered overlayer (0.5 ML) requires substantially higher doses. Finally, at coverages exceeding 0.5 ML, there is an increase in the rate of oxygen uptake as NiO islands begin to form. The surface NiO layers are saturated at 300 K at an oxygen exposure of 300 L, as indicated by our FYNES and XPS results [24,25] as well as by many other previous investigations [19-23]. The oxygen content in the 300 L (saturated) NiO film is equivalent to roughly 2.5 ML of bulk NiO [20,25]. In our study, this saturated NiO layer prepared at 300 K is characterized by a relatively sharp, 12-spot ring LEED pattern, which could be interpreted as two NiO(111) units misoriented by 30 degrees. It should be pointed out that a (7x7) LEED pattern, which has been interpreted as being the result of patches of NiO(111) which are misoriented with respect to one another at the island boundaries, has also been reported for a similar NiO/Ni(100) surface [21].

In Figure 1, we show a set of representative HREELS spectra of O/Ni(100) and NiO/Ni(100) at various O_2 exposures at 300 K. The chemisorption stage of oxygen on Ni(100) is marked by the progressive formation of p(2x2)-O (Figure 1b) and c(2x2)-O (Figure 1c) overlayer structures. The assignment of the vibrational features in Figures 1b and 1c is as follows [17,22]: the 230 cm-1 and 425 cm-1 features (Figure 1b) are due to a surface phonon mode and the v(Ni-O) mode of oxygen in the p(2x2)-O structure, respectively; the 335 cm-1 feature in Figure 1c is related to the v(Ni-O) mode of a c(2x2)-O overlayer structure. At oxygen exposures in excess of 40 L, one starts to observe the formation of NiO islands on the surface, characterized by a v(Ni-O) mode at 515 cm-1. After exposing the Ni(100) surface to 300 L of oxygen at 300 K (Figure 1d), one obtains a saturated NiO/Ni(100) layer, which is characterized by a broad v(Ni-O) mode of NiO at 515 cm-1. Since the LEED observation of this surface shows only the NiO(111) patterns, this 515 cm-1 feature could readily be assigned to the v(Ni-O) stretching mode of NiO(111).

Our XPS and AES results indicate that the 300 L (saturated) NiO(111) layer is thermally unstable and it undergoes a phase transformation at temperatures above 525 K [24,25]. Such a phase separation is also supported by the HREELS spectroscopic changes from Figure 1d to Figure 1e. After heating the NiO/Ni(100) layer to 800 K, the broad v(Ni-O) at 515 cm-1 is replaced by a sharp feature at 565 cm-1 and a broad mode at ~ 300 cm-1. The corresponding LEED observation of this surface reveals patterns corresponding to NiO(100) and c(2x2)-O structures, which allows us to readily assign the 565 and ~300 cm-1 features to the v(Ni-O) modes of NiO(100) and c(2x2)-O, respectively. Thus, both the HREELS and LEED results suggest that the NiO(111) films prepared at 300 K undergo a thermally-induced phase separation to produce a surface with a mixture of crystalline NiO(100) clusters and of chemisorbed c(2x2)-O regions. The additional vibrational feature observed at 1125 cm-1 in Figure 1e is related to the overtone and double loss of the v(Ni-O) mode of the crystalline NiO(100) clusters [17]. The observation of this overtone/double loss feature could also be used as an indication that the NiO(100) clusters are well-ordered [17].

In order to further understand the nature of this thermally-induced phase separation of NiO(111) films, we have carried out a detailed XPS and AES investigation of this surface. Both the XPS O(1s) and the AES O(KLL) features show a significant decrease in intensity (about a factor of 3) upon heating the NiO(111) film to 800 K, with the onset of the decrease starting at about 525 K. Considering the limited escape depth of the O 1s electrons and the AES O(KLL) electrons (~10 Å) [26], the attenuation in the XPS and AES intensities could be attributed to the formation of NiO(100) islands which are sufficiently thick that the escape probability of the XPS and AES electrons is reduced [25].

As mentioned in the Introduction, one of the main advantages of the FYNES method is its capability of detecting species deep into the bulk (~ 2000 Å) [14]. If the thermally-induced changes in XPS and AES are related to the formation of thick NiO islands, one would expect that the FYNES spectra should remain the same before and after annealing the NiO/Ni(100) layer to 800 K, if the NiO clusters are less than 2000 Å in thickness. Such an observation is shown in the upper panel of Figure 2, where the FYNES spectra of a NiO/Ni(100) layer before and after annealing to 800 K are compared. The two most intense oxygen K-edge features in Figure 2 are observed at 533 and 541.5 eV, respectively. These two features were assigned to electronic transitions to the $3e_g$ (Ni 3d + O 2pσ) and the $3a_{1g}$ (Ni 4s + O 2pσ) orbitals of NiO, respectively. The atomic orbitals given in parathenses are those contributing

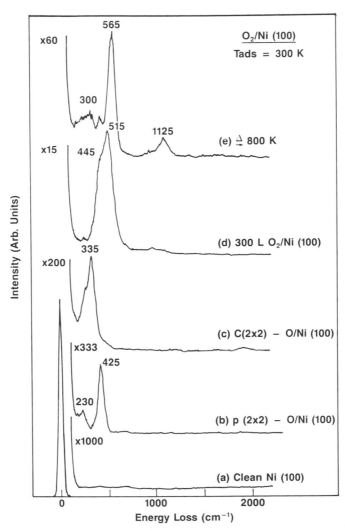

Figure 1.   HREELS spectra of O/Ni(100) and NiO/Ni(100). The O_2 exposures are: (a) 0.0 L; (b) 1.0 L; (c) 40 L; (d) 300 L and (e) 300 L. All spectra were recorded at 300 K.

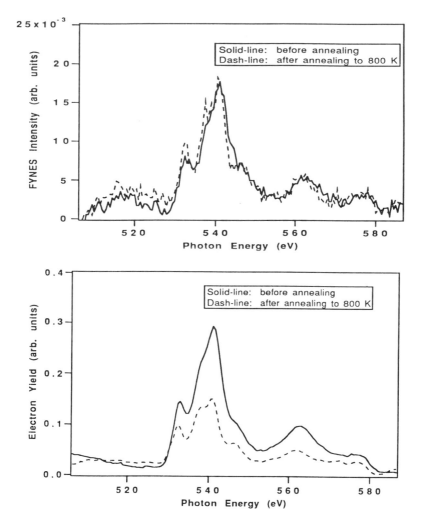

Figure 2.   Upper-panel: Fluorescence-yield near-edge spectra of a NiO/Ni(100) layer measured before and after annealing the surface to 800 K. Lower-panel: Simultaneous measurement of the partial electron-yield of the same NiO/Ni(100) surface.

dominantly to the corresponding molecular orbitals. More details concerning the assignment of other K-edge features in Figure 2 can be found elsewhere [24].

The important conclusion from the FYNES data is that the total oxygen content near the surface region is fairly constant before and after the thermal treatment. The conclusion that the thermally-induced attenuation of the XPS and AES intensities is related to the limited escape depth of either O(1s) electrons or the AES electrons is further confirmed in the lower-panel of Figure 2. These two near-edge spectra were obtained simultaneously with the FYNES spectra by measuring the partial electron-yield of the same NiO/Ni(100) surface [24]. The detection depth of the oxygen content is again limited by the fact that the electron-yield measurments detect only near-surface oxygen. Similar to the XPS and AES measurements, the near-edge spectra by means of measuring the partial electron-yield demonstrate a decrease in the oxygen features upon heating, even though the corresponding FYNES results indicate that the oxygen content near the surface region remains the same. Based on a comparison between an electron scattering model simulation and our XPS and FYNES results, we have estimated that the NiO(100) clusters, produced by heating NiO(111) to 800 K, are roughly 50 Å thick and occupy approximately 6% of the surface [25]. The rest of this annealed surface is occupied by the chemisorbed oxygen with a c(2x2)-O overlayer structure.

We have also investigated the further oxidation of Ni(100) surface by using both FYNES and XPS [25]. Although the NiO/Ni(100) surface is saturated at 300 K after 300 L of O_2, additional NiO can be produced by exposing the annealed, i.e., the [NiO(100)+c(2x2)-O]/Ni(100) surface, to oxygen. This further oxidation occurs on the c(2x2)-O regions of the annealed surface. A thicker, better ordereed NiO(100) film could be obtained by repeated cycles of dosing (300 L O_2 at 300 K) followed by annealing to 800 K. The growth of ordered NiO(100) layers upon multiple-cycles of dosing and annealing is also supported by our LEED observation of a gradual sharpening of the NiO(100) pattern and a gradual decrease of the c(2x2)-O pattern, which eventually disappears after multiple cycles of dosing-annealing. The HREELS results also indicate that after repeated cycles of dosing-annealing, the intensities of the multiple v(Ni-O) losses can be better described by a Poisson distribution, which would be expected if a bulk-like NiO(100) overlayer is produced [17].

We have applied the FTIR method to investigate the surface properties of NiO/Ni(100) films by following the adsorption of CO. A comparison of FTIR spectra of CO on various surfaces is shown in Figure 3. Figure 3a is a typical CO chemisorption spectrum on a clean Ni(100) surface at 300 K. The CO coverage of this surface is about 0.5 ML with all CO molecules occupying the terminal sites on clean Ni(100). The vibrational feature observed at 2032 cm-1 can be easily assigned to the v(CO) stretching mode of the terminal CO species [27]. The saturation adsorption of CO (10 L) on chemisorbed and oxidized O/Ni(100) and NiO surfaces (Figures 3b-3e) was performed at 80 K. Such a low temperature is required since a fraction of CO molecules starts to desorb from these surfaces at about 100 K. Figure 3b was obtained after a saturation adsorption of CO on a c(2x2)-O/Ni(100) surface; the FTIR spectrum is characterized by one terminal v(CO) mode 2095 cm-1. This frequency is blue-shifted by 63 cm-1 from the clean Ni(100) surface. Such a chemisorbed oxygen-induced frequency shift has also been observed for the adsorption of CO on oxygen-covered Ni(111) [28]. It can be readily explained by the fact that the presence of surface oxygen reduces the electron density donation from the metal surfaces to the $2\pi^*$ anti-bonding orbitals of CO, and therefore gives rise to a blue-shift in the v(CO) frequencies.

The FTIR spectrum collected after the adsorption of CO on a saturated NiO(111) film prepared at 300 K (Figure 3c) is significantly different from both clean and c(2x2)-

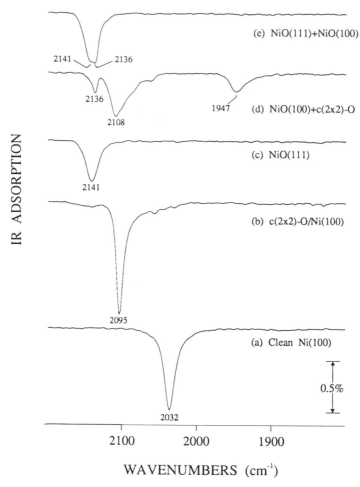

Figure 3. FTIR spectra after a saturation CO adsorption on various surfaces. The adsorption of CO on the clean Ni(100) surface was performed at 300 K. All other adsorption experiments were carried out at 80 K. See text for the preparation of various substrates.

O surfaces. This spectrum is characterized by a single $v(CO)$ feature at 2141 cm^{-1}. Based on previous IR studies on NiO powders [29,30] and on NiO films produced on Ni(111) [30], this feature can be assigned to the terminal CO species adsorbed on a Ni^{2+} site on NiO. The fact that only one $v(CO)$ feature is observed in Figure 3c further confirms that the composition of the top-most layer of this surface is NiO, since there are no $v(CO)$ features related to either bare Ni-sites or sites which are modified by the chemisorbed oxygen. In addition, the thermally-induced phase separation of NiO(111) -> NiO(100) + c(2x2)-O is also confirmed by the comparison of Figures 3c and 3d. Upon the adsorption of CO on a NiO/Ni(100) surface previously annealed to 800 K, the FTIR spectrum (Figure 3d) is characterized by CO features related to the adsorption on the Ni^{2+} sites (2136 cm^{-1}) and on Ni sites that are modified by chemisorbed oxygen (2108 cm^{-1}). One additional $v(CO)$ feature is observed at 1947 cm^{-1} on the annealed surface. This feature is observed on a c(2x2)-O overlayer at low CO coverages (spectrum not shown); it has also been detected by a previous HREELS studies of the adsorption of CO on a c(2x2)-O/Ni(100) surface [31]. At present we tentatively assign this feature to a bridging CO species on the c(2x2)-O/Ni(100) layer.

The further oxidation of this annealed [NiO(100)+c(2x2)-O] layer is also reflected in the FTIR spectrum shown in Figure 3e. The NiO substrate in this spectrum was prepared by an additional exposure of 400 L O$_2$ at 300 K to the annealed surface prior to the adsorption of CO. By comparing Figure 3d and Figure 3e, one can obtain important information concerning the further oxidation of the annealed NiO/Ni(100) surface. In agreement with our XPS data, the further oxidation occurs only on the c(2x2)-O region, as indicated by the disappearance of the c(2x2)-O modified sites (i.e., $v(CO)$ features at 2108 and 1947 cm^{-1}). This conclusion is further supported by the fact that the $v(CO)$ feature due to the Ni^{2+} sites of NiO(100) clusters (2136 cm^{-1}) is unaltered by the reappearance of the 2142 cm^{-1} feature (related to the Ni^{2+} sites of NiO(111) upon the oxidation of c(2x2)-O regions). Figure 3e also indicates that FTIR using CO as a probing molecule is a rather powerful technique since it even allows one to distinguish the Ni^{2+} sites of NiO(111) from those of NiO(100). More details concerning the FTIR characterization of various NiO and K/NiO surfaces will be published in a separate paper [32].

Interaction of Alkali-Metals with NiO/Ni(100). We have investigated the interaction of potassium with the unannealed NiO(111) and the annealed [NiO(100)+c(2x2)-O] films by using HREELS, FTIR, XPS, AES and TPD. Our results indicate that potassium interacts strongly with NiO. An example of such a K-NiO interaction is shown in Figure 4, where the thermal desorption of potassium is compared from a clean Ni(100) surface and from a NiO/Ni(100) surface. The potassium coverage on both surfaces are ~ 2.5 "ML", with 1.0 "ML" corresponding to a K/Ni ratio of 0.38 [32,33]. As shown in Figure 4, the thermal desorption of K from a clean Ni(100) is characterized by a multi-layer peak at ~ 330 K, a second-layer peak at ~ 370 K and a broad peak due to the first monolayer potassium centered at ~ 550 K. All potassium desorbs from Ni(100) as indicated by the AES observation that a clean Ni(100) surface was always obtained after the K/Ni(100) layer was heated to 900 K. However, the thermal desorption of potassium from the NiO/Ni(100) layer is significantly different. Potassium does not desorb from the surface until 700 K, with the desorption peak centered at 760 K, suggesting a strong interaction between potassium and NiO. It is also obvious from Figure 4 that, by comparing the peak areas in the two TPD spectra, a significant amount of K still remains on the surface after the K-doped NiO/Ni(100) layer is heated to 900 K. Our combined HREELS,

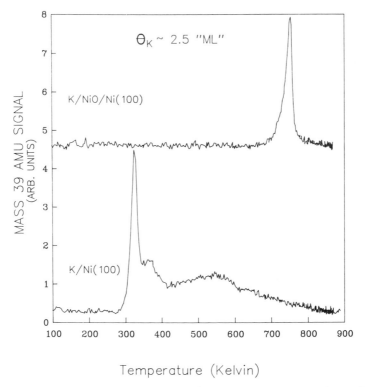

Figure 4. Thermal desorption spectra of K on Ni(100) and on NiO/Ni(100). The NiO/Ni(100) layer was prepared by 300 L O_2 exposure at 300 K followed by annealing to 800 K. The potassium-deposition was carried out at 90 K.

AES and TPD measurements also suggest that some of the potassium penetrates into the NiO clusters at T > ~600 K [32].

The interaction of potassium with NiO by forming K-O and K-Ni bonds has been observed by using HREELS, as shown in Figure 5. Figure 5a is a typical HREELS spectrum of an annealed NiO/Ni(100) surface recorded at 90 K. This spectrum is characterized by the v(Ni-O) mode of the c(2x2)-O phase at 325 cm^{-1}, the v(Ni-O) mode of the crystalline NiO clusters at 570 cm^{-1} and its higher order overtones and multiple losses at 1145 and 1725 cm^{-1}. By depositing ~ 2.5 "ML" K on the surface (Figure 5b), one observes a new low-frequency feature at ~ 210 cm^{-1}. The v(Ni-O) mode of the c(2x2)-O phase is also shifted from 325 to ~400 cm^{-1} upon the deposition of potassium. At higher temperatures (Figures 5c-5e), one observes two sequentially developed low-frequency features at 220-235 cm^{-1} and at ~ 175 cm^{-1}, respectively. Based on the vibrational data for potassium on various transition metal surfaces [34], these two features could be assigned to the stretching motions of v(K-O) and v(K-Ni) modes, respectively. After heating the K-doped NiO/Ni(100) layer to 800 K (Figure 5e), the spectrum is characterized by four vibrational features and the assignment of these features is as follows: the 177 cm^{-1} feature is related to a v(K-Ni) mode and its frequency shifts upward/downward as the oxygen coverage increases/decreases; the 375 cm^{-1} mode is related to the v(Ni-O) mode of regions with low local oxygen coverage resembling the c(2x2)-O phase, but its frequency is shifted upward (as compared with Figure 5a) as a result of the interaction with K; the 573 cm^{-1} feature is due to the v(Ni-O) mode of NiO clusters and the 1133 cm^{-1} mode is related to the overtone and double loss of the v(Ni-O) mode. The 1380-1490 cm^{-1} feature in Figures 5b-5d is most likely a CO or a carbonate feature as a result of CO adsorption (from the residual gas in the UHV chamber) during the course of experiments.

The strong interaction of potassium with oxygen on Ni surfaces is also demonstrated by our experimental observation of a potassium-accelerated oxidation of Ni(100), as shown in Figure 6. Figure 6a is recorded after depositing ~2.0 "ML" of potassium on a clean Ni(100) surface. After exposing 1.0 L of oxygen to the K/Ni(100) layer at 300 K, the HREELS spectrum (Figure 6b) in the low-frequency region is characterized by vibrational features at 255, 425 and 540 cm^{-1}. As discussed earlier in Sec. 3.1, such a low oxygen exposure on a clean Ni(100) surface gives rise only to the chemisorption of a p(2x2)-O overlayer (Figure 1b). By comparing Figure 6b and Figure 1b, one can notice that, on the K/Ni(100) surface, in addition to the formation of the p(2x2)-O overlayer (the 425 cm^{-1} mode), the surface is also characterized by the vibrational features of v(K-O) at 255 cm^{-1} and v(Ni-O) of NiO at 540 cm^{-1}. As shown earlier in Figure 1, on a clean Ni(100) surface, the formation of NiO at 300 K could only be achieved at an oxygen exposure > 40 L. The fact that the NiO formation is observed at an O_2 exposure of 1.0 L on the K/Ni(100) surface clearly indicates that the presence of potassium accelerates the oxidation of the Ni surface. Such an observation is more evident upon heating this surface to higher temperatures (Figure 6c-6d). We should point out that if a p(2x2)-O layer on a clean Ni(100) surface is heated to 800 K, both the HREELS spectrum and LEED pattern remain to be a characteristic p(2x2)-O surface. Heating the p(2x2)-O layer to 800 K does not give rise to the conversion of p(2x2)-O to NiO on the clean Ni(100) surface. This is clearly not the case for the O/K/Ni(100) layer shown in Figure 6. The formation of well-characterized NiO is evident after the surface is heated to 800 K, as indicated by the relatively strong NiO feature at 570 cm^{-1} in Figure 6d. Thus, the results shown in Figure 6 provide a clear experimental evidence that the presence of

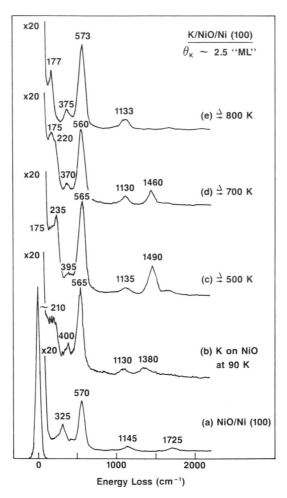

Figure 5. HREELS investigation of the thermal behavior of a K-doped NiO/Ni(100) layer. The NiO/Ni(100) layer was annealed to 800 K prior to the deposition of potassium at 90 K. All HREELS measurements were performed at 90 K.

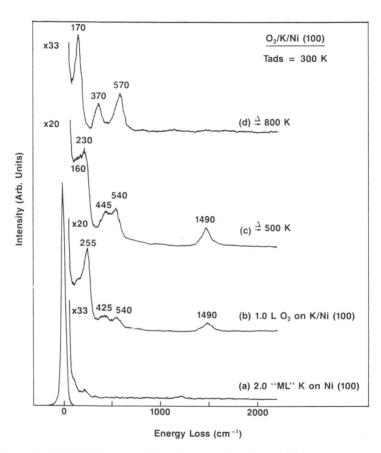

Figure 6. HREELS spectra following the adsorption of 1.0 L oxygen on a K/Ni(100) layer at 300 K. The K/Ni(100) surface was prepared by evaporating ~2 ML of potassium onto a clean Ni(100) surface at 300 K.

potassium on Ni(100) promotes the oxidation of the surface. Similar potassium-promoted oxidation has been previously observed for the oxidation of O/K/Cu(100) [35].

One essential question to be answered for the alkali-promoted activity of NiO is whether the alkali-doping modifies the oxidation states of NiO from Ni^{2+} -> Ni^{3+}. In principle, such a question could be answered by performing the XPS measurements. However, in our XPS experiments of the K/NiO/Ni(100) system, we did not observe any well-resolved Ni^{3+} states. This could be due to the fact that XPS probes both the surface and the near surface region ~10 Å below the surface. For an annealed K/NiO layer, the XPS spectra of the Ni-region could be contributed from the combination of following Ni species on or near the surface region: metallic Ni, Ni modified by the c(2x2)-O, Ni^{2+} of unmodified NiO(100), and the possible Ni^{3+} sites which are produced upon the doping of potassium. Therefore, if the number of Ni^{3+} sites is relatively small compared to other Ni species, it would be very difficult to obtain a distinct Ni^{3+} transition from the XPS measurements. It should be pointed out that a previous XPS study of NiO films and of βNiOOH [36] showed some evidence of the presence of Ni^{3+}, although the XPS features were not well-resolved and could be due to the presence of defects or adsorbed impurities [36].

On the other hand, FTIR using CO as a probe could in principle be applied to investigate the presence/absence of the Ni^{3+} sites, since this method is surface sensitive and since one would expect to observe a distinct ν(CO) frequency if CO molecules are adsorbed on the Ni^{3+} sites. A comparison of FTIR spectra following the adsorption of CO on NiO, K/NiO and Li/NiO is shown in Figure 7. The K/NiO and Li/NiO surfaces were prepared by evaporating ~ 3 ML of alkali metals onto NiO/Ni(100) at 80 K followed by annealing the surface to 800 K. The FTIR spectra in Figure 7 were all recorded at 80 K at a background CO pressure of 1x10^{-3} torr. Such a CO background pressure was required for the FTIR detection of the high-frequency ν(CO) features on the alkali-modified NiO surfaces (the 2160 cm^{-1} feature in Figure 7b and the 2156 cm^{-1} feature in Figure 7c). As shown in Figure 7a, the FTIR spectrum recorded at 10^{-3} torr of CO on NiO/Ni(100) is similar to that recorded under UHV conditions (Figure 3d). The spectrum is again characterized by CO molecules adsorbed on the Ni^{2+} sites of NiO clusters at 2140 cm^{-1}, and by the terminal-CO and bridging-CO on the c(2x2)-O regions at 2106 and 1932 cm^{-1}, respectively.

The FTIR spectrum of CO on the K/NiO/Ni(100) surface (Figure 7b) is characterized by five ν(CO) features. Among them the 2160 cm^{-1} feature could only be detected at a CO background pressure of 1x10^{-3} torr, while the other four features could be observed under UHV conditions after the adsorption of CO. In general, the presence of potassium on clean Ni(100) increases the degree of backdonation of electrons to the $2\pi^*$ antibonding orbitals of CO and therefore gives rise to a red-shift in the ν(CO) frequency. On the other hand, the presence of chemisorbed oxygen on Ni(100) reduces the degree of backdonation and results in a blue-shift in the ν(CO) frequency. We have carried out a detailed FTIR investigation of the complex effect of potassium and chemisorbed oxygen on the adsorption of CO on Ni(100) [32]. In brief, the 2090, 2013 and 1898 cm^{-1} features in Figure 7b are related to the adsorption of CO on Ni sites which are non-uniformly modified by the coadsorbed potassium and atomic oxygen [32]. In addition, the 2134 cm^{-1} feature in Figure 7b can be assigned to the adsorption of CO on the Ni^{2+} sites of NiO clusters that are not modified by the presence of potassium.

The most interesting observation in Figure 7b is the detection of the ν(CO) feature at 2160 cm^{-1}. The fact that it could only be detected at 80 K and at a

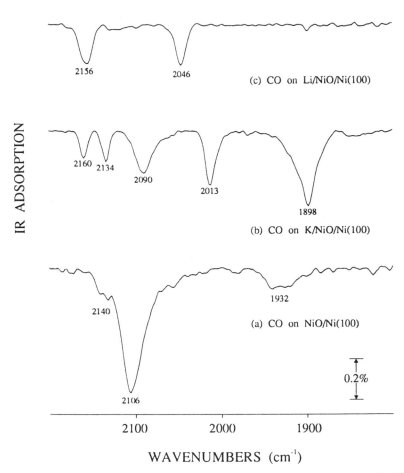

Figure 7. A comparison of CO adsorption on NiO/Ni(100) and K/NiO/Ni(100) surfaces. The FTIR spectra recorded at a CO pressure of 3×10^{-3} torr with the surface temperature at 80 K.

background CO pressure of 10^{-3} torr indicates that the thermal desorption temperature of this weakly adsorbed CO species is lower than 80 K. In general, the $\nu(CO)$ frequency and the corresponding thermal desorption temperature of adsorbed CO species could be used as an indicative of the oxidation states of the Ni sites. It has been observed previously that the Ni-CO bond is contributed dominantly by the backdonation of electrons into the $2\pi^*$ orbitals of CO [30]. Therefore, CO molecules adsorbed on a Ni site with higher oxidation state would give rise to a higher $\nu(CO)$ frequency and a lower thermal desorption temperature. Our combined FTIR and TPD results also demonstrated the following trend between the Ni oxidation states and the corresponding $\nu(CO)$ frequencies and thermal desorption temperatures of adsorbed CO molecules: on Ni^0 of clean Ni(100): 2030 cm^{-1}, 410 K; on $Ni^{\delta+}$ of c(2x2)-O modified Ni: 2095 cm^{-1}, 190 K; on Ni^{2+} of NiO: 2136-2141 cm^{-1}, 115 K; and on the new Ni sites due to the interaction of NiO with potassium: 2160 cm^{-1}, < 80 K. Such a comparison suggests that the new $\nu(CO)$ feature observed on the K/NiO surface could most likely be attributed to the adsorption of CO on Ni^{3+} sites which are produced as a result of the interaction of potassium with NiO. Such an assignment is also in agreement with a previous IR study of the adsorption of CO on NiO powders, where a 2156 cm^{-1} CO feature was assigned to the adsorption of CO on the Ni^{3+} sites [29].

The presence of Ni^{3+} sites by alkali-doping is also supported by the FTIR spectrum of CO on Li-modified NiO/Ni(100) (Figure 7c). Due to a better lattice match between Li and Ni, one would expect that the NiO layer would be more uniformly modified by the presence of Li, as demonstrated in Figure 7c. The high-frequency $\nu(CO)$ region of Figure 7c is characterized by one $\nu(CO)$ feature at 2156 cm^{-1}. By comparing this spectrum region with that of the K/NiO/Ni(100) surface, it is clear that all Ni^{2+} sites of NiO clusters are converted to the Ni^{3+} sites due to the presence of Li. In addition, the c(2x2)-O regions are also more uniformly modified by the presence of Li, as indicated by a single chemisorbed $\nu(CO)$ feature at 2046 cm^{-1}. More details concerning the FTIR investigations of the CO adsorption on various alkali-modified Ni sites can be found elsewhere [32].

Reactivities of NiO and K/NiO with H_2 and Methane.   We have investigated the surface reactivity of the annealed NiO/Ni(100) surface towards H_2 by using EELS, FYNES, XPS and LEED [24,25]. These results indicated that the annealed NiO/Ni(100) surface could be reduced by exposing the surface to H_2 at 600 - 800 K. The removal of oxygen occurs via the formation and subsequent desorption of H_2O. Details concerning the kinetics of the reaction of hydrogen with NiO/Ni(100) by using XPS and FYNES can be found elsewhere [24,25]. In the present paper, we will concentrate on our HREELS results to compare the reactivities of hydrogen with NiO/Ni(100) and K/NiO/Ni(100).

A set of HREELS results following the removal of oxygen by H_2 is shown in Figure 8; the corresponding LEED patterns of each surface after the hydrogen-treatment are given in the Figure Caption. The NiO/Ni(100) layer was prepared by heating a saturated NiO/Ni(100) layer to 800 K. This surface was then exposed to 2.0×10^{-6} torr of H_2 at 800 K. After the surface was treated by H_2 for a certain period of time, the H_2 gas was pumped away and the HREELS spectra were recorded after the surface was cooled back to ~ 300 K. It should also be pointed out that the HREELS spectroscopic changes shown in Figure 8 are not due to thermal effects, as suggested by the absence of any noticeable change in the HREELS spectra after the NiO/Ni(100) surface was annealed *in vacuo* at 800 K for 10 minutes.

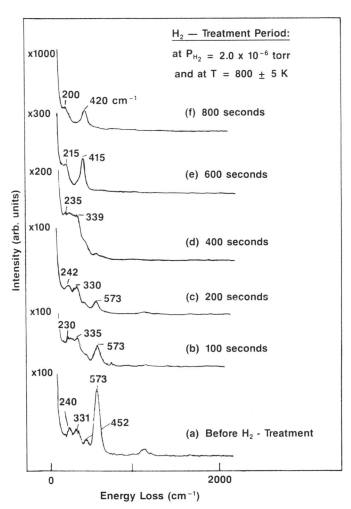

Figure 8. HREELS spectra following the reaction of NiO/Ni(100) with hydrogen. The surface was treated at 800 K at a H$_2$ pressure of 2x10^{-6} torr. All HREELS spectra were collected at 300 K after the reaction with H$_2$. The corresponding LEED patterns are: (a-c) NiO(100) + c(2x2)-O; (d) c(2x2)-O; (e) p(2x2)-O; (f) Ni(100) + very weak p(2x2)-O.

Both the HREELS results shown in Figure 8 and the corresponding LEED investigations indicate that the oxygen species are gradually removed from the surface via the sequence of NiO(100) + c(2x2)-O -> c(2x2)-O -> p(2x2)-O -> Ni(100). The HREELS results indicate that, for a surface with mixtures of c(2x2)-O and NiO(100) clusters, the NiO clusters are preferentially reduced by H_2. In fact, the trend shown in Figure 8 is almost the reverse of the gradual oxidation of Ni(100) as that shown in Figure 1.

For comparison, HREELS results following the reaction of hydrogen with an annealed K/NiO surface under similar experimental conditions are shown in Figure 9. The K/NiO/Ni(100) surface was prepared by evaporating ~2.5 "ML" of potassium onto a saturated NiO/Ni(100) films before the surface was annealed to 800 K. By comparing Figure 9 with Figure 8, one can notice that the presence of potassium actually reduces the reaction rate of NiO with hydrogen. For example, after exposing the surface to 2.0×10^{-6} torr of H_2 for 400 seconds, almost all NiO is removed in the case of NiO/Ni(100) (Figure 8d), while there is still a substantial amount of NiO remaining in the case of K-doped NiO/Ni(100) (Figure 9c). Such a potassium-related decrease of reactivity towards H_2 is also confirmed in our time-resolved XPS measurements, which indicate that the oxygen removal rate is about a factor of 2 slower due to the presence of potassium on NiO.

On the other hand, the presence of potassium on NiO/Ni(100) noticeably increases the reaction rate of NiO with CH_4. Figure 10 shows a comparison of HREELS results following the reaction of CH_4 with NiO and K-doped NiO. The preparations of the NiO and K/NiO layers in Figure 10 are identical to those described in Figure 8 and Figure 9, respectively. As shown in the right-panel of Figure 10, upon the reaction of NiO/Ni(100) with CH_4 at 800 K, the intensity of the v(Ni-O) mode decreases rather slowly. However, a noticeable increase in the oxygen removal rate is observed in the K-doped NiO/Ni(100) surface after reactions with CH_4 under identical reaction conditions (left panel of Figure 10). Assuming that the relative HREELS intensity of the v(Ni-O) mode of NiO at 560-570 cm^{-1} could be used as an indication of the amount of NiO clusters on the surface, such a comparison indicates that the removal of NiO by CH_4 occurs in a faster rate for the K-doped NiO/Ni(100) surface.

The results shown in Figures 8-10 demonstrate that the reactivity of the model NiO/Ni(100) surface can be significantly modified by the presence of potassium. The experimental results indicated that, although K-doped NiO/Ni(100) could be reduced by H_2, the presence of potassium in this case did not give rise to any enhancement in the reactivity. In fact, the rate of oxygen removal was even reduced by a factor of ~2 for the K-doped surface, suggesting that the presence of potassium did not promote the reactivity of NiO towards the H-H bond cleavage. On the other hand, the presence of potassium noticeably increases the reactivity of NiO with CH_4. Therefore, these results would most likely suggest that, as compared to the H-H bond scission, the K-doped NiO selectively promotes the cleavage of the C-H bond.

Conclusions

The results presented above demonstrate that one could obtain fundamental information concerning the structure, oxidation state and reactivities of a model K/NiO/Ni(100) catalyst by using a combination of advanced surface science techniques. The experimental approaches described here could in principle be applied to other model catalysts which are of importance in the selective oxidation of hydrocarbons. These results also demonstrate that it is very important to use a variety of surface techniques to obtain complementary results.

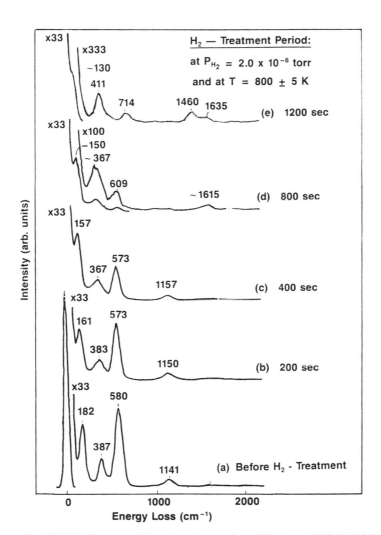

Figure 9. HREELS spectra following the reaction of K-doped NiO/Ni(100) with hydrogen at 800 K. All experimental conditions were identical as those described in Figure 8.

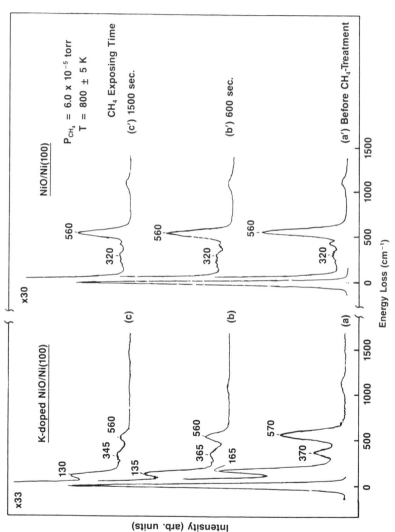

Figure 10. A comparison of the reactivity of CH$_4$ towards clean and K-doped NiO/Ni(100). Both NiO/Ni(100) and K-doped NiO/Ni(100) layers were annealed to 800 K prior to the CH$_4$-treatment. All HREELS measurements were performed at 300 K after the CH$_4$-treatment at 800 K.

Literature Cited

1. Otsuka, K., Lin, Q. and Morikawa, A. , *Inorg. Chimica Acta*, **1986**, *118*, L23.
2. Ungar, R.K., Zhang, X. and Lambert, R.M. *Appl. Catal.,* **1988**, *42*, L1.
3. Driscoll, D.J., Martir, W., Wang, J.-X. and Lunsford, J.H. *J. Am. Chem. Soc.*, **1985**, *107*, 58.
4. Lin, C.-H., Ito, T., Wang, J.-X. and Lunsford, J.H. *J. Am. Chem. Soc.*, **1987**, *109*, 4808.
5. Nelson, P.F. and Cant, N.W. *J. Phys. Chem.,* **1990**, *94*, 3756.
6. Hutchings, G.J., Scurrell, M.S. and Woodhouse, J.. *Chem. Soc. Rev.*, **1989**, *18*, 251.
7. Lunsford, J.H. *Catal. Today*, **1990**, *6*, 235.
8. Jones, C.A., Leonard, J.S. and Sofranko, J.A. *J. Energy and Fuels*, **1987**, *1*, 12.
9. Scurrell, M.S. *Applied Catal.*, **1987**, *32*, 1.
10. Badyal, J.P.S., Zhang, X. and Lambert, R.M., *Surf. Sci.*, **1990**, *225*, L15.
11. Peng, X.D., Richards, D.A. and Stair, P.C., *J. Catalysis*, **1990**, *121*, 99.
12. Moggridge, G.D., Badyal, J.P.S. and Lambert, R.M., *J. Catalysis*, **1991**, *132*, 92.
13. Chen, J.G., Weisel, M.D., Hardenbergh, J.H., Hoffmann, F.M., Mims, C.A. and Hall, R.B., *J. Vac. Sci. Technol. A*, **1991**, *9*, 1684.
14. Fischer, D.A., Colbert, J. and Gland, J.L., *Rev. Sci. Instrum.*, **1989**, *60*, 1596.
15. Hrbek, J., de Paola, R.A. and Hoffmann, F.M., *J. Chem. Phys.*, **1984**, *81*, 2818.
16. Hoffmann, F.M., *J. Chem. Phys.*, **1989**, *90*, 2816.
17. Chen, J.G., Weisel, M.D. and Hall, R.B., *Surf. Sci.*, **1991**, *250*, 159.
18. de Paola, R.A., Hrbek, J. and Hoffmann, F.M., *J. Chem. Phys.*, **1985**, *82*, 2484.
19. Holloway, P.H. *J. Vac. Sci. Technol.*, **1981**, *18*, 653.
20. Norton, P.R., Tapping, R.L. and Goodall, J.W. *Surf. Sci.*, **1977**, *65*, 13.
21. Wang, W.-D., Wu, N.J. and Thiel, P.A. *J. Chem. Phys.*, **1990**, *92*, 2025.
22. Rahman, T.S., Mills, D.L., Black, J.E. Szeftel, J.M., Lehwald, S. and Ibach, H., *Phys. Rev. B*, **1984**, *30*, 589 .
23. Brundle, C.R. *J.Vac.Sci.Technol. A*, **1985**, *3*, 1468.
24. Chen, J.G., Fischer, D.A., Hardenbergh, J.H. and Hall, R.B., *Surf. Sci.*, in press.
25. Hall, R.B., Mims, C.A., Hardenbergh, J.H. and Chen, J.G., in "*Surface Science of Catalysis*", Eds. Dwyer, D.J. and Hoffmann, F.M., ACS Sympossium Series **1991**, *482*, 85.
26. Penn, D.R., *J. Electron Spectrosc. Related Phenom.*, **1976**, *9*, 29.
27. Uvdal, P., Karlsson, P.-A., Nyberg, C., Andersson, S. and Richardson, N.V., *Surf. Sci.*, **1988**, *202*, 167.
28. Xu. Z., Surnev, L., Uram, K. and Yates, Jr. J.T., to be published.
29. Platero, E.E., Colluccia, S. and Zecchina, A., *Surf. Sci.*, **1986**, *171*, 465.
30. Yoshinobu, J., Ballinger, T.H., Xu. Z., Jansch, H.J., Zaki, M.I., Xu, J. and Yates, Jr. J.T., *Surf. Sci.*, **1991**, *255*, 295.
31. Andersson, S., *Solid State Commun.*, **1977**, *24*, 183.
32. Chen, J.G., Weisel, M.D., Hoffmann, F.M., and Hall, R.B., to be published.
33. Campbell, C.T. and Goodman, D.W., *Surf. Sci.*, **1982**, *123*, 413.
34. Paul, J., de Paola, R.A and Hoffmann, F.M., in "*Physics and Chemistry of Alkali Metal Adsorption*", Eds., Bonzel, H.P., Bradshaw, A.M. and Ertl, G., Materials Science Monographs, **1989**, *57*, 213.

35. Hoffmann, F.M., Liu, K.C., Tobin, R.G., Hirshlungl, C.J., G.P. Williams and Dumas, P., *Surf. Sci.*, in press.
36. Carley, A.F., Chalker, P.R. and Roberts, M.W., *Proc. Royal Soc. London*, A, **1985,** *399*, 167.

RECEIVED November 6, 1992

CHARACTERIZATION OF OXIDATION CATALYSTS

Chapter 12

Active Crystal Face of Vanadyl Pyrophosphate for Selective Oxidation of n-Butane

T. Okuhara, K. Inumaru, and M. Misono

Department of Synthetic Chemistry, Faculty of Engineering, The University of Tokyo, Bunkyo-ku, Tokyo 113, Japan

Catalytic properties of the crystal faces of $(VO)_2P_2O_7$ for selective oxidation of butane were examined by using large plate-like crystallites with average size of 5 μm and thickness of 40 nm. The $(VO)_2P_2O_7$ produced maleic anhydride with 60%-selectivity in the butane oxidation at 713 K. When SiO_2 was deposited on the surface of $(VO)_2P_2O_7$ (the atomic ratio of Si to V at the surface was 6) by the reaction of $Si(CH_3)_4 + O_2$ at 773 K, the surface lost the catalytic activity. TEM-EDX observation revealed that SiO_2 was located both on the basal (100) face, in which $V^{4+}(=O)$-O-V^{4+} sites are present, and on the side faces such as (001). The SiO_2-deposited $(VO)_2P_2O_7$ was fractured into small crystallites by applying pressure on it to create the side faces predominantly. It was found that the fractured sample was active mainly for the formation of CO and CO_2 from butane. Therefore, it is concluded that the face which is selective for the formation of maleic anhydride is (100), while the side faces such as (001) are active for non-selective oxidation. When the parent $(VO)_2P_2O_7$ (without SiO_2-deposition) was fractured by the same method, the activity increased a little and the selectivity to maleic anhydride decreased from 60% to 56%. This fact supports the above conclusion.

V-P oxides are known to be efficient for the production of maleic anhydride (abbreviated as MA) from n-butane (*1-3*). A single-phase $(VO)_2P_2O_7$ has been inferred to be the active catalyst phase (*4-6*). In addition, the catalytic properties of $(VO)_2P_2O_7$ varied depending on the microstructure of $(VO)_2P_2O_7$ particles (*7,8*). Some claimed that the

0097–6156/93/0523–0156$06.00/0
© 1993 American Chemical Society

active phase is a combination of $(VO)_2P_2O_7$ and $VOPO_4$ (*9-11*). Volta et al. have proposed that MA is formed on $(VO)_2P_2O_7$ surface with the participation of V^{5+} (*11*). On the other hand, P-rich surface phases have been reported to be active on the basis of XPS analysis (*12,13*).

The active site of $(VO)_2P_2O_7$ for the butane oxidation is thought to be $V^{4+}(=O)-O-V^{4+}$, which is located on the (100) face of $(VO)_2P_2O_7$ (*1*), where XRD lines were indexed following Gorbunova and Linde (*14*). Busca et al. have proposed that these sites are responsible for the first step, i.e., the dehydrogenation of butane to butene adsorbed on the surface (*15*). Some researchers inferred that the defects in the (100) face play an important role in the reaction (*1,5*).

We have previously demonstrated by means of XPS that the surface compositions of various $(VO)_2P_2O_7$ prepared with different methods were close to those of the bulk (P/V = 1) (*16*). This supports that the V^{4+} pair sites are active. However, no direct evidence for the active sites and active faces has been presented yet. The difficulty in determining them is due to the fact that most of efficient $(VO)_2P_2O_7$ consist of non-uniform fine crystallites. In the present study, it has been attempted to differentiate the reactivities of the crystal faces of $(VO)_2P_2O_7$ and assign the effective face by using large plate-like crystallites of similar size.

Experimental

Catalysts.   To obtain large plate-like crystallites of $(VO)_2P_2O_7$, the following method was adopted after several trials and errors. $(VO)_2P_2O_7$ was prepared from a well-defined precursor, $VOHPO_4 \cdot 0.5H_2O$, which was obtained by reduction of $VOPO_4 \cdot 2H_2O$ with 2-butanol by the same method as described in the literature (*17,18*). In our previous papers, this was called C-4 (*7,8*). Powder of V_2O_5 was added to an aqueous solution of H_3PO_4 (40%) (P/V atomic ratio = 7). Then the solution was stirred for several hours at room temperature. The precipitate formed (probably $VOPO_4 \cdot nH_2O$, n> 2) was refluxed at 393 K for 2 h. In order to dehydrate the precipitate to $VOPO_4 \cdot 2H_2O$, it was suspended into an aqueous solution of H_3PO_4 (85%) and the solution was again refluxed at 393 K for 2 h. The obtained crystallites were confirmed to be $VOPO_4 \cdot 2H_2O$ by XRD (*10,19*). By the reduction with 2-butanol at 358 K for 2 h in 2-butanol solution, $VOHPO_4 \cdot 0.5H_2O$ was formed (see below). It was necessary to stir the solution slowly to avoid the destruction of the crystallites during the above procedure. The obtained $VOHPO_4 \cdot 0.5H_2O$ was treated in a flow of N_2 at 823 K for 5 h to form $(VO)_2P_2O_7$.

In order to cover the surface of $(VO)_2P_2O_7$ with the SiO_2 overlayer, $(VO)_2P_2O_7$ was treated with a mixture of $Si(CH_3)_4$ and O_2 as follows.

After $(VO)_2P_2O_7$ was evacuated at 773 K for 1 h in a closed circulation system (250 cm^3), the mixture of $Si(CH_3)_4$ (200 Torr) and O_2 (70 Torr) was introduced into the system at the same temperature. $Si(CH_3)_4$, CO, and CO_2 as well as O_2 were analyzed with an on-line gas chromatograph. The resulting sample is denoted by $SiO_2/(VO)_2P_2O_7$, for which the amount of SiO_2 was evaluated from the pressure decrease of $Si(CH_3)_4$.

The crystallites of $(VO)_2P_2O_7$ and $SiO_2/(VO)_2P_2O_7$ were fractured by applying pressure (630 Kg·cm^{-2}) with a tablet molding machine. The location of Si on the surface of $(VO)_2P_2O_7$ was analyzed by a TEM-EDX (JEM 2000 FXII, JEOL). SEM (TSM-T20, JEOL) and XRD (MXP-3, MAC Science) were used for the characterization of the catalysts.

Catalytic Reactions. Catalytic oxidation of butane was carried out in a flow reactor at 713 K after the catalysts were pretreated in an N_2 flow at 773 K for 2 h. The feed gas consisted of 1.5% butane, 17% O_2, and N_2 (balance) (7). The products were analyzed with an on-line gas chromatograph. W/F (W = catalyst weight/g, F = flow rate of butane/mol·h^{-1}) was changed in the range of 1.1 x 10^2 - 11 x 10^2 mol^{-1}·g·h by controlling the total flow rate.

Results and Discussion

Characterization of Catalysts. Figure 1 shows the XRD patterns of the crystallites of $VOPO_4·2H_2O$ obtained after the reflux in the 85%H_3PO_4 solution (Figure 1a), $VOHPO_4·0.5H_2O$ (Figure 1b), and $(VO)_2P_2O_7$ (Figure 1c). $VOPO_4·2H_2O$ gave the XRD pattern similar to that reported already (*10,19*), in which the main peak at 11.8° (2θ) corresponds to the (002) face. The peaks were very intense and sharp, showing high crystallinity.

The XRD pattern of the sample obtained by the reduction of $VOPO_4·2H_2O$ with 2-butanol was the same as that of $VOHPO_4·0.5H_2O$ (*17,18*) (orthorhombic, a= 7.434 Å, b= 9.620 Å, c= 5.699 Å). There was no evidence for the presence of other impurity phases. The line widths of the XRD peaks of $VOHPO_4·0.5H_2O$ were similar to those of the parent $VOPO_4·2H_2O$, while peak intensities were about one-tenth those of the $VOPO_4·2H_2O$.

$(VO)_2P_2O_7$ obtained by the heat treatment of $VOHPO_4·0.5H_2O$ showed the same XRD pattern (Figure 1c) as reported in the literature (*17,18*). Gorbunova et al. (*14*) determined the crystal structure of $(VO)_2P_2O_7$ by the single crystal method to be orthorhombic (a=7.725 Å, b=16.576 Å, c=9.573 Å), where the peak at 22.8° corresponds to (100) plane. The line width of the peak at 22.8° was about 0.5 ° (2θ), which corresponds to

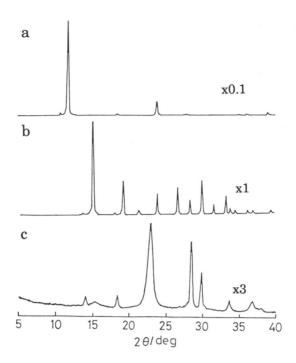

Figure 1. XRD patterns of precursors and catalyst.
(a) $VOPO_4 \cdot 2H_2O$, (b) $VOHPO_4 \cdot 0.5H_2O$ (c) $(VO)_2P_2O_7$

a thickness of about 20 nm in the [100] direction. The thickness estimated in the same way was about 80 nm for $VOHPO_4 \cdot 0.5H_2O$. These values were comparable with those estimated from the surface area (see below).

Figure 2 provides the SEM micrographs of these samples. $VOPO_4 \cdot 2H_2O$ (Figure 2a) consisted of square plate-like crystals with lengths of about 15 μm. $VOHPO_4 \cdot 0.5H_2O$ was also comprised of plate-like crystals having approximate size of 5 μm. The apparent thickness was roughly estimated to be 100 - 150 nm by SEM. As shown in Figures 2b and 2c, only a small difference in length and thickness was observed between $(VO)_2P_2O_7$ and $VOHPO_4 \cdot 0.5H_2O$. This indicates that the dehydration of $VOHPO_4 \cdot 0.5H_2O$ occurs without altering the morphology of crystallites as reported previously (17). However, as stated above, the thickness of the crystallites measured by XRD became small by the dehydration to $(VO)_2P_2O_7$ (from 80 to 20 nm). The thickness of the crystallites measured by SEM probably corresponds to that of the plate formed by stacking of about 10 thin crystallites.

By electron diffraction (*17*), Bordes et al. determined the basal face of the $(VO)_2P_2O_7$, which was prepared by the same method as the present study, to be (100). The same electron diffraction pattern was obtained for the basal plane of $(VO)_2P_2O_7$ in the present study.

The SEM micrographs of the fractured $(VO)_2P_2O_7$ and $SiO_2/$ $(VO)_2P_2O_7$ are given in Figure 3. Crystallites of $(VO)_2P_2O_7$ changed by the fracture into smaller ones with average size of 3 μm (Figure 3a). In this way, the side faces like (001) were newly created. When the crystallites were fractured 5 times, they became much smaller (about 1 μm) as shown in Figure 3b. Figures 3c and 3d clearly showed that $SiO_2/(VO)_2P_2O_7$ had the same crystallite size and shape as the parent $(VO)_2P_2O_7$ (Figure 3c) and the change of the crystallites upon the fracture was also the same as that of $(VO)_2P_2O_7$.

In Table 1, the size of crystallite, surface area, and density of Si on $(VO)_2P_2O_7$ are summarized. From the surface area ($17 \ m^2 \cdot g^{-1}$ for $(VO)_2P_2O_7$), the density ($3.2 \ g \cdot cm^{-3}$) and the size (5 μm in length), the thickness was calculated to be 40 nm, which is comparable with that from the line-width of XRD (20 nm). The surface areas of $(VO)_2P_2O_7$ and $SiO_2/(VO)_2P_2O_7$ changed little upon the fracture, showing that the surface areas of the side faces, which were newly created, were small. This is reasonable because the fraction of the side faces is estimated to 1.6 and 7.4% of the total surface area for the crystallites with the sizes of 5 and 1 μm, respectively, assuming the thickness of 40 nm. If the fracture brought about cleavage along (100) plane of the crystallites to change thickness to half (20 nm), the surface area would become about twice. Thus little change of the surface area indicates that the cleavage along (100) plane upon the fracture is little.

For $SiO_2/(VO)_2P_2O_7$, the density of Si is expressed as the ratio of Si to V atoms present at the surface, Si/V_{surf}, where the density of V_{surf} was evaluated from the crystallographic data (*14*). The value of Si/V_{surf} was 6 (Table 1).

In order to determine the location of SiO_2 on the surface of $(VO)_2P_2O_7$, TEM-EDX was applied. As shown in Figure 4, the electron beam (size; 50 nm x 10 nm) was focused on two positions; position **I** is the surface at the center of the crystal face of (100), position **II** the edge. By EDX analysis, the Si peak was detected on both positions, while the ratio of Si/V peak intensity (Si; 1.739 keV, V; 5.426 keV) for position **II** was about 10 times as large as that for position **I**. This difference does not necessarily correspond to the difference in the density of Si. The amount of V atoms, which are analyzed by EDX, is proportional to the sample volume that the electron beam passed, since the sample volume at position **II** is about half of that at position **I**, i.e., the amount of V detected

Figure 2. SEM micrographs of (a) $VOPO_4 \cdot 2H_2O$, (b) $VOHPO_4 \cdot 0.5H_2O$, (c) $(VO)_2P_2O_7$. Magnification: 2000

Figure 3. SEM micrographs of (a) $(VO)_2P_2O_7$ fractured once, (b) $(VO)_2P_2O_7$ fractured 5 times, (c) $SiO_2/(VO)_2P_2O_7$, and (d) $SiO_2/(VO)_2P_2O_7$ fractured once. Magnification: 2000

Table 1. Crystallite Size, Surface Area, and Si/Vsurf Atomic Ratio of Catalysts

Catalysts	Size[a] (thickness[b])	Surface area[c] $/m^2 \cdot g^{-1}$	Si/Vsurf[d]
$(VO)_2P_2O_7$	5 μm (40 nm)	17	-
$SiO_2/(VO)_2P_2O_7$	5 μm (40 nm)	10	6.0
$SiO_2/(VO)_2P_2O_7$ fractured	3 μm (40 nm)	11	6.0
$(VO)_2P_2O_7$ fractured once	3 μm (40 nm)	17	-
5 times	1 μm (40 nm)	18	-

[a]Average size measured by SEM, [b]Estimated from surface area, [c]Measured by BET method, [d]Atomic ratio of Si to V_{surf}, where V_{surf} is the amount of surface vanadium atoms.

by EDX at position **II** must be half of that at position **I**. On the other hand, the surface area analyzed at position **II** is estimated to about 2.5 times greater than that at position **I**, considering the thickness (40 nm) of the crystallites and the beam size (Figure 4). Thus, if the density of SiO_2 is equal on both (100) and the side face, the ratio of Si/V peak intensity for position **II** becomes 5 times that for position **I**. This value is not far from the experimental value, 10. Therefore, it is likely that the density of SiO_2 is similar on both faces or somewhat higher at the side faces. In any event, it is concluded that SiO_2 was deposited to a similar extent on the basal face and on the side faces.

Catalytic Oxidation of Butane. The dependences of the conversion of butane on W/F for $(VO)_2P_2O_7$ and $SiO_2/(VO)_2P_2O_7$ are given in Figure 5, where W is the catalyst weight and F is the flow rate of butane. The conversion increased with the increase in W/F. The catalytic activity and the selectivity of oxidation of butane are summarized in Table 2. The parent $(VO)_2P_2O_7$ produced MA with the selectivity of about 60% at the conversion of 46%, which is in agreement with the previous data (7,8). $SiO_2/(VO)_2P_2O_7$ showed a very low activity, indicating that the active sites on the surface of $(VO)_2P_2O_7$ were effectively poisoned by SiO_2. When the Si/V_{surf} ratio was less than 6, the conversion increased gradually with time and MA was formed at the later stage of the reaction. In this case, SiO_2 overlayers probably peeled off from the surface of $(VO)_2P_2O_7$ because of the incomplete formation of the SiO_2 overlayers.

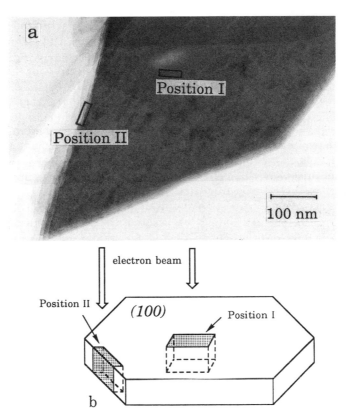

Figure 4. TEM micrograph of and illustration of EDX analysis.
(a) TEM micrographs of $SiO_2/(VO)_2P_2O_7$ (x160000). EDX analysis was
performed at positions **I** and **II**, (b) Illustration of EDX analysis (beam
size; 50 nm x 10 nm)

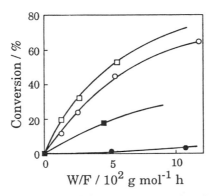

Figure 5. W/F dependencies of the conversion for the butane oxidation
over $(VO)_2P_2O_7$. (◯) $(VO)_2P_2O_7$, (▢) $(VO)_2P_2O_7$ fractured 5 times, (●)
$SiO_2/(VO)_2P_2O_7$, (▪) $SiO_2/(VO)_2P_2O_7$ fractured once.

When the inactive $SiO_2/(VO)_2P_2O_7$ was fractured, it exhibited a certain activity and produced mainly CO and CO_2, with the formation of a small amount of crotonaldehyde (10%). This result clearly indicates that the side faces such as (001) and (210) are non-selective for the formation of MA. Since the parent $(VO)_2P_2O_7$ gave MA with 60%-selectivity, it is deduced that the (100) face, in which $V^{4+}(=O)-O-V^{4+}$ pair sites are located, is selective for the formation of MA. No change in the crystallite size and shape was detected by SEM after the reaction.

Table 1 shows that the surface area of $SiO_2/(VO)_2P_2O_7$ increased by 1 $m^2 \cdot g^{-1}$ by the fracture, which is consistent with the estimated value from the SEM measurement. The change in the conversion with time was small over $SiO_2/(VO)_2P_2O_7$ fractured once. If the increase in the reaction rate (from 0.05×10^{-5} to 0.68×10^{-5} mol·g^{-1}·min^{-1}) by the fracture is attributed to the surface (side faces) created, the activity per unit surface area is 0.63×10^{-5} mol·m^{-2}·min^{-1}. A possible reason of the apparent high activity of the side faces created by fracture is that the fractured faces consisted of not only side faces but also (100), etc., because of the roughness of the fractured surface or peeling of SiO_2 overlayers by the fracture.

The typical time courses of the butane oxidation over $(VO)_2P_2O_7$ and the fractured $(VO)_2P_2O_7$ (SiO_2-free) are shown in Figure 6. The fractured $(VO)_2P_2O_7$ showed high initial activities as compared with the parent $(VO)_2P_2O_7$. The initial activity increased as the fracture was

Table 2. Catalytic Activities and Selectivities for Oxidation of Butane

Catalysts	Catalytic activity[a] / 10^{-5} mol·g^{-1}·min^{-1}	Selectivity to MA / %	(Conversion / %)
$(VO)_2P_2O_7$	1.80 (0.11)	60	(46)
$SiO2/(VO)_2P_2O_7$	0.05 (0.005)	0	(2)
$SiO2/(VO)_2P_2O_7$ fractured	0.68 (0.062)	1.2	(20)[b]
$(VO)_2P_2O_7$ fractured once	1.90 (0.11)	56	(47)
5 times	2.26 (0.13)	56	(52)

[a]Reaction rate: Butane; 1.5%, O_2; 17% at 713 K. The figures in the parentheses are the rates divided by the total surface area; 10^{-5} mol·m^{-2}·min^{-1}, [b]Crotonaldehyde was formed with the selectivity of 10%.

repeated (from 1 to 5). Over $(VO)_2P_2O_7$ fractured 5 times, the %-conversion decreased appreciably and the selectivity increased with time at the initial stages of the reaction. This result shows that the side faces

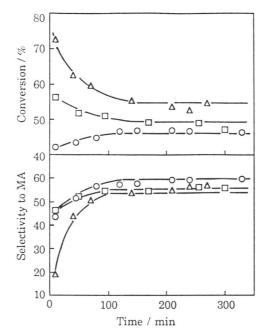

Figure 6. Time courses of butane oxidation over $(VO)_2P_2O_7$ and $(VO)_2P_2O_7$ fractured at 713 K. (○); parent, (□); fractured once, (△); fractured 5 times. The feed gas consisted of butane 1.5%, O_2 17%, and N_2 (balance).

created freshly tend to be deactivated under the reaction conditions. As shown in Table 2, the stationary catalytic activities of the fractured $(VO)_2P_2O_7$ were slightly higher than that of the parent $(VO)_2P_2O_7$. On the other hand, the selectivities of the fractured $(VO)_2P_2O_7$ (without SiO_2 deposition) were slightly lower than that of the parent $(VO)_2P_2O_7$. This result is consistent with the conclusion that the side faces are non-selective.

We reported previously that the addition of H_3PO_4 (probably present as $PO_{2.5}$ on the surface after the calcination) to $(VO)_2P_2O_7$, which was prepared by an organic solvent method (5), enhanced the selectivity to MA at high conversion levels, while the activity decreased (20). Considering the low selectivity of the side faces, it is possible that the H_3PO_4 added preferentially deactivated the side faces to suppress the formation of CO and CO_2 and/or the secondary oxidation of the product MA there.

In conclusion, the crystal faces of $(VO)_2P_2O_7$ had different reactivities for butane oxidation; the (100) face was selective for the formation MA and the side faces such as (001) are active for non-selective oxidation as illustrated in Figure 7.

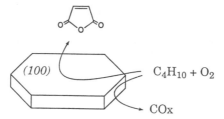

Figure 7.   Selective and non-selective faces of $(VO)_2P_2O_7$ for oxidation of butane.

Acknowledgments

We would like to thank to JEOL for the TEM-EDX measurement. This work was in part supported by Tokuyama Science Foundation and a Grant-in-Aid for Scientific Research (No. 04805081) from the Ministry of Education, Science and Culture of Japan.

Literature Cited
(1) Centi, G.;Trifiro, F.; Ebner, J. B.; Franchetti, V. M. *Chem. Rev.,* **1988,** *88,* 55.
(2) Hodnett, B. K. *Catal. Rev. Sci. Eng.,* **1985,** *27,* 373.
(3) Contractor, R. M.; Bergna, H. E.; Horowitz, H. S.; Blackstone, C. M.; Malone, B.; Torardi, C. C.; Griffiths, B.; Chawdhry, U.; Sleight, A. W. *Catal. Today,* **1987,** *1,* 40.
(4) Shimoda, T.; Okuhara, T.; Misono, M. *Bull. Chem. Soc. Jpn.,* **1985,** *58,* 2163.
(5) Busca, G.; Cavani, F.; Centi, G.; Trifiro, F. *J. Catal.,* **1986,** *99,* 400.
(6) Moser, T. P.; Schrader, G.; *J. Catal.,* **1985,** *92,* 216.
(7) Misono, M.; Miyamoto, K.; Tsuji, K.; Goto, T.; Mizuno, N.; Okuhara, T. *Stud. Surf. Sci. Catal.,* **1990,** *55,* 605.
(8) Okuhara, T.; Tsuji, K.; Igarashi, H.; Misono, M. *Catalytic Science and Technology,* **1990,** *Vol. 1,* Kodansha, Tokyo-VCH, Weinheim, p. 443.
(9) Hodnett, B. K.; Delmon, B. *Appl. Catal.,* **1985,** *15,* 141.
(10) Bordes, E.; Courtine, P. *J. Catal.,* **1977,** *57,* 236.
(11) Harrouch Batis, N.; Batis, H.; Ghorbel, A.; Vedrine, J. C.; Volta, J. C. *J. Catal.,* **1991,** *128,* 248; Abdelouahab, F. B.; Olier, R.; Guilhaume, N.; Lefebvre, F.; Volta, J. C. *J. Catal.,* **1992,** *134,* 151.
(12) Yamazoe, N.; Morishige, H.; Teraoka, Y. *Stud. Surf. Sci. Catal.,* **1989,** *44,* 15.
(13) Cornaglia, L. M.; Caspani, C.; Lombardo, E. *Appl. Catal.,* **1991,** *74,* 15.
(14) Gorbunova,Y. E.; Linde, S. A. *Dokl. Akad. Nauk SSSR,* **1979,** *245,* 584.

(15) Busca, G.; Centi, G.; Trifiro, F.; Lorenzelli, V. *J. Phys. Chem.*, **1986**, *90*, 1337.
(16) Okuhara, T.; Nakama, T.; Misono, M. *Chem. Lett.*, **1990**, 1941.
(17) Bordes, E.; Courtine, P. *J. Solid State Chem.*, **1984**, *55*, 270.
(18) Johnson, J. W.; Johnston, D. C.; Jacobson, A. J.; Brody, J. F. *J. Am. Chem. Soc.*, **1984**, *106*, 8123.
(19) R'kha, C.; Vandenbarre, M. T.; Livage, J. *J. Solid State Chem.*, **1986**, *63*, 202.
(20) Okuhara, T.; Misono, M. *Catal. Today*, in press.

RECEIVED November 25, 1992

Chapter 13

Selective and Nonselective Pathways in Oxidation and Ammoxidation of Methyl-Aromatic Compounds over Vanadia–Titania Catalysts

Fourier Transform Infrared Spectroscopic Studies

G. Busca

Istituto di Chimica, Facoltà di Ingegneria, Università di Genoa, Piazzale J. F. Kennedy, I–16129 Genoa, Italy

FT-IR data show that the alkyl-aromatics toluene, ortho-, meta- and para-xylene are activated by the surface of vanadia-titania catalysts in the form of benzyl species. Surface reaction pathways have been established upon heating with and without gas-phase oxygen, relevant with respect to the mechanisms of heterogeneously catalyzed oxidation and ammoxidation of alkylaromatics. Side-chain oxidation, giving aromatic aldehydes and carboxylic acids, as well as phthalic anhydride from o-xylene, involves successive reaction with the catalyst oxidized surface sites. On the contrary, ring oxidative breaking giving maleic and citraconic anhydrides involves reaction with gas-phase oxygen. Side-chain oxidation and ammoxidation routes are interplayed.

The gas-phase selective oxidation of o-xylene to phthalic anhydride is performed industrially over vanadia-titania-based catalysts (*1-3*). The process operates in the temperature range 620-670 K with 60-70 g/Nm^3 of xylene in air and 0.15 to 0.6 sec. contact times. It allows near 80 % yield in phthalic anhydride. The main by-products are maleic anhydride, that is recovered with yields near 4 %, and carbon oxides. Minor by-products are o-tolualdehyde, o-toluic acid, phthalide, benzoic acid, toluene, benzene, citraconic anhydride. The kinetics and the mechanism of this reaction have been the object of a number of studies (*2-7*). Reaction schemes have been proposed for the selective pathways, but much less is known about by-product formation.

Vanadia-titania (*8*) and other supported vanadia catalysts (*9*) can also be applied for the production of aromatic nitriles by ammoxidation of toluene and of the three xylene isomers: alumina-supported V-Sb-based oxides seem to be the best catalysts (*10*). Detailed kinetic studies of toluene ammoxidation have been reported recently using different vanadia-titania catalysts (*11,12*). Ammonia inhibits toluene conversion, while benzonitrile yields (up to 80 % near 610 K) are mainly limited by

0097–6156/93/0523–0168$06.00/0

combustion of the hydrocarbon and by the partial oxidation of ammonia to N_2 and N_2O. The mechanisms proposed for catalysts like V_2O_5-TiO_2 (anatase) *(11,13)*, V_2O_5-TiO_2 (B) *(12)*, V_2O_5-Al_2O_3 *(9)* and V-Sb-Bi-oxides *(14)* show marked differences.

The purpose of the present paper is to offer a contribute to the understanding of the mechanisms of these reactions by using an IR spectroscopic method and well-characterized "monolayer" type vanadia-titania (anatase) as the catalyst. We will focus our paper in particular on the following subjects: i) the nature of the activation step of the methyl-aromatic hydrocarbon; ii) the mechanism of formation of maleic anhydride as a by-product of o-xylene synthesis; iii) the main routes of formation of carbon oxides upon methyl-aromatic oxidation and ammoxidation; iv) the nature of the first N-containing intermediates in the ammoxidation routes.

Experimental.

The vanadium-titanium oxide catalyst was prepared by dry impregnation of Degussa P25 TiO_2 (surface area 53 m^2/g) with a boiling water solution of ammonium metavanadate, followed by drying in air at 720 K for 3 h. The loaded amount was 10 % as V_2O_5 by weight, higher than that needed to complete the theoretical monolayer *(6)*. The resulting surface area of the catalyst was 48 m^2/g.

For the adsorption and oxidation experiments the catalyst powder was pressed into self-supporting disks and activated by outgassing in the IR cell at 720 K for 2 h. The IR spectra were recorded with a Nicolet MX1 Fourier transorm instrument, using IR cells built to perform measurements in controlled atmospheres.

Results and Discussion.

Characterization of activated hydrocarbon entities. In Figure 1 the IR spectra of the surface species arising from the adsorption of toluene, ortho-, meta- and para-xylene on vanadia-titania at r.t. are reported. In all cases, two different adsorbed species are found, one completely desorbed by outgassing in the temperature range 300-370 K, and the other resistant to evacuation at 370 K. The IR spectra of the former species correspond to those of the adsorbed intact hydrocarbons, with poor vibrational perturbation. The IR spectra of the latter species are similar in the four cases showing that they are due to closely-related entities. They show in the region 1650-1300 cm^{-1} the typical ring vibration modes of mono- and/or di-substituted aromatics *(15)*. The detection of some very typical features (like the relatively strong bands present in all cases near 1350 cm^{-1} and the features in the region 1350-1150 cm^{-1}) rules out their assignment to particular conformers or surface complexes of the intact molecules. The detection in IR of the 8a and 8b ring vibration modes near 1590 cm^{-1} (inactive in the IR spectrum of p-xylene for symmetry reasons) also in the case of the para-xylene derivative indicate that the intermediate is no more centrosymmetric. The strength of the bands near 1590 cm^{-1} in all cases is justified by increased electron density on the aromatic rings *(15)*.

The additional data below allow us to propose an identification of these species:
i) further reaction of these species with the catalyst surface in the presence or in the absence of oxygen gives side-chain oxygenated compounds like benzaldehyde and tolualdehydes, and phthalic anhydride from o-xylene, see below. This shows that further reaction occurs at the side chain, the aromatic ring being and remaining intact.
ii) these spectra do not correspond to those obtained by direct adsorption of oxygenated compounds like phenols, benzyl alcohols, benzaldehyde and toluadehydes. In particular, evidence is found in these spectra neither for phenyl-oxygen bonds nor for C=O or C-O bonds. This indicates that these species do not contain oxygen.
iii) the spectra show medium-strong bands in the region 1300-1150 cm^{-1} that can be assigned to Ar-C stretchings. These bands are found at higher wavenumbers with respect to the values measured for the corresponding methyl aromatics, in good agreement with the values measured for benzyl radical *(16)* and for some metal-benzyl *(17,18)* and methyl-benzyl organometallic complexes *(18)*. For the toluene derivative this mode is found at 1298 cm^{-1}, with respect to 1210 cm^{-1} for liquid toluene.
iv) the very characteristic strong band present in all cases in the region 1365-1350 cm^{-1} is likely due to a CH_2 wagging mode. This mode is in fact very sensitive to the environment of the methylene group and is coupled with the Ar-C stretching modes.

These arguments support the conclusion that these species are benzyl and methyl-benzyl species produced by hydrogen abstraction from one of the methyl groups of the starting hydrocarbons.

Other data can be taken into account to establish the nature of these intermediates:
i) vanadia-titania surface has a medium-high acidic nature, lacking significant basicity *(19,20)*. Also on very basic surfaces alkyl-aromatics acid dissociation is not observed at all, or is limited to few extremely active sites, the IR spectrum of the benzyl anion being not observable *(21)*. Moreover, vanadia-titanias doped by alkali oxides are deactivated with respect to alkyl-aromatic oxidation *(3,22,23)*. Accordingly, a proton abstraction by basic sites giving anionic benzyl species is very unlikely.
ii) both in aerobic *(5,24)* and in anaerobic conditions *(25)* alkyl aromatics reacted over vanadia-based catalysts give rise to methyl-aryl oxidative coupling products, like methyl-diphenyl methanes, antraquinones and methyl-benzophenones from toluene. The nature of these products, their isomer distribution (almost exclusively ortho- and para- isomers) and the nature of the reactants (aromatics activated towards the electrophylic aromatic substitution) strongly suggest that benzyl species react in a cationic form (as $Ar\text{-}CH_2^+$) with the mechanism of aromatic alkylation.
iii) the same products are obtained in solution starting from compounds like the complex ion $[(H_2O)_5Cr(CH_2\text{-}Ar)]^{2+}$, formally containing an anionic benzyl species and Cr^{3+} *(26)*. This means that anionic or covalently bonded benzyl species can react as cations, leaving electrons to the metal center.
iv) the heterogeneously-catalyzed oxidation of alkyl aromatics shows some similarities with the homogeneously catalyzed one, the so-called Amoco "MC" process *(27)*, where toluene is thought to be activated by Co^{3+} complexes with the reaction:

$$Ar\text{-}CH_3 + Co^{3+} = Ar\text{-}CH_2^{\cdot} + Co^{2+} + H^+$$

As a conclusions, we propose the following two-step mechanism for alkyl aromatic activation over vanadia-titania catalysts:

$$Ar\text{-}CH_3 + V^{n+} = Ar\text{-}CH_2^{\cdot} + V^{(n-1)+} + H^+$$

$$Ar\text{-}CH_2^{\cdot} + V^{n+} = Ar\text{-}CH_2^{+} + V^{(n-1)+} \qquad (n = 5 \text{ or } 4)$$

Oxidation of toluene, para-xylene and meta-xylene.   The results of the oxidation experiments of para- and meta-xylene are parallel each other and with those concerning toluene oxidation, reported previously *(13)*. In Figure 2 the spectra relative to the meta- isomer are reported. As already cited, near 373 K the spectrum of the adsorbed intact hydrocarbon is disappeared while the bands due to the m-methylbenzyl species are predominant, although a new band is also observed, weak, near 1635 cm^{-1}. A band at 1635 cm^{-1} grows in the temperature range 373-473 K and later disappears, both in the presence and in the absence of gas phase oxygen, starting both from toluene and from the three xylene isomers. It coincides with the C=O stretching of adsorbed benzaldehyde (see below) and tolualdehydes, and can be assigned to the adsorbed aldehydes produced by oxidation of the benzyl species.

Upon m-xylene oxidation, the above band of m-tolualdehyde disappears near 523 K, while a couple of strong and broad bands is grown near 1530 and 1430 cm^{-1}. These bands are typical of carboxylates and are again observed, with very weak band shifts, upon oxidation of all methyl-benzenes, as well as of the corresponding aromatic aldehydes. They can be assigned predominantly (if not entirely) to benzoate and toluate anions *(28)* (in Figure 2, meta-toluate anions). These bands raise their maximum near 523 K in all cases and suddenly disappear above 673 K, when gas-phase CO_2 begins to be detectable.

These features provide evidence for a reaction path, common to the four methyl-aromatics under study, that involves the following steps: i) activation of the methyl-aromatic in the form of a benzyl species; ii) reaction of this species with the catalyst surface giving the corresponding aldehyde, probably with the intermediacy of benzyloxy- species, as discussed previously *(13)* ; iii) oxidation of the aldehyde by the catalyst surface to give the carboxylate species; iv) decarboxylation of the carboxylate species giving CO_2, and, likely, a demethylated hydrocarbon (toluene from xylenes and benzene from toluene). All these successive processes occur both with and without gas phase oxygen, that, consequently, is not active in them.

However, features belonging to other species not involved in this path are also observed upon m-xylene oxidation in O_2. Near 433 K two other bands are clearly observed at 1710 and 1670 cm^{-1}. A feature near 1700 cm^{-1} persists also near 673 K when a very strong and complex absorption pattern becomes detectable in the 1900-1700 cm^{-1} region. In this region the couples of bands due to symmetric and asymmetric C=O stretchings of the O=C-O-C=O system of cyclic anhydrides typically fall. The

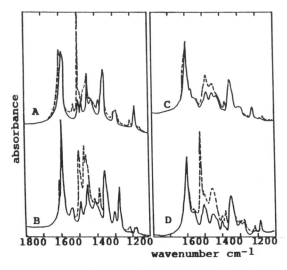

Figure 1. FT-IR spectra of the surface species arising from adsorption of toluene (A), o-xylene (B), m-xylene (C) and p-xylene (D) on vanadia-titania. Full lines: after outgassing at 350 K.

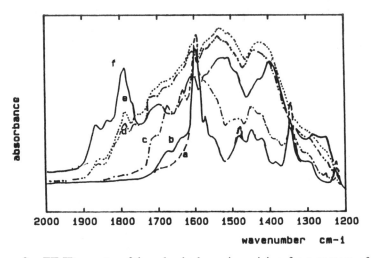

Figure 2. FT-IR spectra of the adsorbed species arising from contact of the catalyst with m-xylene and evacuation at 373 K (a), and further heating in oxygen at 373 K (b), 423 K (c), 473 K (d), 523 K (e) and 573 K (f).

IR spectra of the different cyclic anhydrides (maleic anhydride *(29)*, citraconic anhydride *(30)* and phthalic anhydride *(31)*) differ slightly for the position of these bands, as well as for the position and intensities of the C-O-C stretchings, observed in the 1300-1200 cm^{-1} region. The complexity of the spectra observed after m-xylene oxidation on the catalyst surface (distinct bands are observed at 1870, 1840, 1800, 1790 cm^{-1} and at 1290 and 1260 cm^{-1}) suggests that we are dealing with a mixture of maleic and citraconc anhydrides. Very similar spectra are observed when starting with p-xylene, while only maleic anhydride is observed upon toluene oxidation (bands at 1870, 1790, 1290 and 1260 cm^{-1}). Being maleic and citraconic anhydrides main by-products in toluene and xylenes oxidation, their origin has been investigated further.

Experiments of toluene, meta- and para-xylene oxidation by the catalyst surface in the absence of gas-phase oxygen show the oxidation path at the methyl group, as already cited; however, neither the bands of cyclic anhydrides nor those at 1720 and 1670 cm^{-1} (found in intermediate conditions in O_2) are observed. This suggests that species responsible for these bands are intermediates in the formation of maleic and citraconic anhydrides, and that are formed by reaction of the adsorbed benzyl species with gas-phase oxygen. A similar situation has been observed by studying the oxidation of benzene over vanadia-titania *(32)*, where phenate species where though to react with gas-phase oxygen giving quinone species, likely responsible for the band at 1670 cm^{-1}, successively giving maleic anhdride.

The spectra relative to adsorption and heating of benzaldehyde over vanadia-titania in the presence of gas-phase oxygen are reported in Figure 3. They confirm fully the data discussed above showing adsorbed benzaldehyde (mainly characterized by its C=O stretching at 1635 cm^{-1}, but also by features at 1600, 1582, 1494, 1461 cm^{-1}, ring vibrations, at 1395, CHO deformation, and at 1325, 1238 and 1175 cm^{-1}, ring vibrations *(33)*) that transforms into benzoate species (more intense bands at 1500, 1430 cm^{-1}) in the 423-473 K range. Benzoates later decarboxylate giving CO_2 near 573 K. No bands of maleic anhydride where found starting from the aldheyde. In our experimental conditions benzene is likely formed upon decarboxylation and desorbed as such. However, benzene, if formed, can be oxidized above 573 K over the vanadia-titania surface, giving benzoquinone, maleic anhydride and carbon oxides *(34)*. We believe that this consecutive path is the main one for carbon oxide production and a secondary one for maleic anhydride production. The main way to maleic anhydride should involve the attack of gaseous oxygen to an adsorbed hydrocarbon species, arising from the benzyl species, in a parallel slow way.

So, we can propose the following reaction scheme for the oxidation of toluene, meta- and para-xylene on the surface of vanadia-titania in our experimental conditions:

$$XArCH_3 \rightarrow XArCH_2 \rightarrow XArCH_2\text{-}O\text{-} \rightarrow XArCH=O \rightarrow XArCOOH \rightarrow CO_x \text{ (MA)}$$
$$| \; O_2$$
$$\rightarrow \rightarrow MA \text{ (CA)}$$

where X = H or CH_3; MA = maleic anhydride; CA = citraconic anhydride

This scheme and our data agree with those reported for toluene oxidation over vanadia-titania *(5,8,22,35)*, and meta- and para- xylene oxidation over vanadia-based catalysts *(5,36)*. An excellent correlation exists between the temperature at wich we observe the maximum concentration of the different surface species (aldehydes, carboxylate species, cyclic anhydrides) and those at which selectivity in the corresponding gas-phase products (aldehydes, carboxylic acids and cyclic anhydrides) raises the maximum. For exemple, the main products in toluene oxidation over vanadia-titania catalysts are reported to be benzoic acid and benzaldehyde, with maximum selectivity near 540 K, and maleic anhydride with maximum selectivity near 573 K *(35)*, in good agreement with our data. In particular, our data definitely support that side chain oxidation (giving the aldehyde, carboxylic acids and finally CO_2) and ring oxidative opening (giving maleic and citraconic anhydrides) are mainly competitive paths, probably both starting from adsorbed benzyl species reacting either with the catalyst surface at the side chain or with gas-phase oxygen at the aromatic ring. From our data is reasonable to propose that the rate-limiting step in the selective pathway is the oxidation of the benzyl species (readily formed) by the catalyst surface, according to the opinion of Sanati and Andersson *(12)*.

Oxidation of ortho-xylene. The spectra of the adsorbed species arising from interaction of ortho-xylene with the surface of the vanadia-titania catalyst in the presence of oxygen are shown in Figure 4. The spectra show some parallel features with respect to those discussed above concerning the oxidation of toluene and meta- and para-xylene. Also in this case the o-methyl-benzyl species begins to transform above 373 K, with production of adsorbed o-tolualdehyde (band at 1635 cm^{-1}) and of a quinone derivative (band at 1670 cm^{-1}). Successively bands likely due to o-toluate species (1530, 1420 cm^{-1}) grow first and decrease later with production of CO_2 gas.

However, already near 473 K two strong bands predominate, centered at 1860 and 1790 cm^{-1}, associated to a single band at 1255 cm^{-1}. These features are those of adsorbed phthalic anhydride, as confirmed by the direct adsorption of this compound on the catalyst. Another strong band appears in the same conditions at 1690 cm^{-1}, certainly due to a compound containing a carbonyl or a carboxyl group, very likely phthalide *(15)*. The bands due to phthalic anhydride further grow up to 573 K, while that at 1690 cm^{-1} decreases in intensity. At 593 K all these bands decrease strongly although a more complex weak pattern remains in the 1900-1700 cm^{-1} range, probably due to a mixture of maleic and citraconic anhydrides.

Experiments carried out in the absence of oxygen show that the bands due to phthalic anhydride are detectable, although much weaker, also in the absence of gas-phase oxygen. This indicates that the formation of this compound does not directly involve O_2, but oxidized catalyst surface sites.

In conclusion, we can propose the following reaction scheme for the oxidation of ortho-xylene on the surface of vanadia-titania:

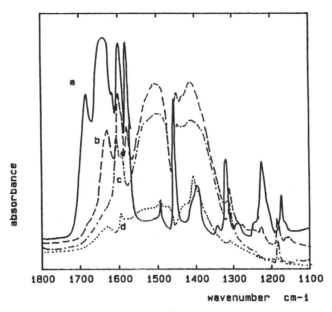

Figure 3. FT-IR spectra of the adsorbed species arisng from contact of the catalyst with benzaldehyde /oxygen at r.t. (a), 423 K (b), 473 K (c) and 573 K (d).

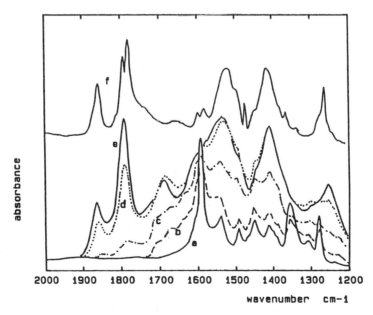

Figure 4. FT-IR spectra of the adsorbed species arising from contact of the catalyst with o-xylene / oxygen at r.t. (a), 373 K (b), 423 K (c), 473 K (d), and 523 K (e). (f): phtalic anhydride adsorbed at r.t., producing also phtalate ions.

$$
\begin{array}{c}
\qquad\qquad\qquad\qquad ----> OHCArCHO \; -> \; CO_x \\
\qquad\qquad\qquad\qquad\quad | \qquad\qquad\qquad\quad | \\
H_3CArCH_3 \; -> \; H_3CArCH_2 \; -> \; H_3CArCH_2\text{-}O\text{-} \; -> \; H_3CArCHO \quad PA \; (\text{-}> MA) \\
\qquad\quad | \qquad\qquad\qquad\qquad\qquad\qquad\qquad\qquad\quad | \qquad\qquad\qquad | \\
\qquad\quad |\; O_2 \qquad\qquad\qquad\qquad\qquad\qquad H_3CArCOOH \; -> \; phthalide \\
\qquad\quad | \qquad\qquad\qquad\qquad\qquad\qquad\qquad\qquad\quad | \\
\qquad\quad \text{-}> \text{-}> MA, CA \; (CO_x) \qquad\qquad CO_x
\end{array}
$$

PA= phthalic anhydride; MA = maleic anhydride; CA = citraconic anhydride

This scheme and our data agree with the product distribution in o-xylene oxidation reported by many authors *(1-7)*, as well as with experiments of oxidation of intermediates *(37)*: o-tolualdehyde and phthalide are observed as the main intermediates in the 523-573 K temperature range, while phthalic anhydride selectivity grows in the 473-573 K range and later only slightly decreases above 600 K when also maleic anhydride appears and conversion is very high.

A relevant feature concerns the main route to maleic anhydride. In fact, different authors disagree on this subject: Wainwright and Foster *(3)* and Wachs and Saleh *(2)* believe that maleic anhydride is mainly formed from phthalic anhydride successive oxidation, while following Bond *(6)* it is formed from "one of the last adsorbed intermediates in the reaction sequence"; finally, in the reaction scheme proposed by Bielanski and Haber *(7)* it is formed through a competitive way. Our data provide evidence for the formation of maleic anhydride near 573 K from all alkyl aromatics. This supports the idea that the main path producing maleic anhydride is competitive (although slow) with that giving phthalic anhydride, and that side-chain oxidation products have a minor role in maleic anhydride production. Our data provide also evidence for the competitivity of the main route to carbon oxides, attributed to decarbonylation of o-toluate anions and combustion of the resulting aryl species.

Ammoxidation of toluene. The IR investigation of toluene ammoxidation *(13)* and of ammonia adsorption *(38)* over vanadia-titania were the subject of previous publications from our reasearch group. We summarize here some relevant conclusions: i) at least two paths to benzonitrile are observed in our IR experiments, one involving reaction of toluene and ammonia giving benzylamine, and the other involving condensation of benzaldehyde and ammonia; ii) the ammonia species active in the former way derives from NH_3 species coordinated over vanadium cation centers *(13)* ; iii) these ammonia species near 400 K can lose an hydrogen, producing active amide species by a redox mechanism *(38)*

A reexamination of the ammonia-toluene coadsorption shows that ammonia prevents the formation of benzyl-species at room temperature: in the presence of ammonia, in fact, toluene only weakly adsorbs at r.t. in a reversible form. This agrees with the strong inhibiting effect of toluene conversion by ammonia due to the competitivity of the two reactants on the same adsorption sites *(11,12)*. The spectrum obtained after coadsorption of toluene and ammonia at 470 K is compared in Figure 5 with those resulting from co-adsorption of benzaldehyde and ammonia at 370 K and of benzylamine in oxygen at 420 K. The spectra show a number of common bands. In particular, bands due to benzylamine are observed after toluene/ammonia coadsorption, showing that this compound is formed by their reaction. From these data, it seems reasonable to propose that benzyl species can be formed in the presence of ammonia over sites from where ammonia is desorbed at 473 K. However, in these conditions also amide species are present on the surface. The production of benzylamine is consequently thougth to be due to a Langmuir-Hinshelwood mechanism involving benzyl- and amide species. The rate limiting step of this process should be the activation of ammonia in the form of the active amide species.

In Figure 6 the spectra of the adsorbed species arising from benzylamine oxidation over the catalyst surface are reported. Adsorbed benzylamine is characterized by bands at 1642 cm^{-1} (NH_2 scissoring), 1610, 1580, 1492, 1450 cm^{-1} (ring vibrations), 1420 cm^{-1}, shifting to 1400 cm^{-1} by decreasing coverage (probably a CH_2 deformation mixed with a NH_2 deformation), and 1218 cm^{-1} (aryl-methylene stretching). Assignments are based on comparison with other amine spectra *(15)*, because we cannot find in the literature a recent and satisfactory description of the IR spectrum of this compound. By heating in oxygen atmosphere at 423 K new bands grow at 1670 cm^{-1} (sharp) and at 1540 cm^{-1}, 1330, 1240, and 1203 cm^{-1}, while also the band at 1642 cm^{-1} apparently increases in intensity. However, although the band at 1670 cm^{-1} at 500 K is almost completely disappeared, the other decrease slowly in intensity by further heating. Near 573 K a small band appears near 2260 cm^{-1}, certainly due to benzonitrile (vCN).

In Figure 7 the spectra of benzonitrile adsorbed and reacted with oxygen on vanadia-titania are reported. Adsorbed benzonitrile is characterized by bands at 2265 cm^{-1} (vCN), 1602, 1495, 1455, 1340, 1295 cm^{-1}, with small perturbations with respect to the pure molecule *(39)*. However, by heating (and also by simple staying on the surface) a transformation occurs, giving strong bands at 1642, 1535 and 1410 cm^{-1}, all rather broad. These bands strongly remind, although slightly shifted, the so-called amide I, amide II and amide III bands of benzamide *(15,40)*, and provide evidence for partial hydrolysis of benzonitrile when adsorbed on the catalyst in the presence of oxygen. At higher temperatures traces of benzoate species can be found together with adsorbed ammonia, showing that complete hydrolysis can take place.

The comparison of the bands observed after benzonitrile and benzylamine transformation over vanadia-titania allows us to suggest that benzamide can also be obtained by oxidation of benzylamine. Moreover, the growth of the absorption near 1640 cm^{-1} and the appearance of bands at 1330 and 1240 cm^{-1} during benzylamine oxidation suggests that benzaldehyde is also formed. A likely assignment for the band at 1670 cm^{-1} is to the stretching of a C=N double bond *(15)*, so being likely indicative of the formation of benzaldimine.

These results strongly suggest that benzylamine can undergo oxy-dehydrogenation in two steps giving successively benzaldimine and benzonitrile. However,

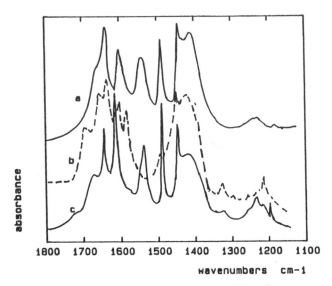

Figure 5. FT-IR spectra of the adsorbed species arising from contact of the catalyst with toluene + NH3 at 473 K (a); benzaldehyde + NH_3 at 373 K (b) and benzylamine + O_2 at 423 K (c).

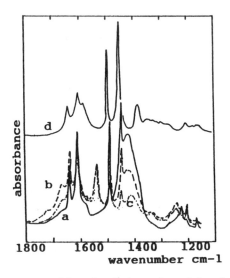

Figure 6. FT-IR spectra of the adsorbed species arising from contact of the catalyst with benzylamine and evacuation at r.t. (a) and further heating in oxygen at 423 K (b), and 473 K (c). (d) liquid benzylamine.

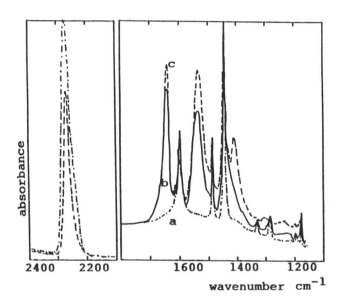

Figure 7. FT-IR spectra of the adsorbed species arising from contact of the catalyst with benzonitrile and evacuation at r.t. (a) and further heating in oxygen at 373 K (b), and 473 K (c).

benzaldimine can also be formed by ammonia condensation with benzaldehyde. This equilibrium could be established also in the gas phase. We have found it, because we can observe the imine starting from the aldehyde and ammonia, and the aldehyde starting from the amine. As reported above, benzonitrile too can undergo hydrolysis giving benzamide first and and benzoate ions later. Also these reactions are reversible, according to Niwa et al. *(9)* that proposed a way to benzonitrile by condensation of benzoate and ammonium ions. Indeed, nitrile is also obtained by reaction of benzoic acid with ammonia over vanadia-titania *(13)*, and the dehydration of amides to nitriles is reported to occur over acid heterogeneous catalysts *(41)*.

So, our data and the known organic chemistry suggest, according to the literature *(8,12)*, that ammoxidation and oxidation mechanisms are interplayed. We can consequently propose the following mechanism for the surface reactions relative to ammoxidation of toluene:

$$NH_3 \rightarrow NH_2 \,\text{------}$$

```
                   |-> ArCH2NH2 -> ArCH=NH ->  ArCN
          |                          | |        H2O| |
          |                        H2O | | NH3      ArCONH2
ArCH3 -> ArCH2 - |                      | |        H2O| | NH3
                   |-> ArCH2O- -> ArCH=O -----> ArCOOH -> COx
```

This picture agrees with the recent kinetic investigations reported by Cavalli et al. *(8)* and by Sanati and Andersson *(12)* concerning anatase- type and (B)-type V_2O_5-TiO_2 catalysts, respectively. In particular, our data support the following conclusions given by Sanati and Andersson:
i) the active hydrocarbon intermediate is common to the oxidation and the ammoxidation routes (the benzyl species).
ii) ammoxidation is faster than oxidation, because the reaction of benzyl species with amide is faster than that with surface oxide species.
iii) the rate-determining step in the ammoxidation is a surface reaction (likely the formation of the amide from coordinated ammonia).

Nevertheless, these conclusions imply that the route via benzylamine is the fastest one, and this agrees with the conclusions of Azimov et al. *(10,14)*.

On the other hand, we definitely demonstrate here that aromatic nitriles are not stable against degradation, as assumed by Sanati and Andersson *(12)*. We consequntly think, according to previous data concerning a related catalytic system *(42)*, that the main combustion ways in ammoxidation imply nitrile and/or imine hydrolysis and benzoate ions combustion. This is not in contrast with the effect of water that is used as an additive to enhance selectivity in nitriles *(10)* because water (that being a product of the ammoxidation reaction is always present in stoichiometric amounts sufficient to perform nitrile hydrolysis), when adsorbed on the catalyst in big amounts poisons the nucleophylic sites (oxide ions) responsible for the attack to the CN multiple bonds of nitrile and imine Accordingly benzonitrile is adsorbed weakly on hydrated surfaces and does not apparently undergo any transformation.

Conclusions.

The IR study performed in static controlled atmospheres in the IR cell allowed us to identify a number of adsorbed intermediate and secondary products, together with the main reaction products in oxidation and ammoxidation of toluene and the three xylene isomers. Surface reactions schemes are proposed that account for most of the mechanistic features of the heterogeneously-catalyzed industrial reactions.

Our data support the following conclusions:

i) activation of methyl-aromatics occurs with a redox mechanism giving nearly neutral benzyl species and reduced vanadium centers.

ii) maleic anhydride, the main by-product besides CO_x in o-xylene oxidation, should be mainly produced by a slow competitive way involving reaction of an hydrocarbon intermediate at the aromatic ring with gas-phase (or weakly adsorbed) oxygen.

iii) the main route of formation of carbon oxides upon methyl-aromatic oxidation and ammoxidation should involve decarboxylation and combustion of benzoic and/or toluic acids.

iv) benzaldimine appears to be a key intermediate in ammoxidation of toluene, and can be obtained by two different routes: dehydrogenation of benzylamine and condensation of ammonia with benzaldehyde. In the first faster mechanism the rate-limiting step should be the activation of ammonia, while in the second one, probably slower, it should be constituted be the side-chain oxidation of the benzyl species.

Acknowledgements.

The author acknowledges financial support from CNR (Rome, Italy), Progetto Finalizzato Chimica Fine, and Prof. F. Trifirò and V. Lorenzelli for helpful discussions.

Literature Cited.

1. Franck, H.G.; Stadelhofer, J.W., *Industrial Aromatic Chemistry*, Springer Verlag, Berlin, **1988**.
2. Wachs, I.E.; Saleh, R.Y.; Chan, S.S.; Chersich, C.; *Chemtech*, **1985**, (12), 756; Saleh, R.Y.; Wachs, I.E.; *Appl. Catal.* **1987**, *31*, 87.
3. Wainwright, M.S.; Foster, N.R., *Catal. Rev. Sci. Technol.* **1979**, *19*, 211.
4. DeLasa, H.; *Canad. J. Chem. Eng.,* **1983**, *61*, 710.
5. Andersson, L.H.S.; *J. Catal.,* **1986**, *98*, 138.
6. Bond, G.C. *J. Catal.,* **1989**, *116*, 531; Bond, G.C.; Bruckman, K.; *Disc. Faraday Soc.,* **1981**, *72*, 235.
7. Bielanski, A.; Haber, J. *Oxygen in Catalysis*, Dekker, New York, **1991**.
8. Cavalli, P.; Cavani, F.; Manenti, I., Trifirò, F., *Catalysis Today*, **1987**, *1*, 245; *Ind. Eng. Chem. Res.* **1987**, *26*, 639.
9. Niwa, M.; Ando, H.; Murakami, Y. *J. Catal.,* **1981**, *70*, 1.
10. Rizayev, R.G.; Mamedov, E.A.; Vislovskii, V.P.; Sheinin, V.E.; *Appl. Catal. A: General,* **1992**, *83*, 103.
11. Cavalli, P.; Cavani, F.; Manenti, I.; Trifirò, F.; ElSawi, M.; *Ind. Eng. Chem. Res.* **1987**, *26*, 804.
12. Sanati, M.; Andersson, A., *Ind. Eng. Chem. Res.,* **1991**, *30*, 312 and 320.
13. Busca, G.; Cavani, F.; Trifirò, F. *J. Catal.,* **1987**, *106*, 471.

14. Azimov, A.B.; Vislovskii, V.P.; Mamedov, E.; Rizayev, R.G.; *J. Catal.* **1991**, *127*, 354.
15. Bellamy, L.J., *The Infrared Spectra of Complex Molecules*, Chapmann and Hall, London, 3rd ed., **1980**.
16. Morrison, V.J.; Laposa, J.D.; *Spectrochim. Acta,* **1976**, *32A,* 1207.
17. Greene, J.H.S.; *Spectrochim. Acta,* **1968**, *24A,* 863.
18. Cattanach, C.J.; Mooney, E.F.; *Spectrochim. Acta,* **1968**, *24A,* 407.
19. Busca, G.; Centi, G.; Marchetti, L.; Trifirò, F.; *Langmuir,* **1986**, *2*, 568.
20. Busca, G; *Langmuir,* **1986**, *2*, 577.
21. Garrone, E.; Giamello, E.; Coluccia, S.; Spoto, G.; Zecchina, A.; *Proc. 9th Int. Congr. Catalysis,* Calgary, **1988**, p. 1577
22. Zhu, J.; Andersson, S.L.T.; *J. Chem. Soc. Faraday Trans. 1,* **1989**, *85,* 3629.
23. Kotter, M.; Li, D.X.; Riekert, L.; in *New Developments in Selective Oxidation,* Centi, G.; Trifirò, F., Eds., Elesevier, Amsterdam, **1990**, pp. 267-274.
24. Germain, J.E.; Laugier, R.; *Bull. Soc. Chim. France,* **1971**, 650.
25. King, S.T.; *J. Catal.,* **1991**, *131,* 215.
26. Kita, P.; Jordan, R.B.; *Inorg. Chem.,* **1989**, *28,* 3489.
27. Partenheimer, W.; In *Catalysis of Organic Reactions,* Blackburn, D.W., Ed.; Dekker, New York, **1990**, pp. 321-346.
28. Greene, J.H.S.; *Spectrochim. Acta,* **1977**, *33A,* 575.
29. Mirone, P.; Chiorboli, P.; *Spectrochim. Acta,* **1962**, *18,* 1425.
30. Rogstad, A.; Cyvin, B.N.; Christensen, D.H.; *Spectrochim. Acta,* **1976**, *32A, 487.*
31. Hase, Y.; Davanzo, C.U.; Kawai, K.; Sala, U.; *J. Mol. Struct.,* **1976**, *30,* 37.
32. Busca, G.; Ramis, G.; Lorenzelli, V.; in *New Developments in Selective Oxidation,* Centi, G.; Trifirò, F., Eds., Elesevier, Amsterdam, **1990**, pp. 825-831.
33. Greene, J.H.S.; Harrison, D.J.; *Spectrochim. Acta,* **1976**, *32A,* 1265.
34. Miyamoto, A.; Mori, K.; Inomata, M.; Murakami, Y.; *Proc. 8th Int. Congr. Catalysis,* Berlin, **1984**, Vol. IV, pp. 285-296.
35. VanHengstum, A.J.; VanOmmen, J.G.; Bosch, H.; Gellings, P.J.; *Appl. Catal.,* **1983**, *8,* 369.
36. Mathur, B.C.; Viswanath, D.S.; *J. Catal.,* **1974**, *32,* 1.
37. Vanhove, D.; Blanchard, M.; *J. Catal.,* **1975**, *36,* 6.
38. Ramis, G.; Busca, G.; Bregani, F.; Forzatti, P.; *Appl. Catal.,* **1990**, *64,* 259.
39. Greene, J.H.S.; *Spectrochim. Acta,* 1976, *32A,* 1265.
40. Knyseley, R.N.; Fassel, V.A.; Farquhar, E.L.; Gray, L.S.; *Spectrochim. Acta,* **1962**, *18,* 1217.
41. Joshi, G.W.; Rajadhyaksha, R.A.; *Chim. Ind.* (London), **1986**, 876.
42. van den Berg, P.J.; van der Wiele, K.; den Ridder, J.J.J.; *Proc. 8th Int. Congr. Catalysis,* Berlin, **1984**, Vol. V, pp. 393-404.

RECEIVED November 6, 1992

Chapter 14

Activation of Silver Powder for Ethylene Epoxidation at Vacuum and Atmospheric Pressures

N. C. Rigas, G. D. Svoboda, and J. T. Gleaves

Department of Chemical Engineering, Washington University, St. Louis, MO 63130

Transient response techniques are used to investigate the activation of silver powder for ethylene epoxidation at vacuum and atmospheric pressures. Results indicate that the activation process is qualitatively the same in both pressure regimes. Numerical simulation of the process indicates that activation involves the concurrent incorporation of oxygen into surface and subsurface sites. The reaction selectivity parallels the incorporation of oxygen into the subsurface.

The heterogeneous selective oxidation of ethylene to ethylene oxide over silver based catalysts has become a significant world wide industry since its development in the 1950's *(1-3)*. Much industrial research has focused on modifying the catalyst composition to maximize catalyst selectivity and activity, and in modern processes ethylene is converted to ethylene oxide with a selectivity of approximately 80%. In spite of these advances, a variety of questions remains regarding the nature of the active silver surface that leads to epoxidation. One important aspect that is not well understood and has received relatively little attention in the literature is the mechanism of the activation process that leads to a catalytic surface. Observing changes in the chemical characteristics of a surface during activation can provide insights into the nature of the active/selective phase that is formed. A clearer understanding of the activation process is also important for producing more selective catalysts. From a fundamental viewpoint, silver's unique character makes it an extremely important catalytic system, and developing a greater understanding of its operation would add to our overall knowledge of oxidation processes.

In silver catalyzed epoxidation, the selective oxygen species preferentially adds across the C=C double bond. However, oxygen covered silver surfaces can also give products similar to those found in metal oxide catalyzed reactions. For example, Madix and coworkers *(4-6)* recently found that butene can be oxidized to butadiene, dihydrofuran, furan, and maleic anhydride by oxygen atoms chemisorbed on silver single crystals. The initial activation of butene involves the abstraction of an acidic allylic hydrogen similar to the activation of propene by bismuth molybdate or the activation of butene by vanadyl pyrophosphate *(7,8)*. In these reactions, oxygen reacts as a Bronsted base and a nucleophile. Atomically adsorbed oxygen on Ag (110) surfaces can also add across the C=C double bond of olefins such as norbornene, styrene, or 2,3-dimethylbutene *(6,9)*. In these cases, the C-H bonds are much more weakly acidic. Thus, chemisorbed oxygen atoms can display decidedly different chemical characteristics depending on the nature of the reacting hydrocarbon.

The reactive nature of an oxygen adspecies may also be influenced by the electronic characteristics and structure of the silver surface. Experimental evidence and theoretical calculations indicate that more than one form of atomically adsorbed oxygen exists on a silver surface *(10,11)*. It has been suggested that subsurface oxygen plays a key role in forming the selective surface for epoxidation *(10,12)*. For example, Van Santen and coworkers *(13)* have proposed that activation of a silver surface initially involves the filling of subsurface sites and that ethylene oxide production does not occur until after the subsurface is fully populated. Recently, Bukhtiyarov and coworkers *(14)* investigated ethylene epoxidation on polycrystalline silver foils and determined that two forms of atomically adsorbed oxygen are present on surfaces active for ethylene epoxidation. In addition, they observed a strong promoting effect of carbon during the formation of subsurface oxygen *(15)* and suggested that carbon stabilizes surface defects *(14)* that are necessary for ethylene oxide production. These findings are consistent with studies by Grant and Lambert *(16)* that indicate that treatment of a Ag (111) surface with only oxygen is not sufficient for ethylene epoxidation, but that exposure to a reaction medium is necessary.

Previous reaction studies have tended to focus on the reactivity of equilibrated silver catalysts of fixed composition. In this paper, ethylene oxidation is studied under nonsteady-state conditions in an effort to unravel aspects of silver activation and the role of subsurface oxygen in forming a selective catalyst. Results from a series of steady-flow and transient studies over polycrystalline silver metal powder are presented. These studies were carried out using a modified TAP reactor system *(17)* that is capable of performing steady-flow and transient experiments at vacuum conditions and at atmospheric pressures. The results provide some important insights into the mechanism of activation and the nature of the active catalytic surface. In addition, the results demonstrate a new method of monitoring the chemical changes in a catalyst surface during nonsteady-state processes.

Modified TAP Reactor System

The reactor system used in this study is a modified version of the TAP reactor system *(17-19)* developed by Gleaves and Ebner. A schematic of the system is shown in Figure 1. The principal modification is a movable high-pressure sealing assembly that permits operation from 10^{-7} to 2500 torr. When the sealing assembly is engaged, reactor effluent is split between a vacuum bleed and an external vent that is connected to a back-pressure regulator. The back-pressure regulator is used to control the reactor pressure in the range of 100 to 2500 torr. Effluent from the vacuum bleed is detected by a quadrupole mass spectrometer (QMS), and flow from the back pressure regulator is monitored by a gas chromatograph. When the high-pressure assembly is disengaged, the reactor is continuously evacuated, and all of the reactor effluent is sent to the QMS. Switching between the high-pressure (100-2500 torr) and vacuum (<100 torr) mode of operation can be accomplished without disturbing the catalyst sample or venting it to atmosphere. A detailed schematic of the microreactor and valve manifold, showing high-pressure and vacuum operation, is presented in Figure 2.

In addition to the high-pressure assembly, the modified system incorporates a new real-time data collection system coupled with a PC based computer. Experimental parameters, such as the valve firing sequence and the reactor temperature-control program, can be set from the computer. Reactants are introduced through two high-speed pulse valves or two continuous feed valves that are fed by mass flow controllers. In high-speed transient response experiments, the QMS is set at a particular mass value and the intensity variation as a function of time is obtained. In steady-flow experiments,

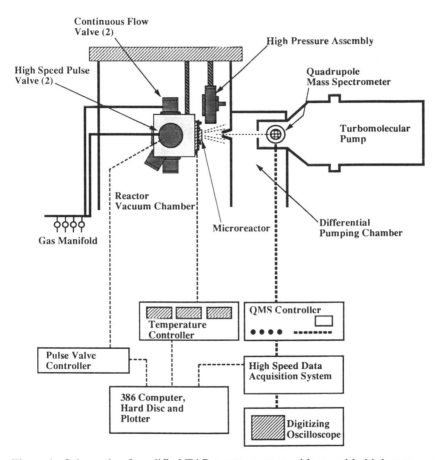

Figure 1. Schematic of modified TAP reactor system with movable high pressure assembly disengaged for vacuum operation.

Atmospheric Pressure Operation

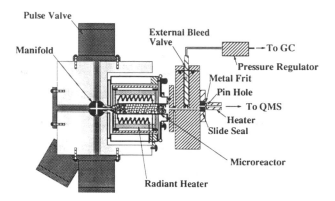

Vacuum Operation

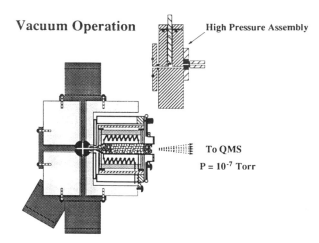

Figure 2. Schematic of valve manifold assembly, microreactor, radiant heater, and high pressure assembly depicting vacuum and high pressure operation.

the entire mass spectrum is collected. In temperature-programmed experiments, both single-mass and multiple-mass spectra may be obtained.

Experimental

To eliminate support effects, polycrystalline silver powder was used as the catalytic substrate. Its activation was studied at vacuum and atmospheric conditions using both steady flow and transient experiments. A typical reactor charge contained 0.45-0.50 grams of polycrystalline Ag powder. Powder samples were obtained from Aesar Johnson-Matthey (99.995% purity) and sieved to give 300-350 μm particles. Reactant gases were obtained from Matheson, MDM Scientific, and Isotec and were used as received without further purification. Reactant gases included Matheson ultra high purity helium (99.999%), Matheson high purity ethylene (99.7%), Matheson high purity argon (99.7%), Matheson ultra high purity oxygen-16 (99.8%), MDM Scientific & Chemical ethylene-d4 (99%), and Isotec oxygen-18 (99%).

Reaction studies were performed with pure gases and gas blends containing 15% argon. Argon was used as an inert reference to monitor changes in the bed transport characteristics. In high-pressure (800 torr) experiments, the reactants were introduced from one of the continuous valves or from the high-speed pulse valves as a step transient. In the step transient experiments, helium was used as an inert carrier gas and introduced from one of the continuous valves. Step transients were produced by pulsing one of the high-speed pulse valves at 6 pps. In vacuum experiments, a carrier gas was not employed and the reactants were introduced in a pump-probe format by alternately pulsing oxygen-16 from one high-speed valve and ethylene-d4 from the other. Pulse intensities were determined by monitoring the pressure drop in the oxygen and ethylene feed reservoirs for a known number of pulses. Temperature-programmed experiments (TPD and TPSR) were performed by heating the microreactor in a linear fashion at a ramp rate of 10 or 20° C per minute.

Prior to each activation experiment, the silver samples were initially preconditioned for eight hours by alternately oxidizing and reducing the surface following the procedure described by Czanderna *et al. (17)*. The pretreatment procedure was performed with the high-pressure assembly engaged. A cycle consisted of first oxidizing with a 25 cc/min oxygen flow at 800 torr and 525 K for 30 minutes and then reducing with a 25 cc/min hydrogen flow at 800 torr and 625 K for 30 minutes. The sequence was repeated a total of eight times. The surface area was determined before and after pretreatment by the BET method using krypton adsorption and was typically $0.11 - 0.12$ m^2/g. In some experimental sequences, the same sample was used a number of times. After each reaction, these samples were restored to their pretreatment state by exposure to a 25 cc/min oxygen flow at 800 torr and 525 K for 30 minutes followed by a 25 cc/min hydrogen flow at 800 torr and 625 K for 30 minutes.

In addition to experiments performed with pretreated silver, experiments were also conducted with preoxidized silver samples. The preoxidized samples were prepared by exposing a clean silver sample to 650 Torr of oxygen at 473 K for 1 hour. Oxygen adsorption was determined by measuring the reactor pressure drop over the 1 hour exposure period. For a typical sample, the pressure drop corresponded to an oxygen uptake of $\approx 1 \times 10^{19}$ O-atoms/gram of silver.

Results

Temperature-Programmed Surface Reaction (TPSR) Experiments at 800 Torr.

Pretreated and preoxidized silver exhibited no reactivity toward an ethylene/argon mixture at reaction temperatures (443 - 543 K) and atmospheric pressures (750-800 torr). The desorption spectrum of a pretreated sample showed no evidence of oxygen desorption when the sample was heated *in vacuo* to 673 K. These results are consistent with those of other workers *(13,20-28)* using powders and single crystals.

Pretreated silver became active when ethylene/argon and oxygen/argon mixtures were simultaneously fed over the silver at 800 torr and at temperatures above 423 K. Both ethylene oxide and carbon dioxide were produced. Figure 3 shows a typical three dimensional temperature-intensity-mass number plot collected during a TPSR experiment. In this example, the oxygen-16 to ethylene feed ratio was 3:1 and the total flow rate was 25 cc/min Mass spectra were collected every 2° C beginning at 373 K and stored in memory along with the collection temperature. Similar spectra were obtained using other oxygen to ethylene feed ratios.

The TPSR spectrum has reactant peaks at m/e = 32, 16, 28, 27, 26, and 40 corresponding to the parent and fragment peaks of oxygen, ethylene, and argon, respectively. A product peak at m/e=44 begins to appear at temperatures above 423 K and corresponds to the parent ion of both ethylene oxide and carbon dioxide. Changing the feed to ethylene-d4 gives rise to new product peaks at m/e = 48 and m/e = 46 as well as the peak at m/e = 44. The former two peaks are indicative of ethylene-d4 oxide.

Step Transient Experiments at 800 Torr. Activation of pretreated silver was performed under isothermal conditions at 493, 523, and 543 K and 800 torr using a step transient format. A typical spectrum collected at 493 K, obtained by simultaneously pulsing ethylene-d4 and oxygen-18 from separate pulse valves into a continuous helium flow, is plotted in Figure 4. In this example, the oxygen to ethylene ratio was 2:1. As observed in the steady-flow TPSR experiments, the pretreated sample is readily activated, while the preoxidized samples remain inactive.

The integrated transient responses of ethylene-d4 (m/e = 32), oxygen-18 (m/e = 36), ethylene-d4 oxide-18 (m/e = 50), and carbon dioxide-18 (m/e = 48) from the spectra shown in Figure 4 are displayed in Figures 5a and 5b. For reference, the transient response of argon taken from a separate experiment under identical conditions is included. The delayed response of the oxygen step transient indicates oxygen chemisorption. The total uptake, which includes the oxygen incorporated in reaction products, corresponds to 7.2×10^{18} molecules. Ethylene conversion plateaus approximately 20 seconds into the step transient and remains essentially constant. The maximum ethylene conversion is approximately 1%.

The transient responses of $C_2D_4{}^{18}O$ and $C^{18}O_2$ are delayed relative to the ethylene response. The rise time in the $C^{18}O_2$ response is more rapid than the $C_2D_4{}^{18}O$ response, indicating that during the initial stages of oxygen incorporation carbon dioxide production is favored. The $C^{18}O_2$ response mirrors that of the ethylene input and reaches a maximum at the point that oxygen breakthrough begins. This indicates that carbon dioxide production is not effected by oxygen that is incorporated during the rise portion of the oxygen transient response. In contrast, the $C_2D_4{}^{18}O$ response resembles the oxygen response and begins to plateau when the oxygen uptake is approximately 90% complete. The lag in the $C_2D_4{}^{18}O$ response indicates that the ethylene oxide producing sites form more slowly than the nonselective CO_2 producing sites.

Pump-probe Experiments at Vacuum Conditions. Consistent with the high-pressure activation studies, pretreated surfaces could be activated under vacuum conditions, while preoxidized surfaces could not. Figures 6a-c show a typical set of the transient responses for ethylene-d4, oxygen-16, carbon dioxide, and ethylene-d4 oxide over a pretreated silver surface using a pump-probe format. As previously explained, an alternating sequence of oxygen and ethylene-d4 pulses were introduced into the reactor starting with an oxygen pulse. Figure 6a shows the transient responses at m/e = 32 when constant intensity pulses of oxygen and ethylene-d4 were introduced at an oxygen to ethylene ratio of 2:1. Figures 6b and c show the CO_2 and C_2D_4O responses

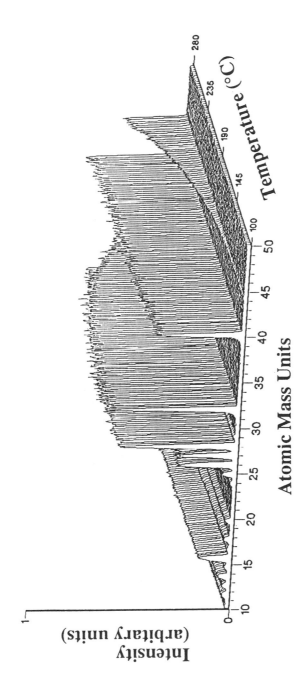

Figure 3. TPSR spectra of the activation process over pretreated silver powder at 800 torr with a 3:1 oxygen/ethylene feed ratio.

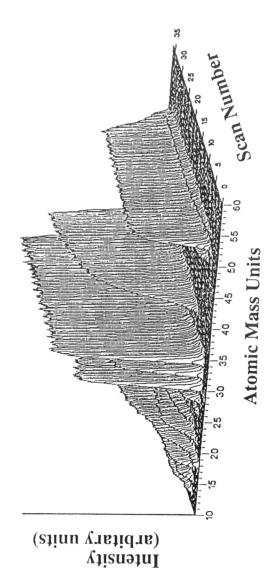

Figure 4. Transient response spectra of the activation process over pretreated silver powder at 800 torr for step inputs of oxygen and ethylene.

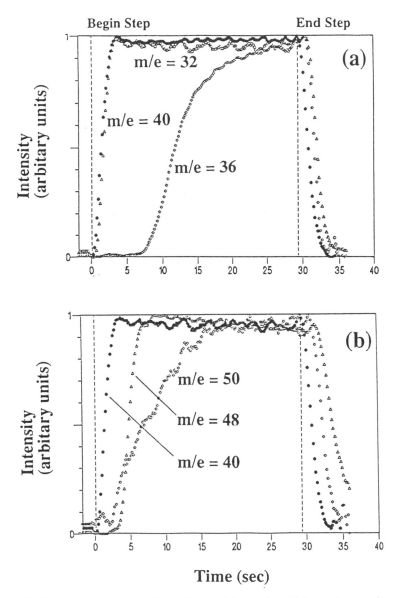

Figure 5. Transient responses of step inputs: (a) ● - m/e = 40 (argon), Δ - m/e = 32 (ethylene-d4), ◊ - m/e = 36 (oxygen-18), (b) ● - m/e = 40 (argon), Δ - m/e = 48, (carbon dioxide-18), ◊ - m/e = 50 (ethylene-d4 oxide-18).

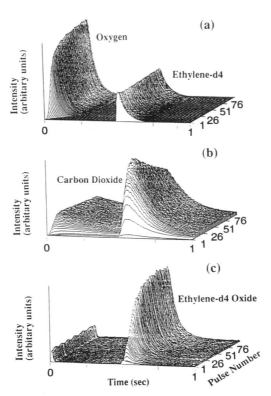

Figure 6. Transient responses of pump-probe inputs: (a) Oxygen and ethylene-d4 inputs (m/e = 32) separated by ≈ 0.25 seconds; (b) carbon dioxide (m/e = 44) product response; (c) ethylene-d4 oxide (m/e = 48) product response (Note: peak at t = 0.0 is minor contaminant in oxygen feed).

respectively. An argon response was also collected, but is not shown. The argon pulse shape and pulse intensity was invariant with time, indicating that the bed transport characteristics and valve performances were constant.

As observed in the high-pressure transient and steady-flow experiments, oxygen is readily chemisorbed by a pretreated silver surface under high-vacuum conditions. The oxygen uptake, corrected for oxygen incorporated in reaction products (i.e., ethylene oxide, carbon dioxide and water), is plotted in Figure 7 along with the carbon dioxide production. Oxygen is reversibly adsorbed at reaction temperatures and some desorption will occur between oxygen pulses. This is not accounted for in the oxygen uptake calculations. However, at temperatures below 493 K the desorption rate is relatively slow compared to the pulsing rate, and oxygen loss due to desorption is negligible.

Carbon dioxide is produced upon introduction of the second set of reactant pulses and occurs during both the oxygen and ethylene-d4 inputs. Since gas-phase ethylene-d4 is not present during the oxygen pulse (the pulse separation between the ethylene-d4 pulse and the next oxygen pulse is 2.0 seconds), carbon dioxide production must result from the reaction of oxygen with a carbon containing adspecies. Carbon dioxide production during the oxygen pulse reaches a maximum within 65 oxygen pulses and remains constant, indicating that the surface carbon reaches a steady state. The pulse width of the CO_2 pulse formed during the oxygen input is essentially constant, while the pulse width formed during the ethylene-d4 input varies with oxygen coverage. In the latter case, the CO_2 pulse width decreases with increasing coverage. The shape difference between the two sets of CO_2 pulses indicates that the formation mechanisms are not equivalent. The CO_2 pulse formed during the oxygen input is the broader, indicating the formation mechanism is slower.

As indicated in Figure 7, the CO_2 production initially increases with oxygen uptake, goes through a maximum, and then rapidly decreases. Ethylene-d4 oxide production, plotted in Figure 8, also increases and then decreases with oxygen uptake, but more slowly than CO_2. As a result, the reaction selectivity shown in Figure 9 increases with increasing oxygen uptake.

Oxygen Incorporation

During the initial stages of an oxygen incorporation experiment, oxygen is unevenly distributed across the length of the catalyst bed. The surface and subsurface sites at the front of the catalyst bed are filled first and those at the reactor exit more slowly. Low-pressure oxygen transient response experiments provide a means of investigating the mechanism of this process and provide some interesting details related to the relative rates of oxygen incorporation into the surface and subsurface, as well as its relationship to product selectivity.

The oxygen incorporation process can be described by a mathematical model similar to that reported by Svoboda *et al.* (29) with modifications to include subsurface sites and nonlinear adsorption kinetics. Assuming oxygen chemisorption is dissociative (30), and oxygen migration into the subsurface is first order (31), equations 1 - 3 below can be used to model oxygen incorporation, provided: 1) Knudsen diffusion is the predominate mode of gas transport, 2) oxygen desorption is negligible, 3) the subsurface concentration is uniform, 4) diffusion from the subsurface to the surface is negligible, 5) the catalyst bed is isothermal, 6) the silver powder is nonporous, 7) changes in the number of surface sites are negligible, and 8) reaction is negligible.

$$\varepsilon_b \frac{\partial C}{\partial t} = D_e \frac{\partial^2 C}{\partial z^2} - \rho_s(1-\varepsilon_b)\frac{1}{2}\left(\frac{\partial \theta}{\partial t} + \frac{\partial \theta_s}{\partial t}\right) \tag{1}$$

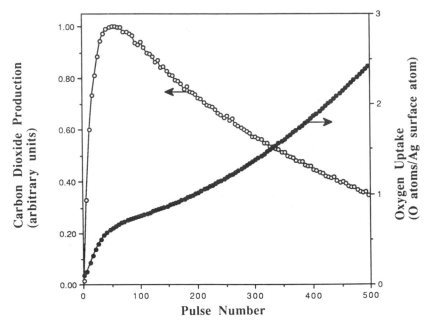

Figure 7. Carbon dioxide production and oxygen uptake as a function of pulse number.

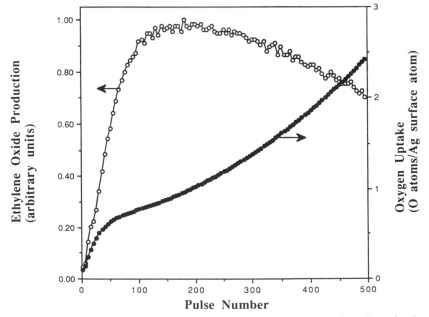

Figure 8. Ethylene-d4 oxide production and oxygen uptake as a function of pulse number.

$$\frac{\partial \theta}{\partial t} = 2k_a C(1-\theta)^2 - \frac{\partial \theta_s}{\partial t} \tag{2}$$

$$\frac{\partial \theta_s}{\partial t} = k_s \theta \left(1 - \frac{\theta_s}{\theta_{ss}}\right) \tag{3}$$

Equation 1 represents the gas-phase oxygen mass balance, and equations 2 and 3 the surface and subsurface mass balances, respectively. C is the gas-phase oxygen concentration, θ is the fractional surface coverage, θ_s is the fractional subsurface concentration normalized to the surface coverage, θ_{ss} is the subsurface saturation concentration normalized to the surface coverage, D_e is the Knudsen diffusion coefficient, k_a is the adsorption rate constant, k_s is the subsurface incorporation rate constant, ρ_s is the saturation oxygen concentration per volume silver, and ε_b is the reactor bed void fraction. With appropriate boundary conditions *(29)*, equations 1-3 can be solved to give the time dependent concentration profile of oxygen in the reactor voids, on the silver surface, and in the silver subsurface with respect to the reactor axial coordinate. The oxygen output response obtained at the QMS, $Y(t)$, is determined by evaluating the flux at the reactor exit as follows:

$$Y(t) = -D_e \left(\frac{\partial C}{\partial z}\right)_{z=L} \tag{4}$$

The equations are solved for an assumed set of parameters, $\mathbf{P} = [\varepsilon_b, D_e, \rho_s, k_a, k_s, \theta_{ss}]$, using finite difference equations, which are described in standard texts on the subject *(32)*. The vector of unknown parameters $\mathbf{P}$ is determined by minimizing the mean square relative error between the model-predicted and experimental breakthrough curves. Minimization of the mean-square relative errors was obtained using Marquardt's method *(33)*.

Figure 10 shows the result obtained for a typical oxygen breakthrough curve. The experimental data points were taken from the pump-probe data presented above at 240 °C. The total measured oxygen uptake was 1.6×10^{19} O-atoms/gram of silver. The solid line is the model fit. D_e and ε_b were determined in separate experiments so that only four unknown parameters had to be determined by the model. The model gave values of $\rho_s = 1 \times 10^{-4}$ mole cm^{-3}, $k_a = 2 \times 10^8$ cm^{3}mole^{-1}sec^{-1}, $k_s = 8 \times 10^{-2}$ sec^{-1}, and $\theta_{ss} = 2$. The ratio of subsurface to surface oxygen at saturation (θ_{ss}) indicates that $\approx 33\%$ of the oxygen present at saturation occupies surface sites. This is similar to the 50% value reported by Van Santen for oxygen incorporation in polycrystalline silver powders at 200 °C *(22)*.

The value obtained for k_a can be used to calculate the sticking coefficient s_o for oxygen on clean silver powder at 240 °C using the relationship *(34)*:

$$s_o = \frac{k_a}{\left(\dfrac{RT}{2\pi M}\right)^{\frac{1}{2}} \sigma} \tag{5}$$

where $\sigma = $ (density of Ag)(measured surface area)$/\rho_s$.

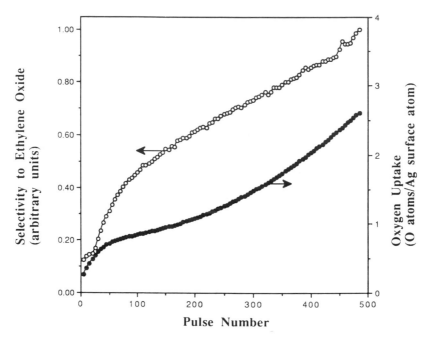

Figure 9. Selectivity to ethylene-d4 oxide and oxygen uptake as a function of pulse number.

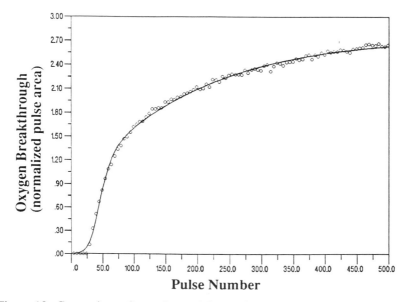

Figure 10. Comparison of experimental (open circles) and model-predicted (solid line) oxygen breakthrough curves.

Substituting the experimentally determined value for k_a and the experimentally determined value for σ of 3 x 10^8 cm^2/mole into equation 5, gives $s_o = 4$ x 10^{-5}. This value lies within the range of literature values for the initial dissociative sticking coefficient on the (110) face ($s_o = 5 - 30$ x 10^{-4}) at 200 °C *(30)* and the (111) face ($s_o \approx 1$ x 10^{-6}) at 217 °C *(35)* of silver single crystals.

The modeling results yield values for the sticking coefficient, and the ratio of surface to subsurface oxygen that are consistent with those of other workers. The results indicate that the activation process involves a concurrent filling of surface and subsurface sites, but that the latter sites are filled much more slowly. This is shown graphically in Figure 11.

Discussion

The results of the high and low-pressure transient activation studies indicate that the activation process follows a similar pattern in both pressure regimes. In both cases, oxidation products are formed during the initial incorporation of oxygen. Results of the low-pressure pump-probe experiments indicate that CO_2 and C_2D_4O production begins after the first oxygen pulse, when total oxygen adsorption is below .05 monolayers. Also, in both the high and low-pressure studies, the surface is less selective during the initial stages of the activation process, and selectivity increases with oxygen uptake.

At vacuum pressures and reaction temperatures, the predominant oxygen adspecies is atomic oxygen *(28,36)*. Molecularly adsorbed oxygen has a significantly shorter surface lifetime and will either desorb or chemisorb dissociatively *(25)*. In pump-probe experiments, the concentration of molecularly adsorbed oxygen is negligible during an ethylene pulse since the oxygen and ethylene inputs are separated in time. As a result, molecularly adsorbed oxygen is not expected to play a significant role in ethylene conversion. Previous TAP reactor studies using oxygen isotopes *(25)*, as well as a variety of other studies *(10,26)*, have indicated that atomically adsorbed oxygen, and not molecularly adsorbed oxygen, is the active species in ethylene epoxidation. Results of the pump-probe experiments presented in this paper are consistent with those findings.

In addition to reacting with ethylene, atomically adsorbed oxygen can migrate to the subsurface where it is blocked from directly participating in an oxidation reaction but can modify the electronic properties of the surface. Specifically, reaction of subsurface oxygen with silver surface atoms will withdraw electron density from the surface. This in turn will increase the electrophilic character of silver surface atoms*(10)*. Recent theoretical calculations by van den Hoek *et al.* *(37)* on Ag(110) clusters indicate that subsurface oxygen may also reduce the bond energy between silver and adsorbed oxygen and convert the repulsive interaction between adsorbed oxygen and ethylene into an attractive interaction. As a result, the barrier for epoxide formation disappears.

Results from the oxygen incorporation modeling study presented in this work indicate that the reaction selectivity is strongly influenced by the incorporation of subsurface oxygen. Figure 12a plots the reaction selectivity, together with the model-predicted average subsurface concentration, as a function of oxygen pulse number. Figure 12b plots CO_2 production, together with the model-predicted average surface concentration, as a function of pulse number. Comparison of the selectivity curve with the incorporation curves indicates that the rise in C_2D_4O selectivity parallels the increase in subsurface oxygen, but not the increase in surface oxygen. In contrast, the maximum CO_2 production occurs when the surface coverage begins to plateau, and CO_2 production rapidly decreases with increasing subsurface concentration. These results indicate that incorporation of subsurface oxygen is crucial to the formation of a selective surface. They are also consistent with the supposition that subsurface oxygen modifies the reactive characteristics of the oxygen adspecies.

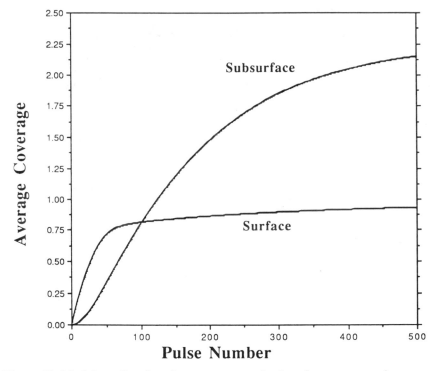

Figure 11. Model-predicted surface coverage and subsurface concentration as a function of pulse number.

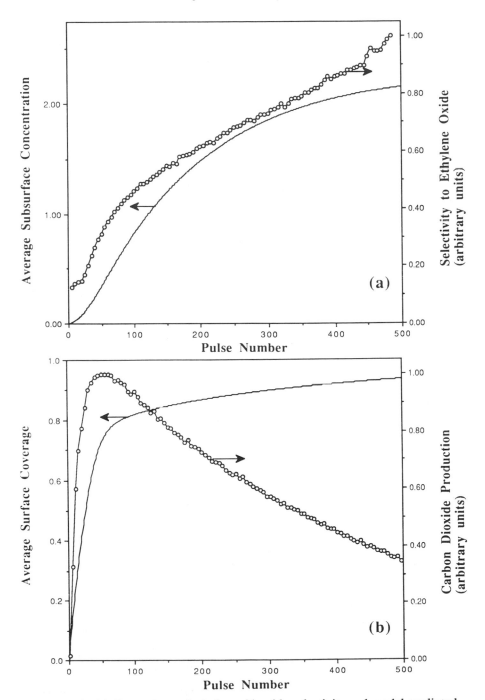

Figure 12. (a) Comparison of ethylene-d4 oxide selectivity and model-predicted average subsurface oxygen concentration as a function of pulse number. (b) Comparison of carbon dioxide production and model-predicted average oxygen surface coverage as a function of pulse number.

Another significant feature of the pump-probe experiments is the detection of CO_2 during the oxygen pulse. Since ethylene is not present in the oxygen feed, oxygen must be reacting with carbon adspecies deposited during the ethylene input. Unlike the CO_2 produced during the ethylene pulse, the intensity of the oxygen-generated CO_2 rises to a plateau, but does not decrease with increasing oxygen incorporation. In addition, the pulse width is broader for the oxygen-generated CO_2 pulse than the ethylene generated pulse, and the former does not narrow with pulse number. This behavior indicates that the pulse width is controlled by the surface reaction rate, and not CO_2 desorption.

Taken in conjunction with previous studies that indicate that subsurface oxygen (10-14) is essential to the formation of ethylene oxide, and that surface carbon (14-16) plays a key role in the formation of an active surface, the current results can be interpreted in terms of the multistep mechanism presented in Figure 13. According to this mechanism, the activation process begins with the adsorption of oxygen and the formation of partially oxidized silver sites. Step two involves the adsorption of ethylene and the formation of carbon adspecies. The initial adsorption of ethylene may involve the formation of a π-complex between ethylene and a partially oxidized silver atom bound to an adsorbed oxygen atom. The production of carbon dioxide during the oxygen pulse is clear evidence that carbon adspecies are formed. In step three, the carbon adspecies promote the rearrangement of the silver surface, the incorporation of subsurface oxygen, and the formation of stable defect sites. Step three follows the proposals of Bukhtiyarov and coworkers (14) that carbon has a strong promoting effect on the formation of subsurface oxygen and stabilizes surface defects. Step four involves the production of ethylene oxide at an active site and the continued incorporation of subsurface oxygen.

The evidence presented in this study suggests that the selective catalytic site is comprised of both surface and subsurface oxygen species, and a partially oxidized silver atom. In addition, the site may include one or more carbon adspecies. As well as stabilizing the defect site, the carbon adspecies may promote site isolation and prevent over oxidation of adsorbed ethylene molecules. Deactivation of the silver surface occurs when the surface is completely oxidized, the surface carbon is depleted, and no silver sites are available for ethylene adsorption. In this regard, it is important to note that the amount of oxygen incorporated during both the high and low pressure activation processes, and in the oxygen pretreatment experiments was $\approx 10^{19}$ O-atoms/gram of silver. Since the active catalyst surface contains approximately the same quantity of oxygen as an inactive surface, it follows that the former is not simply a reduced state of the latter. Rather, the active surface must contain unique sites that are absent in the inactive surface. This result provides additional support for the supposition that carbon plays a role in creating and stabilizing ethylene epoxidation sites.

Conclusion

In the present work, a novel modification of the TAP reactor system has been described which permits transient response experiments to be performed at vacuum and atmospheric pressures. This system was used to investigate the activation of silver powder for ethylene epoxidation. The experimental results indicate that activation at high and low pressures is qualitatively the same. Pretreated and preoxidized silver exhibited no reactivity toward an ethylene/argon mixture at reaction temperatures. Pretreated silver became active when exposed to an ethylene/oxygen feed, but preoxidized surfaces remained inactive. In both pressure regimes, CO_2 production predominates at low oxygen coverages, but ethylene oxide selectivity increases with oxygen coverage. Low-pressure pump-probe experiments indicated that carbon adspecies are formed on the silver surface.

Oxygen incorporation at low pressures was simulated using a simple transport-kinetics model, and kinetic parameters were determined. Results of the transport-

1. Oxygen adsorption.

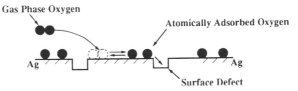

2. Ethylene adsorption.

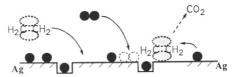

3. Buildup of Surface Carbon and Subsurface Oxygen and Formation of Stabilized Defect Sites.

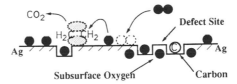

4. Ethylene Oxide Production at Defect Site.

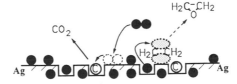

5. Surface Deactivation via Formation of Oxide Film.

Figure 13. Proposed mechanism for activation of silver metal powder for ethylene epoxidation.

kinetics model indicate that the activation process involves a concurrent filling of surface and subsurface sites, but that the latter sites are filled much more slowly. The model-predicted surface coverage and subsurface concentration was compared with the ethylene oxide selectivity and CO_2 production. The results were consistent with current models of ethylene epoxidation that indicate that subsurface oxygen is necessary for adsorbed oxygen to react to epoxide. A multistep mechanism was proposed to explain the activation process in which carbon adspecies play a role in the formation and stabilization of catalytic sites active for ethylene epoxidation.

Literature Cited

1. Voge, H.H.; Adams, C.R. *Adv. Catal.***1967**, *17*, 151.
2. Zomerdijk, J.; Hall, M. Catal. Rev. *Sci. Eng.* **1981**, *23*, 163.
3. Detwiler, H. R.; Barker, A.; Richarz, W. *Helv. Chim. Acta.* **1979**, *62*, 1689.
4. Roberts, J.T.; Capote, A. J.; Madix, R.J. *Surf. Sci.* **1991**, *253*, 13.
5. Roberts, J.T.; Capote, A. J.; Madix, R.J. *J. Am. Chem. Soc.* **1991**, *113*, 9848.
6. Roberts, J.T.; Madix, R.J. *J. Am. Chem. Soc.* **1988**, *110*, 8540.
7. Grasselli, R.K.; Burrington, J.D. *Adv. Catal.* **1981**, *30*, 133.
8. Centi, G; Trifiro, F; Ebner, J.R.; Franchetti, V. *Chem. Rev.* **1989**, *28*, 400.
9. Hawker, S.; Mukoid, C.; Badyal, J.P.S.; Lambert, R.M. *Surf. Sci.* **1989**, *219*, L615.
10. van Santen, R. A.; Kuipers, H. P. C. E. *Adv. Catal.* **1987**, *35*, 265.
11. Carter, E. A.; Goddard, W. A. III *J. Catal.* **1988**, *112*, 80.
12. Grant, R. B.; Lambert, R. M. *J. Chem. Soc., Chem. Commun.* **1983**, 662.
13. van Santen, R. A.; DeGroot, C. P. M. *J. Catal.* **1986**, *98*, 530.
14. Bukhtiyarov, V. I.; Boronin, A. I.; Savchenko, V. I. *Surf. Sci. Lett.* **1990**, *232*, L205.
15. Boronin, A. I.; Bukhtiyarov, V. I.; Vishnevskii, A. L.; Boreskov, G. K.; Savchenko, V. I. *Surf. Sci.* **1988**, *201*, 195.
16. Grant, R. B.; Lambert, R.M. *J. Catal.* **1985**, *92*, 364.
17. Gleaves, J. T.; Ebner, J. R.; Kuechler, T. C. *Cat. Rev. Sci.* **1988**, *30*, 49.
18. Ebner, J. R.; Gleaves, J. T., U.S. Patent 4626412, Dec. **1986**.
19. Gleaves, J. T.; Harkins, P. T., U. S. Patent 5,039,489 **1991**.
20. Czanderna, A. W. *J. Phys. Chem.* **1966**, *70*, 2120.
21. Barteau, M. A.; Bowker, M.; Madix, R. J. *J. Catal.* **1981**, *67*, 118.
22. Backx, C.; Moolhuysen, J.; Geenen, P.; Van Santen, R. A. *J. Catal.* **1981**, *72*, 364.
23. Campbell, C. T.; Paffett, M. T. *Surf. Sci.* **1984**, *143*, 517.
24. Rigas, N. C.; Gleaves, J. T.; Mills, P. L. *New Developments in Selective Oxidation*, Elsevier (Amsterdam) **1990**; pp. 707 (1990)
25. Gleaves, J. T.; Sault, A. G.; Madix, R. J.; Ebner, J. R. *J. Catal.* **1990**, *121*, 202.
26. Grant, R.B.; Lambert, R. M. *J. Chem. Soc., Chem. Commun.* **1983**, 58.
27. Force, E. L.; Bell, A. T. *J. Catal.* **1975**, *40*, 356.
28. Bowker, M.; Barteau, M.A.; Madix, R. J. *Surf. Sci.* **1980**, *92*, 528.
29. Svoboda, G. D.; Gleaves, J. T.; Mills, P.L. *Ind. Eng. Chem. Res.* **1992**, *31*, 19.
30. Engelhardt A.; Menzel D. *Surf. Sci.* **1976**, *57*, 591.
31. Borg, R. J.; Dienes G. J.*The Physical Chemistry of Solids*, Academic Press, New York, N.Y.,**1992**.
32. Jenson, V. G.; Jeffreys, G. V. *Mathematical Methods in Chemical Engineering*, Academic Press, New York, N.Y., **1977**.
33. Seinfeld, J. H.; Lapidus, L. *Mathematical Methods in Chemical Engineering, Vol. 3, Process Modeling, Estimation, and Identification,* Prentice-Hall, Englewood Cliffs, NJ, **1974**.

34. Rieck, J. S.; Bell, A. T. *J. Catal.* **1984**, *85*, 143.
35. Campbell, C. T. *Surf. Sci.* **1985**, *157,* 43.
36. Barteau, M. A.; Madix, R. J. *Surf. Sci.* **1980**, *97*, 101.
37. van den Hoek, P.J.; Baerends, E.J.; van Santen R. A. *J. Phys. Chem.* **1989**, *93*, 6469.

RECEIVED November 6, 1992

Chapter 15

Structure and Reactivity of Tin Oxide-Supported Vanadium Oxide Catalysts

B. Mahipal Reddy

Catalysis Section, Indian Institute of Chemical Technology, Hyderabad 500 007, India

The surface structure and reactivity of vanadium oxide monolayer catalysts supported on tin oxide were investigated by various physico-chemical characterization techniques. In this study a series of tin oxide supported vanadium oxide catalysts with various vanadia loadings ranging from 0.5 to 6.4 wt.% have been prepared and were characterized by means of X-ray diffraction, oxygen chemisorption at -78°C, solid state ^{51}V and ^{1}H nuclear magnetic resonance and electron spin resonance techniques. Reactivities of these catalysts were also evaluated for partial oxidation of methanol to formaldehyde at atmospheric pressure. Oxygen uptake results suggest the formation of V-oxide 'monolayer' at about 3.2 wt.% vanadia loading; where a maximum activity for methanol oxidation was also observed. Solid-state ^{51}V NMR results show the presence of two types of vanadium oxide species, one due to a dispersed phase and the other due to the crystalline vanadia phase, supporting oxygen chemisorption results. ESR results also reveal V-oxide in highly dispersed state at the same 3.2 wt.% loading. A direct relationship was also noted between the oxygen uptake and MeOH oxidation activity of the catalysts.

Supported vanadium oxides represent one of the technologically most important class of solid catalysts. These catalysts are useful for partial oxidation of various hydrocarbons (1), ammoxidation of alkyl substituted N-heteroaromatic compounds (2) and most recently for NO_x reduction (3). For a catalyst to be a successful one in industry, it should exhibit high activity with maximum selectivity, thermal and mechanical stability and long life etc. For getting some of these functionalities, the active component has to be dispersed uniformly on a support material.

The V_2O_5-SnO_2 mixed oxide combination is often used for the oxidation of benzene, naphthalene and various other organic compounds.

0097–6156/93/0523–0204$06.00/0
© 1993 American Chemical Society

A good number of patents are reported in literature (4). However, very little attention has been paid towards investigation of the role of SnO_2 as promoter in vanadia-tin oxide catalysts. These catalysts are normally prepared by coprecipitation or mixing of individual metal oxides together. Therefore, it is hard to understand the active sites and their relevance to the catalytic reaction. The best method to overcome this difficulty is perhaps to study a vanadia monolayer on tin oxide support, because only then vanadium oxide interacting directly with the SnO_2 basal plane is exposed on the surface. Ready formation of vanadium oxide monolayers on various supports such as Al_2O_3, TiO_2, ZrO_2 and CeO_2 has already been established in previous investigations (5). Hence the primary purpose of this investigation was to prepare a vanadia monolayer on tin oxide support and additionally to identify the active sites available for methanol oxidation.

Experimental

Catalyst Preparation.   Tin oxide support was prepared from stannic chloride by hydrolysis with dilute ammonia solution. The resulting stannic hydroxide precipitate, washed several times with deionized water till it was free from chloride ions, was dried at 120°C for 16 h and calcined at 600°C for 6 h in air. The tin oxide support thus obtained had a N_2 BET surface area of 30 $m^2 g^{-1}$. The V_2O_5/SnO_2 catalysts with various vanadia loadings ranging from 0.5 to 6.4 wt.% were prepared by the standard wet impregnation method. Tin oxide support (0.5 mm average particle size) was added to a stoichiometric aqueous ammonium metavanadate solution and excess water was evaporated on a hot plate with continuous stirring. The impregnated samples were further oven dried at 120°C for 12 h and calcined at 500°C for 5 h in an air circulation furnace.

Oxygen Uptake Measurements.   Oxygen chemisorption measurements were made at -78°C on a standard static volumetric all-glass, high-vacuum system equipped with a mercury diffusion pump and an in-line liquid nitrogen cold trap (6). The standard procedure employed for oxygen uptake measurement was reduction of catalyst sample for 5 h at 500°C followed by evacuation for 2 h (1 x 10^{-6} torr) at the same temperature. Before admitting oxygen the system was further evacuated for 1 h at the temperature of chemisorption (-78°C) and then purified oxygen was let in from a storage bulb into the catalyst chamber. The first adsorption isotherm was obtained representing the sum of physisorbed and chemisorbed oxygen. The physisorbed oxygen was then removed by evacuating (1 x 10^{-6} torr) for 1 h at the same temperature and soon after, a second isotherm representing only the physisorbed oxygen was generated in an identical manner. From these two isotherms, which are parallel in the pressure range studied (100-300 torr), the volume of the chemisorbed oxygen was determined (7). The BET surface area of the catalyst was obtained by the N_2 physisorption at -196°C using 0.162 nm^2 as the area of cross-section of the nitrogen molecule.

X-Ray Diffraction.   X-ray powder diffraction patterns were recorded on a Philips PW 1051 diffractometer with nickel-filtered CuK_α radiation (λ = 1.54187 Å).

Solid State NMR Measurements. The solid-state ^{51}V NMR spectra have been recorded on a Bruker CXP-300 NMR spectrometer at a frequency of 78-86 MHz in the frequency range of 150 kHz, using 1 μs radio frequency pulses with repetition rate of 10 Hz. Chemical shifts were measured relative to $VOCl_3$ as an external reference. The ^1H NMR spectra with magic angle spinning technique (MAS) were obtained on the same instrument at a frequency of 300 MHz. The frequency range was 50 kHz, (Π/2) pulse duration was 5 μs, and the pulse repetition frequency was 1 Hz. The chemical shifts were measured relative to tetramethylsilane as an external standard. Surface OH groups were quantitatively estimated by measuring the area under the peak with reference to a known standard sample (8).

Electron Spin Resonance. The ESR spectra of reduced and unreduced catalysts were recorded on a Bruker ER 200D-SRC X-band spectrometer with 100 kHz modulation at ambient temperature. Reduced catalysts for ESR study were prepared according to the procedure described elsewhere (9). After hydrogen reduction at 500°C for 4 h the sample was evacuated at the same temperature for 2 h and sealed off under vacuum.

Activity Measurements. To test catalytic properties of various samples partial oxidation of methanol to formaldehyde was studied in a flow micro-reactor operating under normal atmospheric pressure (10). For each run about 0.2 g of catalyst sample was used and the activities were measured at 175°C in the absence of any diffusional effects. The feed gas consisted of 72, 24 and 4% by volume of nitrogen, oxygen and methanol vapor respectively. Reaction products were analysed with a 10% Carbowax 20 M column (2m long) maintained at 60°C oven temperature.

Results and Discussion

X-ray powder diffraction patterns of SnO_2, V_2O_5 and V_2O_5/SnO_2 samples are presented in Figure 1. In all the samples studied only the lines assignable to SnO_2 support (ASTM 21-1250) were obtained and no lines due to the crystalline V_2O_5 or V_2O_5-SnO_2 intermediate compounds were detected. The absence of characteristic V_2O_5 XRD lines may be either due to the absence of crystalline vanadia phase or the crystallites formed are less than the detection capability of the XRD technique (i.e., < 4 nm size). In the case of bulk V_2O_5-SnO_2 catalysts prepared by a precipitation method only SnO_2 phase at $Sn/(V+Sn) > 0.7$ and V_2O_5 at $Sn/(V+Sn) < 0.3$ were reported (11). Nevertheless, the conventional XRD technique fails to provide any further information regarding the nature of V-oxide phase on SnO_2 support.

Oxygen uptakes obtained at -78°C on various V_2O_5/SnO_2 catalysts are shown in Table 1. The tin oxide support was also found to chemisorb some small amount of oxygen under the experimental conditions employed in this study. Therefore, the contribution of pure support was substracted from the uptake results. The amount of chemisorbed oxygen on V_2O_5 catalyst increased linearly as a function of reduction temperature in the range 300 to 500°C and then levelled off. Therefore, 500°C was chosen as the standard temperature of reduction by hydrogen. The factor for the conversion of unit volume of chemisorbed oxygen to the corresponding active vanadia area was determined by the method applied

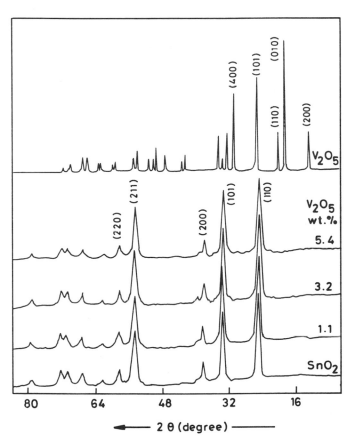

Figure 1.   X-ray powder diffraction patterns of SnO_2, V_2O_5 and V_2O_5/SnO_2 catalysts.

Table 1 : Oxygen Chemisorption and Methanol Oxidation Activity Results on Various V_2O_5/SnO_2 Catalysts

Composition[a] V_2O_5 (wt.%)	O_2 uptake[a] (μ mol g⁻¹ cat.)	BET SA (m^2g^{-1})	Active vanadia area[b] (m^2g^{-1})	Surface coverage[c]	Active site density[d] (nm^{-2})	Dispersion[e] (%)	MeOH conversion (%)	Selectivity (%) HCHO	DME
0.5	4.1	29.0	1.2	4.1	0.17	15.0	10.2	97.1	0.9
1.1	8.9	28.0	2.6	9.3	0.38	14.8	10.6	97.5	0.9
2.2	14.0	25.9	4.1	15.8	0.65	11.7	18.0	95.8	1.8
3.2	17.4	28.7	5.2	18.1	0.73	10.0	21.8	96.8	1.3
4.3	10.8	28.0	3.2	11.4	0.46	4.6	15.8	97.6	2.0
5.4	4.5	25.2	1.6	6.3	0.22	1.5	3.8	96.3	3.1

[a] Balance is SnO_2; [b] Oxygen uptake (μ mol/g cat.) x 0.296 m^2/μ mol O_2.

[c] Defined as : 100 x (active vanadia area/BET surface area of reduced catalyst).

[d] This is equal to the number of oxygen atoms chemisorbed per unit area of the reduced catalyst.

[e] Fraction of vanadium atoms at the surface, assuming $O/V_2O_5 = 1$.

by Parekh and Weller (12).   For this purpose pure V_2O_5 was reduced at 500°C and oxygen was chemisorbed at -78°C.   The conversion factor (0.296 m^2 per μ mole O_2 chemisorbed) was obtained from the ratio of the BET surface area of this reduced sample and the volume of chemisorbed oxygen.   The active vanadia areas and surface coverages of supported catalysts calculated using this conversion factor are listed in Table 1.   It is seen that the oxygen uptake and surface coverage increase as a function of vanadia content upto 3.2 wt.% after which there is a decrease with further loading.   This behaviour is attributed to the formation of V-oxide monolayer due to carrier-catalyst interaction in the lower loading region and the formation of the microcrystallites by multilayer deposition of V-oxide on the monolayer at higher loadings (6, 13, 14).   The decline in the reducibility of V-oxide in the 'post monolayer' region stems from the fact that in the bulky crystallites all the V-oxide units are not accessible to the reducing gas.   On the other hand, in the monolayer region the maximum number of such units are available for the reduction due to high dispersion.

It has been suggested by Bond et al (15) that theoretically a load of 0.145 wt.% V_2O_5 per m^2 surface is required to cover the surface of a titanium dioxide support with a compact single lamella of the vanadium pentoxide structure.   Accordingly, for a 30 $m^2 g^{-1}$ surface area tin dioxide support, the required amount of vanadia is about 4.35 wt.%.   However, the observed oxygen uptake results suggest that the completion of monolayer coverage is at about 3.2 wt.%.   Which is equivalent to about 74% of the theoretical monolayer capacity based on the structure of vanadium pentoxide.   In fact, Bond and coworkers (15) found that the application of surface specific preparative methods such as the reaction of vanadyl triisobutoxide with surface hydroxyl groups of the support resulted in about 0.1 wt.% V_2O_5 per m^2 anatase surface which correspond to about 70% of the theoretical monolayer.   In view of these results an empirical definition of a monolayer was proposed, which is approximate to 70% of the theoretical monolayer capacity.   Baiker et al (16) also reported similar observations by reacting vanadyl triisobutoxide with hydroxyl groups of the titania surface.   In agreement with those findings the present oxygen uptake results also reveal the formation of a monolayer at about 3.2 wt.% V_2O_5 on the tin oxide support.

Dispersion, defined as the percent of V-oxide units available for reduction and subsequent oxygen uptake, can be estimated from the total number of V-oxide units present in a sample and the number of oxygen atoms chemisorbed.   It is observed that (Table 1) dispersion also varied with vanadia loading and is maximum at the lowest loading level.   From the uptake and the BET surface area measurements it is possible to estimate the active site density on the surface.   The active sites are envisaged as the vacancies created by the removal of labile oxygen atoms that take part in the redox processes in oxidation reactions.   Site densities have been calculated from the BET surface areas of the reduced catalysts by assuming a dissociative chemisorption of oxygen on these vacancies.   The active site density also increases in the monolayer level and decreases in the post monolayer (Table 1).   Thus, it is seen that the simple technique of oxygen chemisorption at -78°C gives more meaningful information and was successfully employed to characterize series of vanadia catalysts supported on various carriers such as alumina (6), silica (10, 13), and zirconia (17) etc.

It may be mentioned that though the information obtained was valid, the numerical values obtained were found to be much less than expected. Recently, Oyama et al (18) have proposed that if the temperature of oxygen chemisorption is 370°C, with a prereduction of the catalyst at the same temperature, the results would give much more meaningful information than the results generated at -78°C with a prereduction of the sample at 500°C temperature. Interestingly, it has been observed that the numerical values of the oxygen uptakes on pure unsupported V_2O_5 sample obtained via two different methods are the same (6, 18). This observation clearly signifies the validity of the conclusions drawn from either of the methods.

Solid-state ^{51}V NMR spectra of V_2O_5/SnO_2 catalysts and pure V_2O_5 alone are shown in Figure 2. Two types of distinct signals, with varying intensities depending on the vanadia content on tin oxide support, are the main features of these spectra. Unsupported and crystalline pure V_2O_5 exhibits a line with an axial anisotropy of the chemical shift tensor ($\delta_\perp$ = -310 ppm and $\delta_\parallel$ = -1270 ppm) with small peaks due to the first order quadrupole effects (19). Accordingly, the species around -310 ppm with a shoulder at -1270 ppm was assigned to the microcrystalline vanadia phase and the broad spectrum at -485 ppm to the dispersed vanadia species on the support with a distorted tetrahedral local environment (20). In agreement with oxygen uptake results no crystalline vanadia phase can be seen upto 3.2 wt.% loading indicating that vanadium oxide is in a highly dispersed state. Beyond this loading the formation of crystalline vanadia phase with increased peak intensities can be noted. Similarly, the ^1H MAS NMR results also indicated the formation of a monolayer of vanadium-oxide on SnO_2 support (20). The total number of OH groups of the tin oxide support decreased with increase in vanadia content (Figure 3). This considerable decrease in the concentration of a particular OH groups of the support upon impregnation with an active component is an indication of a strong interaction between the support and the active component (8). The behaviour of levelling off at a certain loading is also a strong indication of the completion of a monolayer coverage of the active component on the surface of the support material (21).

The technique of solid-state ^{51}V NMR used to characterize supported vanadium oxide catalysts has been recently identified as a powerful tool (22, 23). NMR is well suited for the structural analysis of disordered systems, such as the two-dimensional surface vanadium-oxygen complexes to be present on the surfaces, since only the local environment of the nucleus under study is probed by this method. The ^{51}V nucleus is very amenable to solid-state NMR investigations, because of its natural abundance (99.76%) and favourable relaxation characteristics. A good amount of work has already been reported on this technique (19, 20, 22, 23). Similarly, the development of MAS technique has made ^1H NMR an another powerful tool for characterizing Brönsted acidity of zeolites and related catalysts. In addition to the structural information provided by this method direct proportionality of the signal intensity to the number of contributing nuclei makes it a very useful technique for quantitative studies.

ESR spectra of reduced V_2O_5/SnO_2 catalysts are shown in Figure 4. Similar spectra were obtained for unreduced catalysts, but with less intensity of the peaks. Well resolved spectra with hyperfine splittings observed are an indication of the presence of isolated V^{4+} (or VO^{2+})

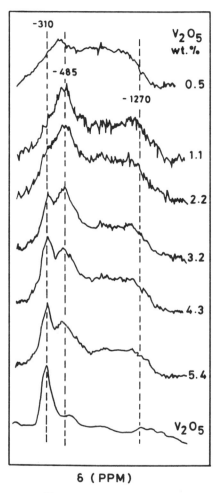

6 (PPM)

Figure 2. Solid-state ^{51}V NMR spectra of V_2O_5 and V_2O_5/SnO_2 catalysts (Reproduced with permission from ref. 20. Copyright 1990 Taylor & Francis, Ltd.)

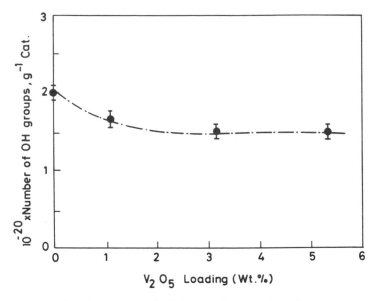

Figure 3. Total number of SnO_2 surface hydroxyl groups plotted as a function of V_2O_5 loading (Reproduced with permission from ref. 20. Copyright 1990 Taylor & Francis, Ltd.)

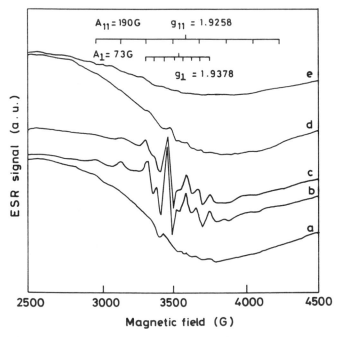

Figure 4. ESR spectra of reduced V_2O_5/SnO_2 catalysts obtained at ambient temperature : a, 1.1; b, 2.2; c, 3.2; d, 5.4; e, 6.4 wt.% V_2O_5 respectively.

species in distorted oxygen environment because the spin of the unpaired electron associated with V^{4+} ($3d^1$) interacts with the nuclear magnetic moment of ^{51}V nuclei (I = 7/2) to give rise to eight parallel and eight perpendicular components of hyperfine splittings, with its intensity being proportional to the quantity of the species (9).   Calculated spectral parameters obtained here indicate that the resultant structures of the surface V-oxide species after reduction are slightly different from the structures of unreduced catalysts.   The presence of hyperfine splittings with high intensities for 2.2 and 3.2 wt.% catalysts suggest that these catalysts have a higher number of contributing nuclei (V^{4+} or VO^{2+}) than the other compositions.   This can be the result of strong interaction between V-oxide units and SnO_2 support or partial substitution of VO_x into SnO_2 lattice.   The enhanced intensity of the hyperfine spectrum at 3.2 wt.% loading is again in agreement with the other results.   The absence of high intensity hyperfine spectra at higher loadings may be due to the presence of weakly interacting microcrystalline vanadia which on reduction yields V^{3+}, an ESR inactive species.

The ESR g values ($g_{\parallel}$ = 1.9258 and $g_{\perp}$ = 1.9378) for the reduced 3.2 wt.% catalyst suggest that V-oxide structure is in between the distorted tetrahedral or square pyramidal (where $g_{\perp} > g_{\parallel}$) (24) geometry. Many authors studied the ESR spectra of V-oxide species supported on a variety of carriers (9, 24, 25) and indicated that the surface VO_x structure is highly sensitive to V_2O_5 loading, nature and structure of the support oxide, and to the variety of pretreatments.   Our ESR results on V_2O_5/SnO_2 catalysts are also in line with the published literature.

Partial oxidation of methanol was used as a test reaction to study the oxidation activity of the catalysts in relation to vanadia loadings. These studies were carried out at a low temperature (175°C) in order to minimise the formation of higher order products like   dimethoxy methane, methyl formate etc. The activity and selectivity results obtained using various catalysts are presented in Table 1. Here again, the conversion of methanol increased with vanadia loading upto 3.2 wt.% and then declined with further increase in the vanadia content. The total conversion of methanol plotted as a function of oxygen uptake is shown in Figure 5.   It clearly demonstrates that methanol oxidation activity is directly proportional to the amount of oxygen chemisorbed at -78°C (26). Similar results were also obtained in our earlier work with vanadia catalysts supported on alumina, zirconia etc. (6, 10, 17, 26).   Hence, based on these observations it was proposed that the coordinatively unsaturated sites of the reduced vanadium oxide patches are the locations for the dissociative chemisorption of oxygen at -78°C which also happen to be the active sites for the catalytic oxidation activity.

Recently, Feil et al (27) and Chung and coworkers (28) proposed mechanisms for the oxidation of methanol, which involves the dissociative adsorption of CH_3OH (CH_3O^- + H^+) to form surface methoxide ions. The next step in the catalytic cycle is the abstraction of a methyl hydrogen by a surface oxygen. This is followed by a rapid intramolecular rearrangement and the desorption of formaldehyde as well as other products. This mechanism has also been widely accepted (29). A correlation as observed in Figure 5 indicates clearly that the CUS are the exact locations for the initial dissociative adsorption of methanol, where oxygen too adsorbs selectively at -78°C.   Hence, a direct correlation between oxygen uptake and the total activity for the partial oxidation of methanol is observed.   This is in agreement with the observation

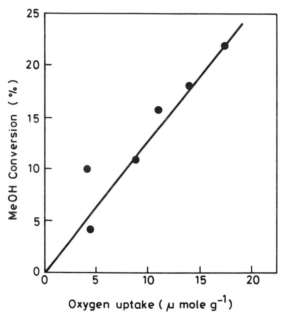

Figure 5.   Methanol oxidation activity plotted as a function of O_2 uptake.

of Chung et al (28) that oxygen vacancies are the active sites and the 'methoxy' species are reaction intermediates. It may be mentioned that active sites (oxygen vacancies) are greatly influenced by geometric and electronic factors and are responsible for the formation of different products. Formaldehyde, and CO are produced mainly from methoxyl species adsorbed on terminal oxygen vacancy ($M=O$) sites and the bridged oxygen vacancy sites are responsible for the formation of high order products. It is further stated that during reaction, the terminal oxygen vacancy is more abundant than the bridged oxygen vacancy, because the bridged oxygen vacancy is reoxidized more easily than the terminal oxygen vacancy.

Conclusions

Monolayer coverage of vanadium oxide on tin oxide support was determined by a simple method of low temperature oxygen chemisorption and was supported by solid-state NMR and ESR techniques. These results clearly indicate the completion of a monolayer formation at about 3.2 wt.% V_2O_5 on tin oxide support (30 m^2g^{-1} surface area). The oxygen uptake capacity of the catalysts directly correlates with their catalytic activity for the partial oxidation of methanol confirming that the sites responsible for oxygen chemisorption and oxidation activity are one and the same. The monolayer catalysts are the best partial oxidation catalysts.

Acknowledgements

The author wishes to thank the Institute of Catalysis, Novosibirsk, Russia for providing solid-state NMR results and Dr. A.V. Rama Rao, Director, IICT, Hyderabad, India for giving permission to publish this work.

Literature Cited

(1)   Hucknall, D.J.; Selective Oxidation of Hydrocarbons, Academic Press, London, 1974.
(2)   Reddy, B.N.; Reddy, B.M.; Subrahmanyam, M.; J. Chem. Soc. Faraday Trans. 1991, 87, 1649 and references therein.
(3)   Janssen, F.J.J.G.; Van den Kerkhof, F.M.G.; Bosch, H.; Ross, J.R.H.; J. Phys. Chem. 1987, 91, 5921.
(4)   Japan 41,273, 1974; Japan 34,138, 1973; Ger. 1,135,883, 1962; Dutch 64,824, 1949; Brit. 882,089, 1959.
(5)   Bond, G.C.; Tahir, S.F. Appl. Catal. 1991, 71, 1 and references therein.
(6)   Nag, N.K.; Chary, K.V.R.; Reddy, B.M.; Rao, B.R.; Subrahmanyam, V.S.; Appl. Catal. 1984, 9, 225.
(7)   Reddy B.M.; Chary, K.V.R.; Subrahmanyam, V.S.; Nag, N.K.; J. Chem. Soc. Faraday Trans. 1985, 81, 1655.
(8)   Reddy, B.M.; Rao, K.S.P.; Mastikhin, V.M.; J. Catal. 1988, 113, 556.
(9)   Chary, K.V.R.; Reddy, B.M.; Nag, N.K.; Subrahmanyam, V.S.; Sunandana, C.S.; J. Phys. Chem. 1984, 88, 2622.
(10) Reddy, B.M.; Narsimha, K.; Rao, P.K.; Mastikhin, V.M.; J. Catal. 1989, 118, 22.
(11) Okada, F.; Satsuma, A.; Furuta, A.; Miyamoto, A.; Hattori, T.; Murakami, Y.; J. Phys. Chem. 1990, 94, 5900.

(12)   Parekh, B.S.; Weller, S.W.; J. Catal. **1977**, 47, 100.
(13)   Nag, N.K.; Chary, K.V.R.; Rao, B.R.; Subrahmanyam, V.S.; Appl.
       Catal. **1987**, 31, 73.
(14)   Narsimha, K.; Sivaraj, Ch.; Reddy, B.M.; Rao, P.K.; Indian J.
       Chem. **1989**, 28A, 157.
(15)   Bond, G.C.; Zurita, J.P.; Flamerz, S.; Gellings, P.J.; Bosch, H.;
       Van Ommen, J.G.; Kip, B.J.; Appl. Catal. **1986**, 22, 361.
(16)   Baiker, A.; Dollenmeier, P.; Glinski, M.; Reller, A.; Appl. Catal.
       1987, 35, 351.
(17)   Chary, K.V.R.; Rao, B.R.; Subrahmanyam, V.S.; Appl. Catal.
       1991, 74, 1.
(18)   Oyama, S.T.; Went, G.T.; Lewis, K.B.; Bell, A.T.; Somorjai, G.A.;
       J. Phys. Chem. **1989**, 93, 6786.
(19)   Sobalik, Z.; Lapina, O.B.; Novogorodova, O.N.; Mastikhin, V.M.;
       Appl. Catal. **1990**, 63, 191.
(20)   Narsimha, K.; Reddy, B.M.; Rao, P.K.; Mastikhin, V.M.; J. Phys.
       Chem. **1990**, 94, 7336.
(21)   Reddy, B.M.; Mastikhin, V.M.; In Proceedings of the 9th Inter-
       national Congress on Catalysis; Phillips, M.J.; Ternan, M.; Eds.;
       Chemical Institutes of Canada : Ottawa, 1988, Vol. 1, p. 82.
(22)   Eckert, H.; Wachs, I.E.; J. Phys. Chem. **1989**, 93, 6796.
(23)   Le Costumer, L.J.; Taouk, B.; Le Meur, M.; Payen, E.; Guelton,
       M.; Grimblot, J.; J. Phys. Chem. **1988**, 22, 1230.
(24)   Cavani, F.; Centi, G.; Foresti, E.; Trifiro, F.; J. Chem. Soc.
       Faraday Trans. **1988**, 84, 237.
(25)   Sharma, V.K.; Wokaun, A.; Baiker, A.; J. Phys. Chem. **1986**,
       90, 2175.
(26)   Reddy, B.M.; Narsimha, K.; Sivaraj, C.; Rao, P.K.; Appl. Catal.
       1989, 55, 1.
(27)   Feil, F.S.; Van Ommen, J.G.; Ross, J.R.H.; Langmuir, **1987**, 3, 668.
(28)   Chung, J.S.; Miranda, R.; Bennett, C.O.; J. Catal. **1988**, 114,
       398.
(29)   Yang, T.J.; Lunsford, J.H.; J. Catal. **1987**, 103, 55.

RECEIVED November 6, 1992

Chapter 16

V–P–O Catalysts in *n*-Butane Oxidation to Maleic Anhydride

Study Using an In Situ Raman Cell

J. C. Volta[1], K. Bere[1], Y. J. Zhang[1], and R. Olier[2]

[1]Institut de Recherches sur la Catalyse, Centre National de la Recherche Scientifique, 2 avenue Albert Einstein, 69626 Villeurbanne Cédex, France
[2]Laboratoire de Physicochimie des Interfaces, Centre National de la Recherche Scientifique, Ecole Centrale de Lyon, 36 avenue Guy de Collongue, B.P. 163, 69131 Ecully Cédex, France

The evolution of the structure of four vanadyl phosphate hemihydrates has been studied and then used as catalysts for n-butane oxidation using an <u>in-situ</u> Raman cell and ^{31}P MAS-NMR. The catalytic performance for maleic anhydride formation can be explained by their transformation into α_{II} and δ $VOPO_4$ on the $(VO)_2P_2O_7$ matrix as evidenced by Raman spectroscopy during the course of the reaction. The best catalytic results correspond to a limited number of V^{5+} sites forming small domains situated on and strongly interacting with the (100) $(VO)_2P_2O_7$ crystalline face. This is therefore another example of a Structure Sensitive Reaction.

It is now considered, by most groups working in this area, that vanadyl pyrophosphate $(VO)_2P_2O_7$ is the central phase of the Vanadium Phosphate system for butane oxidation to maleic anhydride (*1*). However the local structure of the catalytic sites is still a subject of discussion since, up to now, it has not been possible to study the characteristics of the catalyst under reaction conditions. Correlations have been attempted between catalytic performances obtained at variable temperature (380-430°C) in steady state conditions and physicochemical characterization obtained at room temperature after the catalytic test, sometimes after some deactivation of the catalyst. As a consequence, this has led to some confusion as to the nature of the active phase and of the effective sites. $(VO)_2P_2O_7$, V (IV) is mainly detected by X-Ray Diffraction.

0097–6156/93/0523–0217$06.00/0

However, the use of techniques which analyze both the short and the long range orders of the VPO materials, like RED of X-Rays (2), ^{31}P and ^{51}V MAS-NMR (3,4), showed the possible participation of some V (V) structures to the reaction. Previously, there has been some ambiguity insofar as these structures should be a consequence of the reoxidation of the starting $VOHPO_4$, 0.5 H_2O precursor or of the basic $(VO)_2P_2O_7$ matrix and hence do not intervene directly in the reaction mechanism of butane oxidation to maleic anhydride. The possible role of a limited amount of V(V) sites to control the selectivity to maleic anhydride was previously postulated (5).

With the aim to study the characteristics of VPO catalysts in the course of butane oxidation to maleic anhydride together with a simultaneous evaluation of the catalytic performance, we have used Raman spectroscopy which is a very sensitive probe for determining the presence of $VOPO_4$-like entities together with $(VO)_2P_2O_7$. An *in situ* Laser Raman Spectroscopy (LRS) cell was built in our laboratory (6). In the corresponding publication (6), the preparation and the characterization by XRD, ^{31}P and ^{51}V NMR of the different VPO phases, $(VO)_2P_2O_7$, α_{II}, β, γ and δ $VOPO_4$ has been described. The LRS spectra were registered up to 430°C in butane/air atmosphere.

In this communication, we compare VPO catalysts which differ by their conditions of preparation, considering both their LRS spectra registered under reaction conditions and the corresponding catalytic results. LRS data are discussed in relation with results for n-butane oxidation to maleic anhydride.

Experimental

A schematic diagram of the *in situ* cell, built at the Institut de Recherches sur la Catalyse and used for the Laser Raman study of the materials has been described elsewhere (6). It is presented in Figure 1. It is made of three parts of stainless steel with different functions. The solid to be examined under reaction conditions was placed in the lower part on a glass sintered disc. For the present study, 1.5 g of VPO catalyst was used for each run. The temperature was controlled by a thermocouple placed at the centre of the catalyst powder. The reaction gases flowed through the powder in the middle of this lower part which was heated by three thermoregulated fingers. The upper part held a glass window transparent to the laser beam. The gaseous effluent from the cell was analyzed by gas-chromatography. The tightness of the cell was ensured by two gold rings. The composition and flow rate of the reacting gases (2.4% butane/air) were controlled by two flow meters. Experiments were done with a flow

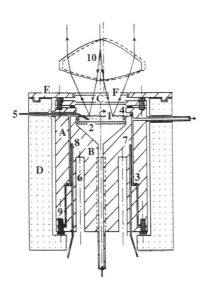

Figure 1 : Λ/Chamber upper part B/Chamber lower part C/Glass window
D/ Isolating shell E/Thermal screen F/Anticaloric filter

1. Catalyst 2. Sintered glass 3 and 4. Tightening rings 5. Temperature controller 6. Heating finger lodging 7. Chamber heating controller 8. Thermal security 9. Mounting bolts 10. Rotating lens.

rate of 3.6 l. h^{-1}. Detection of evolved gases was done by a FID detector. CO and CO_2 amounts were determined by conversion to CH_4 on a $Pt/\gamma Al_2O_3$ catalyst working at 300°C. It was thus possible to analyze all the gases with the only FID detector. Two columns were used in parallel: a 1m 1/4 in. Porapak Q column to separate CO and CO_2 which were further transformed into CH_4, whereas butane, acetic and acrylic acids and maleic anhydride were separated on a 3m 1/8 in. Lac 2R (13%)/H_3PO_4 (2.5%) on Gas ChromQ column. The two columns were heated at 140°C. The tube connecting the cell to the chromatograph was heated at 120°C in order to avoid any condensation of the reaction products. Helium was the carrier gas.

Raman spectra were recorded on a DILOR OMARS 89 spectrophotometer equipped with an intensified photodiode array detector. The emission line at 514.5 nm from Ar^+ ion laser (SPECTRA PHYSICS, Model 164) was used for excitation. The power of incident beam on the sample was 36 mW. Time of acquisition was adjusted according to the intensity of the Raman scattering. LRS spectra were recorded during the activation of the different $VOHPO_4$, 0.5 H_2O precursors up to 440°C in the butane/air atmosphere, with a simultaneous measurement of the butane conversion and selectivity to maleic anhydride. Temperature was maintained at 420°C for 16 hours. 1000 spectra were accumulated for these three periods in order to improve signal to noise ratio. The wavenumber values obtained from the spectra were accurate to within about 2 cm^{-1}. In order to reduce both thermal and photodegradation of samples, the laser beam was scanned on the sample surface by means of a rotating lens in the same way as described in ref. (7). The scattered light was collected in the back scattering geometry. The presence of the different VPO structures in catalytic conditions was determined from the LRS spectra of the pure VPO reference phases recorded at the same temperature and published elsewhere (6).

The ^{31}P spectra were recorded on a Brucker MSL-300 spectrometer operating at 121.4 MHz s. The ^{31}P NMR spectra were obtained under MAS conditions by use of a double bearing probehead. A single pulse sequence was used in all cases and the delays were chosen allowing the obtention of quantitative spectra (typically the pulse width was 2 ms (10°) and the delay was 10 to 100 s. The number of scans was 10 to 100. Spectra were refered to external H_3PO_4 (85%).

Four different $VOHPO_4$, 0.5 H_2O precursors were studied. They were prepared in organic medium according to the EXXON method with isobutanol as reducing reagent (8). They differed by the starting vanadium material (Guilhaume N. and Volta J.C., unpublished results). After the LRS cell examination, the sample powder was stored under dry argon atmosphere to avoid any hydration, prior to a XRD and ^{31}P NMR examination.

Results and Discussion

Figure 2 shows the LRS spectra (1000 accumulations) of the four VPO catalysts recorded at 420°C under reaction conditions with the corresponding catalytic results at stationary state obtained in the LRS cell . It is noteworthy that both butane conversion and MA selectivity are lower than those obtained on the same catalysts in our laboratory reactor for butane oxidation. Note that the thermal characteristics of the LRS cell are highly different from those of a classical tubular reactor for which the catalytic bed is made of cylindrical pellets and is settled in a salt bed. It is likely that there should be a large thermal gradient in the perpendicular direction of the cell from the thermoregulated fingers up to the glass window, so that, in spite of the thinness of the catalytic bed, the sample powder may work at temperature lower than 420°C, as measured. These reasons may explain the poor results obtained in the LRS cell.

It appears that the four VPO catalysts differ in the LRS spectra by the relative distribution of the three $(VO)_2P_2O_7$, α_{II} and δ $VOPO_4$ phases. Raman spectroscopy is highly sensitive to the detection of the $VOPO_4$ structures which are unambiguously observed at 994 cm^{-1} and 1015 cm^{-1} for α_{II} and δ $VOPO_4$ respectively (6). Their presence is confirmed by the XRD spectra (Figure 3) and the ^{31}P NMR spectra (Figure 4) of the catalysts recorded at room temperature after the LRS cell run. The signal observed at - 20.3 ppm is characteristic of α_{II}, while the signal at - 13.3 ppm in the absence of β $VOPO_4$ (Figure 3) is characteristic of δ $VOPO_4$ (6). Note the absence of any broad signal which is indicative of a low V^{5+}/V^{4+} interaction between the large $VOPO_4$ domains and the $(VO)_2P_2O_7$ structure. For the presence of α_{II} and δ $VOPO_4$, there is a very good agreement between the LRS informations obtained at 420°C and the XRD and ^{31}P NMR informations obtained at room temperature. The coexistence of the two α_{II} and δ $VOPO_4$ phases at 420°C is logical insofar as it was previously observed that δ $VOPO_4$ is partially transformed into α_{II} $VOPO_4$ in catalytic conditions (6).

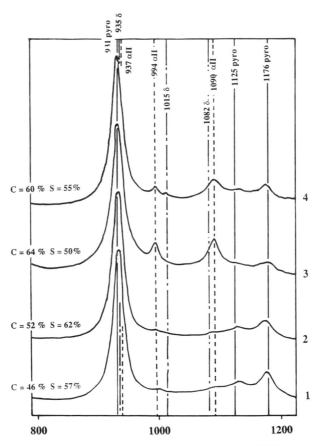

Figure 2 : LRS spectra of the VPO catalysts recorded at 420°C in the *in situ* cell under steady state conditions.

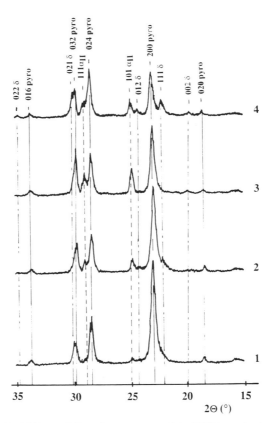

Figure 3 : X-Ray diffraction patterns of the VPO catalysts at room
temperature after the *in situ* Raman cell run.

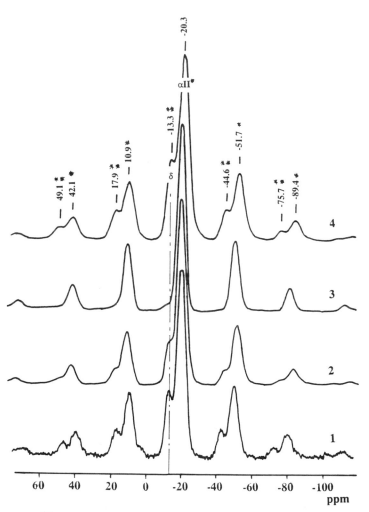

Figure 4 : ^{31}P NMR spectra of the VPO catalysts at room temperature after the *in situ* Raman cell run.

* α_{II} VOPO$_4$; ** δ VOPO$_4$

From Figure 2, it appears that higher selectivity for MA at 420°C is observed for catalysts 1 and 2 for which $(VO)_2P_2O_7$ is principally observed with a small extent of α_{II} $VOPO_4$. Catalysts 3 and 4 which highly develop α_{II} and δ $VOPO_4$ phases together with $(VO)_2P_2O_7$ are less selective but more active. The lower relative intensity of the (200) line (Figure 3) is associated with a higher development of the α_{II} and δ lines, so that we can postulate that nucleation of these two crystalline phases is not associated with the basal (100) $(VO)_2P_2O_7$ face but is more in relation with the lateral faces of the corresponding crystals.

In order to study the influence of the atmosphere of activation both on the catalytic results and on the physicochemical characteristics of the VPO catalyst, a "$(VO)_2P_2O_7$" catalyst (600 mg) was prepared by dehydration of a classical $VOHPO_4$, 0.5 H_2O precursor under an oxygen-free argon atmosphere at 440°C during 26 hours until the stationary state was reached. This catalyst was then tested in a classical reactor with 2.4% butane/air from 300°C up to 450°C for two consecutive runs. Catalytic performances are given in Figure 5. It is clear that both butane conversion and MA selectivity are improved in the second run when compared to the first run. After the second run, catalytic performances approach those of the VPO catalyst prepared from the same $VOHPO_4$, 0.5 H_2O precursor treated directly in the 2.4% butane/air atmosphere. Figure 6 shows the evolution of the ^{31}P NMR spectra before catalysis (Figure 6a), after the first run (Figure 6b) and after the second run (Figure 6c). The first run implies both an increase of the number of the V^{5+} ions interacting with V^{4+} and an increase of the size of their domains (decrease of the band width) (Figure 6b), while the second run implies a redispersion of the V^{5+} ions and a decrease of their number (Figure 6c). This evolution is correlated with the modification of the X-Rays spectra of the catalysts examined before and after the two consecutive runs (Zhang, Y.J., Sneeden, R. and Volta, J.C., Catalysis Today, in press). The main feature is a changing of the profile of the (200) $(VO)_2P_2O_7$ line at around 23° 2Θ, the width of which diminishes from spectra of "$(VO)_2P_2O_7$" to spectra of "$(VO)_2P_2O_7$"(1st run) and spectra of "$(VO)_2P_2O_7$" (2nd run). It is difficult to give a clear explanation to this evolution but we consider that it could be the result of an increase of the organization

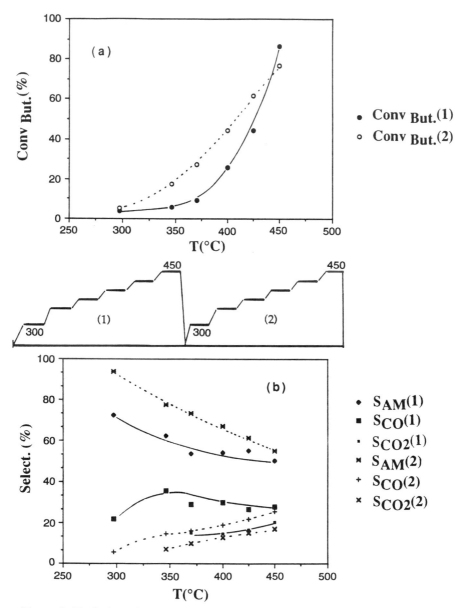

Figure 5 : Evolution of the performances of the "$(VO)_2P_2O_7$" catalyst activated under argon atmosphere after two succesive catalytic runs (300-450°C).

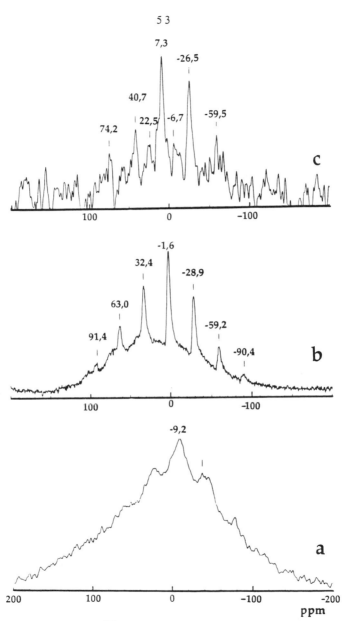

Figure 6 : Evolution of the ^{31}P MAS-NMR spectra of the "$(VO)_2P_2O_7$" catalyst activated under argon atmosphere after two successive catalytic runs (300-450°C).

of the $(VO)_2P_2O_7$ structure (a progressive disappearance of the (201) $(VO)_2P_2O_7$ line is simultaneously observed) which could be connected with the modification of the dispersion of the V^{5+} ions observed by ^{31}P NMR.

The fact that only the profile of the (200) $(VO)_2P_2O_7$ line is perturbated may be explained by suggesting that V^{5+} species principally affect the corresponding (100) basal plane of $(VO)_2P_2O_7$. The improvement of the catalytic performances may be thus associated with a subsequent variation of the V^{5+}/V^{4+} local interaction : the number of the V^{5+} sites interacting locally with V^{4+} sites changes depending on conditions of activation so that there should be a specific ratio for the best catalyst corresponding to small V^{5+} domains on $(VO)_2P_2O_7$ with low interaction.

Conclusions

This study has resulted in interesting informations concerning the active sites of the VPO catalysts for n-butane oxidation to maleic anhydride being obtained. The study of VPO catalysts in the course of n-butane oxidation by an *in-situ* Raman cell has shown that catalytic performances can be explained by the presence of the α_{II} and δ $VOPO_4$ phases on the $(VO)_2P_2O_7$ matrix. However, if $(VO)_2P_2O_7$ is the basic phase for this reaction, the present study confirms the participation of V^{5+} entities detected by ^{31}P NMR. Superficial V^{4+}/V^{5+} distribution is determined by the atmosphere of treatment of the catalysts. This was evidenced both by the catalytic and the physicochemical evolution of a $(VO)_2P_2O_7$ catalyst from the activation of the $VOHPO_4, 0.5 H_2O$ precursor under an oxygen-free argon atmosphere of calcination to the butane/air atmosphere of catalysis. Best catalytic results correspond to a limited number of V^{5+} sites forming small domains with a strong interaction with the $(VO)_2P_2O_7$ matrix. From the evolution of the X-Ray diffraction spectra, it can be postulated that these domains affect principally the basal (100) crystal face. We previously observed that the V^{4+}/V^{5+} distribution depended also on the morphology of the $VOHPO_4, 0.5 H_2O$ precursor which could be determined by the conditions of its preparation (9).

The oxidation of butane to maleic anhydride appears as another example of a Structure Sensitive Reaction (10). The formation of maleic anhydride (MA) could occur on the basal (100) $(VO)_2P_2O_7$ face as it was previously proposed (11), but with a participation of a suitable number of V^{5+} entities. The local superficial V^{5+}/V^{4+}

distribution in this face should control the catalytic results. Side faces of $(VO)_2P_2O_7$ located in the <100> direction which appear to be faces for nucleation and growing of α_{II} and δ VOPO$_4$ phases, should be less selective for MA. The presence of α_{II} VOPO$_4$ is partially a consequence of the transformation of δ VOPO$_4$ under the reaction conditions (*6*). Depending on these ones, a redispersion of the VOPO$_4$ structures should occur from the side faces to the basal (100) $(VO)_2P_2O_7$ face. A model of the catalyst is proposed in Figure 7.

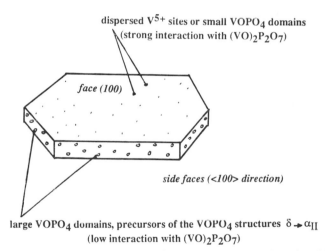

Figure 7 : Model for the transformation of n-butane into maleic anhydride on the "$(VO)_2P_2O_7$" catalyst

Acknowledgments

Authors are indebted to Dr. F. Lefebvre for ^{31}P NMR experiments and for the interpretation of the corresponding results. They thank Dr. J.C. Védrine for fruitfull discussions.

Literature cited

(1) Centi, G., Trifiro, F., Ebner, J.R. and Franchetti, V.M., Chem. Rev., 88, 55, (1988).
(2) Bergeret, G., Broyer, J.P., Gallezot, P., Hecquet, G. and Volta, J.C. J. Chem. Soc., Chem. Comm., 825, (1986).
(3) David, M., Lefebvre, F. and Volta, J.C., 11th Iberoamericano Symposium on Catalysis, Guanajuato, 1988, IMP Edit, p. 365.
(4) Harrouch-Batis, N., Batis, H., Ghorbel, A., Védrine, J.C. and Volta, J.C., J. Catal, 128, 248, (1991).
(5) Cavani, F., Centi, G., Trifiro, F., Grasselli, R.K., Preprints ACS Symposium of Division of Petroleum Chemistry, "Hydrocarbon Oxidation", New Orleans Meeting, Sept. 1987.
(6) Ben Abdelouahab, F., Olier, R., Guilhaume, N., Lefebvre, F. and Volta, J.C., J. Catal., 134, 151, (1992).
(7) Zimmerer, N., Kiefer, W., Appl. Spectrosc., 28, 279, (1974).
(8) Yang, T.C., Rao, K.K., Der Huang, I., Exxon Res. Eng. Co., US. Patent, 4.392.986, (1987).
(9) Guilhaume, N., Roullet, M., Pajonk, G., Grzybowska, B. and Volta, J.C., " New Developments in Selective Oxidation by Heterogeneous Catalysis, Stud. in Surf. Sci. and Catal., P. Ruiz and B. Delmon Ed., Elsevier, Amsterdam, Vol. 72, p.255, (1992).
(10) Volta, J.C. and Moraweck, B., J. Chem. Soc., Chem. Commun., 338, (1980).
(11) Ziolkowski, J., Bordes, E. and Courtine, P., J. Catal., 122, 126, (1990).

RECEIVED November 6, 1992

Chapter 17

Vanadium Catalysts Supported on Titanium (Anatase) Characterized by Ammonia and Low-Temperature Oxygen Chemisorption

P. Kanta Rao and K. Narasimha[1]

Catalysis Section, Physical and Inorganic Chemistry Division, Indian Institute of Chemical Technology, Hyderabad 500 007, India

Being one of the reactants in some of the important selective oxidation and ammoxidation reactions, NH_3 can be a simple probe molecule to investigate the chemical nature of V_2O_5-TiO_2 catalysts. Catalysts with different V_2O_5 loadings on anatase were prepared by aqueous impregnation. The catalyst with 6.5 wt% V_2O_5 has shown maximum NH_3 chemisorption capacity at 150°C. Oxygen uptake at -78°C is also maximum for 6.5 wt% V_2O_5-TiO_2. The activity of these catalysts for the vapour-phase oxidation of methanol to formaldehyde correlates with NH_3 and O_2 uptakes. From these results it appears that NH_3 at 150°C and O_2 at -78°C are chemisorbed on identical active sites. Thus, NH_3 chemisorption can be used as a simple probe technique to characterise V_2O_5-TiO_2 (anatase) catalysts.

Vanadium oxide dispersed on supporting oxides (SiO_2, Al_2O_3, TiO_2, etc.) are frequently employed as catalysts in reactions like partial oxidation and ammoxidation of hydrocarbons, and NO_x reduction. The modifications induced on the reactive properties of transition metal oxides like V_2O_5 when they are supported on an oxide carrier has been the subject matter of recent study. There is much evidence showing that the properties of a thin layer of a transition metal oxide interacting with the support are strongly modified as compared to the properties of the bulk oxide (1-3). In the recent past, increasing attention has been focussed

---

[1]Current address: The Leverhulme Center for Innovative Catalysis, University of Liverpool, Liverpool L69 3BX, United Kingdom

0097–6156/93/0523–0231$06.00/0

towards the study of the physicochemical and catalytic properties of V_2O_5-TiO_2 system. V_2O_5 when supported on TiO_2 (anatase) exhibits much superior activity and selectivity compared to bulk V_2O_5 in the partial oxidation of hydrocarbons (4-10). The modifying effect of TiO_2 (anatase) on the supported V_2O_5 phase has been ascribed to the formation of a monolayer of a surface vanadia species coordinated to the TiO_2 support (6-12). Vanadium oxide in excess of the monolayer coverage is converted to disordered V_2O_5 phase (9,13) and, with increased V_2O_5 content, to crystalline V_2O_5 (14). The surface oxovanadium species are the active sites in the V_2O_5-TiO_2 (anatase) catalysts for the selective oxidation of o-xylene to phthalic anhydride (7,9). A moderate amount of crystalline V_2O_5 in excess of monolayer does not affect their catalytic performance in the o-xylene (9) or toluene (8) oxidation, but perhaps serves as a source of VO_x to maintain the monolayer.

The V_2O_5-TiO_2 system has been extensively investigated by a variety of techniques such as extended X-ray absorption fine structure (EXAFS) (9), high resolution electron microscopy (HREM) and X-ray absorption near edge structure (XANES) (12), X-ray photoelectron spectroscopy (XPS) (15-19), Laser Raman Spectroscopy (LRS) (20,21), electron spin resonance (ESR) (22,23) and oxygen chemisorption (4,24) to characterise surface VO_x species. Based on these studies it has been concluded that the vanadia monolayer on TiO_2 (anatase) is composed of tetrahedral vanadate species involving two terminal double-bonded and two bridging oxygen atoms (9,11,12,25,26). However, recent studies have shown that the terminal oxygens are hydroxylated, permitting formation of more than one monolayer with the grafting method (13,26,27). Vanadia monolayer, disordered V_2O_5 phase consisting of V=O...V and small crystallites, in the V_2O_5-TiO_2 (anatase) system are reduced at lower temperatures (450°C) to V^{3+} (11,14) and the reoxidation is also faster and complete at low temperatures (14). There is, thus, a general agreement that the optimal catalytic performance is shown by a catalyst where the amount of vanadium present corresponds to that necessary to form a monolayer of vanadium oxide on the support surface.

In this communication, the results of a systematic study of ammonia chemisorption on V_2O_5-TiO_2 (anatase) catalysts of different vanadia loading is reported. Low temperature oxygen chemisorption is also utilized to determine the monolayer loading of V_2O_5 on TiO_2 (anatase). Partial oxidation of methanol to formaldehyde is studied as a model reaction on these catalysts and the activities of the catalysts are correlated with NH_3 and O_2 uptakes.

EXPERIMENTAL

Catalyst Preparation. About 500 ml of anhydrous titanium tetra chloride (Fluka, 99.9%) was first stabilized in 1.5 l of concentrated hydrochloric acid (AR, Loba, 30% wt/wt). The resulting liquid was mixed in 10 l of distilled water. To this clear colourless solution dilute ammonia was added slowly dropwise up to pH 6.5-7.0. The precipitated hydroxide was filtered and washed several times with distilled water to remove chlorides. The resulting white precipitate was dried at 120°C for 24 h and then calcined at 450°C for 6 h. The X-ray data showed that the calcined TiO_2 is in anatase form. The surface area of the TiO_2 (anatase) determined by N_2 adsorption at liquid N_2 temperature was found to be 92.6 $m^2 g^{-1}$.

Catalysts of different V_2O_5 loadings from 3.5 to 8.0 wt% were prepared by adding requisite amounts of aqueous ammonium meta vanadate solution to calculated amounts of titania (anatase) support. Excess water was evaporated on a water bath and then oven dried at 120°C for 16 h. All the samples were finally calcined at 480°C for 6 h in air.

Ammonia Chemisorption. Ammonia gas has been widely employed as a basic adsorbate to count the number and strength of acid sites on various solid surfaces (28,29). The nature and strength of these sites may relate the activity and selectivity character of the catalysts. Further, all acid sites on catalyst surface are easily accessible to the small molecules of NH_3 (kinetic dia 0.26 nm) and these molecules also selectively adsorb in the presence of sites of different strengths (30). Ammonia being a reactant in selective catalytic reduction of NO_x gases, in ammoxidation reactions and a probe molecule for acidic sites, it can be an inexpensive alternative to investigate the chemical nature of surface of TiO_2 (anatase) supported V_2O_5 catalysts and its change due to the loading by different amounts of V_2O_5.

Ammonia chemisorption experiments were carried out at different temperatures ranging from 25 to 400°C to establish conditions for the adsorption studies on the catalysts. The NH_3 chemisorption experiments were performed on an all glass high vacuum system according to the procedure described by Kanta Rao et al. (31). In a typical experiment, about 0.3 g of the catalyst sample was placed in a glass adsorption cell and evacuated (10^{-6} Torr) at 150°C for 2 h. The first adsorption isotherm representing both reversible and irreversible ammonia adsorption, was generated, allowing 20 min equilibration time at each pressure. Then the catalyst was evacuated at the temperature of adsorption for 1 h to remove reversibly adsorbed ammonia. After this, a fresh second isotherm representing only the reversibly adsorbed ammonia was generated in an identical manner. From the difference between the first and second adsorption isotherms, the irreversibly chemisorbed ammonia was calculated.

Low Temperature Oxygen Chemisorption. The same volumetric high vacuum system used for NH_3 chemisorption with the facility for reducing the samples in situ by flowing hydrogen, was used for the study of oxygen chemisorption. The quantity of chemisorbed oxygen was determined as the difference between two oxygen adsorption isotherms at -78°C according to the procedure described by Nag et al. (32). Prior to the first isotherm, the catalyst sample (about 0.5 g) was reduced for 5 h at 480°C in flowing purified hydrogen (35 cm^3/min), pumped out for 1 h at 480°C and then cooled to -78°C under vacuum (10^{-6} Torr). Between the first and second oxygen adsorption isotherms the sample was evacuated for 1 h at -78°C to remove the physisorbed oxygen from the catalyst surface. Then the BET surface area of the catalyst was determined by N_2 adsorption at -196°C by taking 0.162 nm^2 as the area of cross section of N_2.

XRD. X-ray diffraction patterns were recorded on a Phillips diffractometer using Ni-filtered CuK$_\alpha$ radiation.

Activity Measurements. A microcatalytic reactor interfaced by a six-way gas sampling valve with a gas chromatograph was used to study the vapor phase oxidation of methanol at 150°C. The feed gas consisted of 72, 24 and 4% by volume of nitrogen, oxygen and methanol vapour, respectively. For each run about 0.3 g of catalyst was used and the products were analysed with a 10% carbowax 20M column (2m long). The observed major products were formaldehyde, dimethyl ether and very minute quantities of CO and CO_2.

RESULTS AND DISCUSSIONS

The XRD data of calcined catalysts indicated the presence of the diffraction maxima relative to anatase phase of titania, together with that of V_2O_5 in 8.0 wt% V_2O_5/TiO_2 catalyst. The reflections of V_2O_5 were not found at V_2O_5 loadings upto 6.5 wt%. The other important observation from XRD data is the absence of rutile polymorph transformation from anatase in all loadings of V_2O_5 studied. The absence of characteristic V_2O_5 peaks at vanadia loadings upto 6.5 wt% can be taken as an indication of high dispersion of V_2O_5 on TiO_2 (anatase). This observation agrees well with published work (33). The XRD data confirm the results of low temperature oxygen chemisorption (LTOC) and NH_3 chemisorption results of V_2O_5/TiO_2 (anatase) catalysts.

The effect of chemisorption temperature on the ammonia uptake capacity of 6.5 wt% V_2O_5/TiO_2 is shown in Fig. 1. Ammonia chemisorption capacities increase with temperature upto 150°C and then decrease with further increase up to 400°C. It is worth noting that there is considerable NH_3 uptake even at 400°C. These results are in accordance with the reported literature. A number of studies have been reported on the acidic character of supported transition-metal oxides (22,34-38). Ammonia on V_2O_5 can be either adsorbed in the form of NH_4^+ species on Bronsted acid sites or coordinatively bonded to vanadium ions on Lewis acid sites (39,40). The latter species were observed up to 250°C, however, evidence exists for their occurrence even at higher temperatures of up to 350°C (23). From an FTIR study of adsorbed ammonia on V_2O_5/TiO_2 monolayer catalysts (41) and IR and Raman study of surface acidity of submonolayer and supramonolayer V_2O_5-TiO_2 catalysts (42) it was concluded that both Lewis acid and Bronsted acid sites are present, and are identified as coordinatively unsaturated VO^{2+} ions and V-OH groups, respectively. Otamiri et al. (43) proposes that one ammonia molecule can adsorb on one active site. Dehydroxylation of V_2O_5 and V_2O_5-TiO_2 occurs at temperatures below 277°C and the reduction of vanadium site by dehydroxylation was proposed (23). The increase in NH_3 uptake by V_2O_5-TiO_2 (anatase) with temperature up to 150°C, in the present study, can be attributed to the increase in adsorption sites upon dehydroxylation with temperature. Adsorption of NH_3 on TiO_2 (anatase) supported vanadia and reduced vanadia surface was observed at temperatures of adsorption ranging from 25 to 400°C (44). The reduced catalysts exhibits only Lewis acidity (42). The decrease in NH_3 uptake beyond 150°C may be explained as due to the reaction of some of the vanadia species with NH_3 according to the reaction (44):

$$V_2O_5 + \frac{x}{2} NH_3 \longrightarrow V_2O_{5-x} + \frac{x}{4} N_2O + \frac{3x}{4} H_2O$$ and the possible blocking of the sites by the products N_2O and H_2O.

From Table 1, it can be observed that NH_3 uptake at 150°C is increasing with vanadia loading up to 6.5 wt% and then declines marginally with further increase upto 8.0 wt%. This is in accordance with the findings that the unsupported V_2O_5 exhibits only Bronsted acidity (45) and as the vanadia loading is increased on TiO_2 (anatase) the Lewis and Bronsted acidity is present in varying degree depending on the vanadia loading (42). From XRD, we found that crystalline V_2O_5 was present in 8.0 wt% V_2O_5. Increase of NH_3 uptake with vanadia loading up to monolayer may also be explained as due to the fact that ammonia is believed to be dissociatively adsorbed forming NH_2^- converting an adjacent V=O group to a new acidic V-OH which reacted further with ammonia to form NH_4^+ (46). Another important point to be noted from Table 1 is that the NH_3 uptake on TiO_2 (anatase) is significantly lower compared to the supported catalysts. Ammonia adsorption on TiO_2 is known to be weak and becomes negligible at 200°C although it is appreciable at 100°C (44). Addition of V_2O_5 to TiO_2 (anatase) enhances the acidity of the catalyst significantly.

The conversion of methanol is plotted as a function of NH_3 uptake in Fig. 2. From this figure it appears that ammonia uptake at 150°C reasonably correlates with methanol conversion at 150°C indicating that ammonia uptake is a measure of active sites upon which the reaction takes place. This finding supports the observation of Ai (47) that the oxidation of hydrocarbon is derived

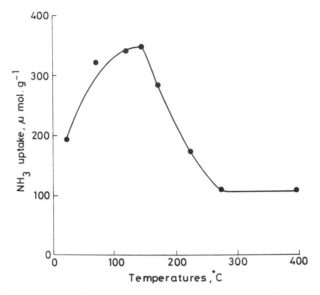

Figure 1. Effect of temperature on ammonia uptake capacity of 6.5 wt.% V_2O_5/TiO_2(A)

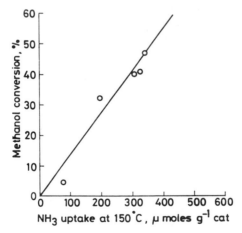

Figure 2. Methanol oxidation activity at 150°C as a function of ammonia uptake at 150°C on V_2O_5/TiO_2(A)

Table 1. Composition, Ammonia and Oxygen Uptakes, Surface Coverage, Active Site Density, Methanol Conversion and Product Selectivities of Catalysts

Catalyst[a]	NH_3 sorb. at 150°C, μ mol/g	O_2 sorb. at -78°C, μ mol/g	BET Surf., m^2/g	% Surf. Cover.[b]	Act.Site Density[c], nm^{-2}	CH_3OH Conv., %	Selectivity, % HCHO	$(CH_3)_2O$
TiO_2	81	-	92.6	-	-	3.7	-	100
3.5 wt%	201	48.1	92.4	15.40	0.627	35.5	96.7	3.3
5.0 wt%	313	84.1	74.6	33.36	1.358	40.1	92.1	7.9
6.5 wt%	344	96.7	73.5	38.93	1.585	46.7	95.1	4.9
8.0 wt%	327	92.6	77.3	35.37	1.422	40.3	97.3	2.7

a Catalysts listed as wt% V_2O_5 on TiO_2 (anatase)

b (Active surface area/BET surface area of reduced catalyst) X 100

c Number of oxygen atoms chemisorbed per unit area of the catalyst from a cooperative action of two functions of a catalyst: an oxidising function and an acidic function. A direct correlation has also been proposed earlier between acid-base properties of titania-vanadia catalysts and their activity and selectivity to the o-xylene oxidation to phthalic anhydride (48).

The chemisorbed ammonia on uncovered surface of TiO_2 is not high as adsorbed ammonia at higher temperatures desorbs with evacuation (44). The active sites for ammonia chemisorption can thus be expected to be surface VO_x species. The supported catalyst has both reducible and unreducible VO_x species as isolated surface VO_x species of coordinatively unsaturated and unreducible V_2O_5 crystallites on the support respectively. The active sites are formulated as a tetrahedral oxohydroxy vanadium complex which shows both Lewis and Bronsted acidity (13). Therefore, it appears that for oxidation and ammoxidation activity the surface of a catalyst should possess both Lewis and Bronsted acidic sites. The active surface of TiO_2 supported V_2O_5 catalysts can be considered as possessing both Lewis and Bronsted acid centres (V=O and V-OH) with electron accepting character. From the results on ammonia chemisorption at 150°C on V_2O_5-TiO_2 (anatase) catalysts it can be inferred that the sites for ammonia chemisorption are reducible and unreducible VO_x species and that the chemisorbed ammonia is directly related to the active site density of VO_x species. The increase of ammonia uptake is an indication of the presence of increased quantity of active sites on the catalyst surface. On the other hand lower value of ammonia uptake for 8 wt% V_2O_5 indicates the presence of lesser number of active sites. It is interesting to note that both ammonia and oxygen uptake values are maximum on catalyst with 6.5 wt% V_2O_5-TiO_2 (anatase).

Thus ammonia chemisorption technique can be used for the estimation of active sites as well as monolayer capacity of TiO_2 (anatase) supported vanadia catalysts.

Ammonia uptake at 150°C, oxygen chemisorption capacities at -78°C along with methanol conversions at 150°C are plotted as a function of vanadia loading in Fig. 3. Ammonia uptake increases with increase in vanadia loading up to 6.5 wt% and then decreases marginally when the vanadia loading increased to 8.0 wt%. Similar trends can be seen with respect to oxygen uptake and methanol conversion values when V_2O_5 loading is increased from 3.5 to 8.0 wt%. Maximum value of O_2 uptake and NH_3 uptake at 6.5 wt% V_2O_5 loading is an indication of surface with maximum number of active sites or complete coverage of active surface area of the support surface by VO_x species. This indicates the formation of vanadium oxide monolayer due to carrier-catalyst interaction (40) in the lower loading region and the formation of crystallites by multilayer deposition of V-oxide on the monolayer at higher V_2O_5 loading.

The O_2 uptake values, BET surface areas of reduced catalysts, the derived active site densities and vanadia surface coverage have been compiled in the Table 1. Oxygen uptake

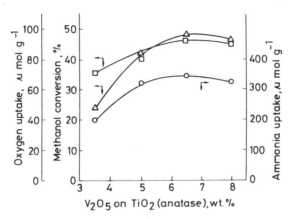

Figure 3. Effect of V_2O_5 loading on ammonia uptake, oxygen uptake and methanol oxidation activity.
(⬤) NH_3 uptake at 150°C, (▲) O_2 uptake at –78°C, (◻) Methanol oxidation activity at 150°C.

increases as a function of V_2O_5 loading up to 6.5 wt% and then decreases slightly with further V_2O_5 loading. The decline in the reducibility of vanadium oxide in post monolayer region (6.5 wt%) stems from the fact that in the bulky crystallites not all the vanadium oxide units are accessible to reducing gas whereas in the monolayer region the maximum number of such units are available for reduction due to high dispersion. Active sites are the vacancies created by the removal of labile oxygen atoms that take part in the redox processes upon which the dissociative chemisorption of oxygen takes place. The active site density presented in Table 1 increases with increase up to monolayer region and then decreases in the post monolayer region. Vanadia surface coverage data indicate that maximum coverage of V_2O_5 is observed at 6.5 wt% V_2O_5. This chemical interaction is possible on a fraction of the total surface of the support, i.e. active surface of the support. Further addition of V_2O_5 leads to formation of crystallites.

It can be seen from Table 1 that on TiO_2 (anatase) support the product of the reaction of methanol is only dimethyl ether. Impregnation of vanadia has drastically reduced the dehydration product, dimethyl ether, with selectivity to formaldehyde (dehydrogenation product) being 92 to 97% at all the loadings studied. The V_2O_5-TiO_2 (anatase) catalysts are clearly highly selective in the selective oxidation of methanol to formaldehyde. It has been reported that impregnation of different supports (Al_2O_3, SiO_2, MgO and TiO_2) with V^{5+} layers leads to an increase in oxidising character of the catalyst, whereas the dehydration activity is lowered in the methanol and n-heptanol oxidation (27). While the observed reactions for methanol are the selective oxidation to formaldehyde and intermolecular dehydration forming dimethyl ether, those with n-heptanol are selective oxidation to n-heptanal, and inter- and intramolecular dehydration leading to n-heptene and di-n-heptyl ether, respectively. It has been observed that on V_2O_5-TiO_2 catalysts the rate of acetone formation is faster than of propene formation in the decomposition of isopropanol (28). Our results on methanol oxidation on V_2O_5-TiO_2 (anatase) support these observations.

The information obtained from oxygen chemisorption confirms the conclusions drawn from the NH_3 chemisorption data.

LITERATURE CITED

1.   Bosh. H., Jansenn, F., Catal. Today **1988**, 2, 1.
2.   Haber, J., Pure and Appl. Chem. **1984**, 56, 1663.
3.   Volta, J.C., Portefaix, J.L., Appl. Catal. **1985**, 18, 1.
4.   Grabowski, R., Grzybowska, B., Haber, J., Sloczynski, J., React. Kinet. Catal. Lett. **1975**, 2, 81.
5.   Bond, G.C., Sarkany, A., Parfitt, G.D., J. Catal. **1979**, 57, 476.
6.   Bond, G.C., Brueckman, K., Faraday Discuss. Chem. Soc. **1981**, 72, 235.

7.   Bond, G.C., Konigs P., J. Catal. **1982**, 77, 309.
8.   van Hengstrum, A.J., van Ommen, J.G., Bosch, H., Gellings, P.J., Appl. Catal. **1983**, 8, 369.
9.   Wachs, I.E., Saleh, R.Y., Chan, S., Chersich, C.C., Appl. Catal. **1985**, 15, 339.
10. Gasior, M., Gasior, I., Grazybowska, B., Appl. Catal. **1984**, 10, 87.
11. Roozeboom, F., Mittelmeijer-Hazeleger, M.C., Moulijn, J.A., de Beer, V.H.J., Gellings, P.J., J. Phys. Chem. **1980**, 84, 2783.
12. Kozlowski, R., Pettifer, R., Thomas, J.M., J. Phys. Chem. **1983**, 87, 5172.
13. Bond G.C., Flamerz, S., Shukri, R., Faraday Discuss, Chem. Soc. **1989**, 87, 65.
14. Bond, G.C., Perez-Zurita, J., Flamerz, S., Gellings, P.J., Bosh H., van Ommen, J.G., Kip, B.J., Appl. Catal. **1986**, 22, 361.
15. Haber, J., Kozlowski, A., Kozlowski, R., J. Catal. **1986**,102, 52.
16. Kang, Z.C., Bao, Q.X., Appl. Catal. **1986**, 26, 251.
17. Machej, T., Remy, M., Ruiz, P., Delmon, B., J. Chem. Soc. Faraday Trans. **1990**, 86, 715.
18. Gil-Llambis, F.J., Eseudey, A.M., Fierro, J.L.G., Lopez Audo, A., J. Catal. **1985**, 95, 520.
19. Bond, G.C., Perez Zurita, J., Flamerz S., Appl. Catal. **1986**, 27, 353.
20. Anderson, S.L.T., J. Chem. Soc. Faraday, Trans.I. **1979**, 25, 1356.
21. Saleh, R.Y., Wachs, I.E., Chan. S., Chersich, C.C., J. Catal. **1986**, 98, 102.
22. Inomata, M., Miyamoto, A., Murakami, Y., J. Catal. **1980**, 62, 140.
23. Busca, G., Marchetti, L., Centi, G., Triffiro, F., J. Chem. Soc. Faraday, Trans. I. **1985**, 81, 1003.
24. Oyama, S.T., Went, G.T., Bell, A.T., Somorjai, G.A., J. Phys. Chem. **1989**, 93, 6786.
25. Haber, J., Proc. 8th Int. Congr. Catal. Dechema, Frankfurt-am Main 1984, 1, 85.
26. van Hengstrum, A.J., van Ommen, J.G., Bosh, H., Gellings, P.J., Appl. Catal. **1983**, 5, 207.
27. Kijenski, J., Baiker, A., Glinski, M., Dollenmeier, P., Wokaun, A., J. Catal. **1986**, 101, 1.
28. Ai, M., J. Catal. **1978**, 54, 223.
29. Sivaraj, Ch., Srinivas, S.T., Nageswara Rao, V., Kanta Rao, P., J. Mol. Catal. **1990**, 60, L23.
30. Shakhtakntinskaya, A.T., Mamedova, Z.M., Mutallibova, Sh. F., Alieva, S.Z., Mardzhanova, R.G., React. Kinet. Catal. Lett. **1989**, 39, 137.
31. Kanta Rao, P., Sivaraj, Ch., Srinivas, S.T., Nageswar Rao, V., In "Catalysis of Organic Reactions", Pascoe, W.E., Ed., Marcel Dekker Inc., New YOrk, U.S.A., 1992, pp 193-204.
32. Nag, N.K., Chary, K.V.R., Reddy, B.M., Rama Rao, B., Subrahmanyam, V.S., Appl. Catal. **1984**, 9, 225.

33.   Chary, K.V.R., J. Chem. Soc. Chem. Commun. **1989**, 104.
34.   Miyata, H., Mukai, T., Ono, T., Kubokawa, Y., J. Chem. Soc. Faraday Trans. I. **1988**, 84, 4137.
35.   Chan, S.S., Wachs, I.E., Murrel, L.L., Wang, L., Hall, W.K., J. Phys. Chem. **1984**, 88, 5831.
36.   Miyata, H., Fujii, K., Ono, T., J. Chem. Soc. Faraday Trans.I. **1988**, 84, 3121.
37.   Kataoka, T., Dumesic, A., J. Catal. **1988**, 112, 66.
38.   Inomata, M., Mori, K., Miyamoto, A., Ui, T., Murakami, Y., J. Phys. Chem. **1983**, 87, 754.
39.   Belokopytov, Yu. V., Kholyanvenko, K.M., Gerei, S.V., J. Catal. **1979**, 60, 1.
40.   Takagi-Kawai, M., Soma, M., Onishi, T., Tamaru, K., Can. J. Chem. **1980**, 88, 2132.
41.   Busca, G., Langmuir. **1986**, 2, 577.
42.   Dines, T.J., Rochester, C.H., Ward, A.M., J. Chem. Soc. Faraday Trans. **1991**, 87, 1611.
43.   Otamiri, J.C., Anderson, A., Catal. Today. **1988**, 3, 211.
44.   Odriozola, J.A., Heinemann, H., Somorjai, G.A., Garcia De La Banda, J.F., Pereira, P., J. Catal. **1989**, 119, 71.
45.   Takagi, M., Kawai, T., Soma, M., Onishi, T., Tamaru, K., J. Phys. Chem. **1976**, 80, 430; J. Catal. **1977**, 50, 441.
46.   Miyata, H., Nakagawa, Y., Ono, T., Kubokawa, Y., J. Chem. Soc. Faraday Trans.I. **1983**, 79, 2343.
47.   Ai, M., J. Catal. **1984**, 85, 324.
48.   Grazybowska-Swierkosz, B., In "Catalysis by Acids and Bases", Studies in Surface Science and Catalysis, Naccache, C., Coudurier, G., Ben Taarit, Y., Verdine, J.C., Ed., Elsevier, Amsterdam, The Netherlands, 1985, pp 45.

RECEIVED November 6, 1992

SYNTHESIS AND REACTIVITY OF NEW MATERIALS

Chapter 18

Hydrocarbon Partial Oxidation Catalysts Prepared by the High-Temperature Aerosol Decomposition Process
Crystal and Catalytic Chemistry

William R. Moser

Department of Chemical Engineering, Worcester Polytechnic Institute, Worcester, MA 01609

The advantages of aerosol processes for the synthesis of complex metal oxide and supported metal catalysts are mainly in the high degree of homogeneity of the finished catalyst and their ease of preparation. The process demonstrated an unusual capability to synthesize ion modified catalysts over a wide range of modifying ion concentrations while maintaining homogeneous solid solutions structures. Catalytic investigations using these complex metal oxide catalysts showed that smooth changes in catalytic performance could be realized as functions of the concentration of the modifying ions. The method led to materials of superior catalytic properties for propylene oxidations by bismuth molybdates and iron modified bismuth molybdates, and methane activation catalysts in the perovskite series. A new phase was obtained from the synthesis of P-V-O catalysts for butane oxidation. Platinum-iridium and silver metal supported catalysts on a high surface area α-alumina displayed an exceptionally high degree of homogeneity.

A fundamental objective of modern catalyst synthesis is to produce solid state catalysts that are single phase and homogeneous solid solutions. This is especially important at the stage of catalyst discovery research where one usually attempts to establish linear relationships between catalyst composition *vs* chemical performance. It is also important in the fabrication of commercial scale catalysts to ensure reproducible catalytic performance in the process reactor. Classical methods of preparation such as co-precipitation, thermal fusion, spray drying, and freeze drying usually result in some multiple phase materials, a condition that becomes more severe as the concentrations of the modifying ions are increased. These methods are sometimes unreliable with regard to reproducibility. In addition, the synthesis of ion substituted homogeneous solid solution catalysts such as perovskites, scheelites, and spinels are difficult to obtain as single phase materials if a moderate to high surface area material is required.

0097–6156/93/0523–0244$06.00/0

This report will describe an aerosol process that was developed in these laboratories over the past few years for the synthesis of complex metal oxide catalysts. Providing that the aerosol reactor has sufficient capabilities for process variability, the method was found in these laboratories to produce a wide range of catalysts not only for hydrocarbon partial oxidation, but also catalysts for a wide range of commodity and fuels chemical processes. Several years ago, we reported (1) the development of a down flow reactor configuration that was found effective for the synthesis of a variety of supported and unsupported catalysts, novel ceramics and superconductors. In the interim, the process was modified to an up flow configuration to afford an even greater latitude in increasing the residence time of the aerosol droplets within the heated zone of the reactor, but the up flow configuration also affords a more even decomposition of the aerosol. This configuration now permits the formation of metal oxide catalysts, ceramics and superconductors that could not be synthesized using the previously described down flow reactor configuration (1). The principal advantage of aerosol processes for metal oxide synthesis is that the process starts with an ideal precursor, an aqueous metal salt solution. This is converted to an aerosol in the gas phase followed by a rapid decomposition at high temperature over 1-5 seconds residence time to produce high surface area homogeneous metal oxide catalysts. The process has demonstrated capabilities superior to all other techniques for the synthesis of phase pure, high surface area catalysts in a reproducible way. The most recent configuration of the process schematic is illustrated in Figure 1, and the continuous process is amenable to the preparation of industrial scale catalysts. The origin of such aerosol processes is found in early patents (2) by Ebner in 1939. Similar processes were described in the mid 1970's by Roy and co-workers (3) for the synthesis of alumina and calcium aluminate. The contribution of this laboratory to aerosol synthesis was to design a process for controlling the injection and decomposition of the aerosol in a well controlled way using facilities which provide the process variability required for a general metal oxide synthesis. The processes which evolved is described as the High Temperature Aerosol Decomposition (HTAD) process; other similar processes are known as spray pyrolysis, and evaporative decomposition of solutions (EDS).

The aerosol is formed as shown in Figure 1 by either an ultrasonic aerosol droplet generator or by a forced air spray nozzle. The residence time within the furnace may be controlled by make-up air fed into the bottom of the reactor to control the contact time for decomposition between 1 to 15 seconds. The temperature is regulated by a three zone furnace which may be controlled between ambient and 1350 °C, either with or without an axial gradient. The process operates under a slight vacuum, and may use a pulsed injection depending on the material to be synthesized. Several process parameters may be used to purposely adjust the surface areas of the desired catalysts. Usual surface areas are in the range of 10-80 m^2/g, although other complex metal oxides, spinels, were synthesized in areas as high as 250 m^2/g.

Studies in these laboratories have resulted in the synthesis and catalytic evaluations on a wide range of perovskites and ion modified homogeneous solid solutions for Fischer-Tropsch catalysis, copper modified spinels for higher alcohol synthesis, ion substituted perovskites for methane activation, alkali modified metal

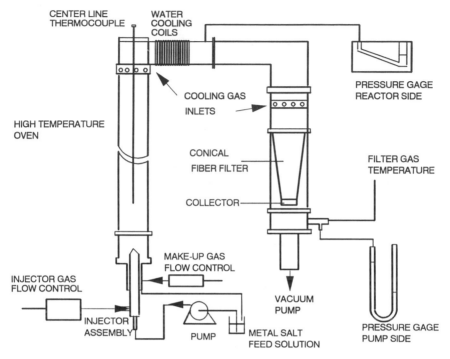

Figure 1. Aerosol process configuration for the continuous preparation of metal oxides by the High Temperature Aerosol Decomposition (HTAD) Process.

oxides for methane activation, methanol synthesis catalysts, propylene oxidation catalysts, silver on alpha-alumina supported catalysts for hydrocarbon oxidation, metallic and bimetallic catalysts for selective dehydrogenations, and high surface area, non-acidic, alpha-alumina. The process also resulted in a wide variety of ceramics of exceptionally high phase purity, nano-phase materials, and the direct synthesis of the orthorhombic structure of the Y-Ba-Cu superconductor and its metallic silver modifications.

This report will concentrate on the synthesis of a wide variety of hydrocarbon partial oxidation catalysts using the HTAD process. The objective is to provide an understanding of the range of advantages of the HTAD process for such catalyst fabrication for widely different oxidation catalysts rather than describing details for one specific system. The synthesis, characterization, and catalytic properties of HTAD oxidation catalysts will be described in the following systems: 1.) bismuth molybdates for the oxidation of propylene to acrolein; 2.) iron modified bismuth molybdate catalysts for propylene oxidation in the substitutional series, $Bi_{(2-2x)}Fe_{2x}Mo_3O_{12}$ where x was varied from 0.0 to 1.0; 3.) ytterbium modified strontium cerium oxide perovskites for the oxidative dimerization of methane to ethane/ethylene; 4.) phosphorus modified vanadium oxides for butane oxidation to maleic anhydride; 5.) platinum, silver, and other noble metals for hydrocarbon oxidation; and Sr and Ca modified $La_{(1-x)}M_xFeO_{(3-y)}$, $La_{(1-x)}M_xCoO_{(3-y)}$, and other perovskites for CO oxidation.

Experimental

Although this report describes the preparation of a wide variety of oxidation catalysts, only the iron modified bismuth molybdates will be described in detail. Other preparations are described in this section through an indication of the starting soluble salts and their synthesis temperatures which are the key process parameters. Other aspects of their HTAD preparations are similar to that described for the iron bismuth molybdates which follows.

The iron substitution series, $Bi_{(2-2x)}Fe_{2x}Mo_3O_{12}$, was synthesized in the HTAD process using the up flow configuration shown in Figure 1 as well as the down flow configuration described before(1). The up flow configuration was found optimum for the synthesis of this type of compound and only those results will be presented here. Although a variety of process parameters were investigated resulting in different morphologies and surface areas, the optimum conditions for synthesis were as follows: Feed solution, Total metal ion molarity of 0.6 M in de-ionized water containing 10% v/v of nitric acid consisting of $Bi(NO_3)_3 \cdot 5H_2O$, $Fe(NO_3)_3 \cdot 9H_2O$, and $(NH_4)_6Mo_7O_{24} \cdot 4H_2O$; Furnace temperature of 900°C and pre-heated make-up air temperature of 200°C flowing at 3-4 ft^3/min into a mullite tube which was 2 3/4 inches inside diameter and 4 ft long. A spraying nozzle was used with a head pressure of 20 psig, and the solution was introduced to the reactor using a pulsed injection of 2 seconds on and 4 seconds off; the product was collected using either a Pall, porous stainless steel filter or a fiber filter. The process was run under 1-5 in. H_2O vacuum to ensure smooth flow of materials through the reactor section.

All solid materials were analyzed by XRD, SEM, infrared, BET surface area, and inductive coupled arc plasma. All up flow samples and co-precipitation samples resulted in acceptable elemental analysis when analyzed for Bi, Fe, and Mo.

The catalysis studies for propylene oxidations were carried out in a heated, well insulated micro-reactor consisting of a 1/2 inch i.d. stainless steel reactor section. Gases were fed by mass flow controllers, and water vapor was fed to the reacting stream by passing the air feed stream through a temperature controlled water saturator. All products were analyzed using an on-line GC. The upper portion of the reactor was filled with 20 mesh silanized quartz, and the catalyst bed consisted of 2.0 mL of the catalyst powder calcined at 450°C and mixed with an equal volume of the silanized quartz. Feed compositions and reactor conditions used were: propylene/O_2/N_2/H_2O = 5.4/14.5/55.7/25.4; The reaction was studied at atmospheric pressures using contact times of 10 seconds. The GC analysis determined effluent concentrations of CO, CO_2, oxygen and hydrocarbons using a back flushed, dual column configuration and thermal conductivity detection for the permanent gases and FID for hydrocarbon and oxygenate analysis.

The methane oxidation studies used a schematic similar to the above except that the reactor section was fabricated entirely from quartz and reaction temperatures from 600 to 1000°C were examined. The precise conditions for the experiments discussed are indicated in the figure captions.

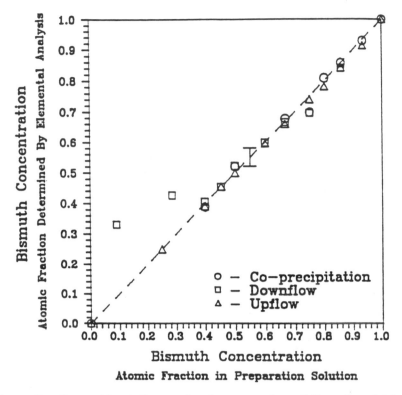

Figure 2. Compositional diagram for the preparation of bismuth molybdate catalysts using the HTAD process configuration shown in Figure 1 at 900°C using air as make up gas. Plot is of concentrations of bismuth used in the reacting solution *vs* arc plasma analyzed concentrations of the finished catalysts directly from HTAD reactor: Circles: Co-precipitation prepared materials. Triangles: Up flow prepared aerosol materials. Squares: Down flow prepared aerosol materials.

Results

Synthesis and Characterization
Bismuth Molybdates And Iron Substituted Bismuth Molybdates

The aerosol syntheses of the binary bismuth molybdate compounds using stoichiometries evenly distributed over a bismuth atom fraction from 0 to 1.0 and the iron modified series $Bi_{(2-2x)}Fe_{2x}Mo_3O_{12}$, x=0.0 to 1.0 were carried out at 900°C. To illustrate that the aerosol synthesis provides finished catalysts having predictable metal ion concentration based on the feedstock compositions, Figure 2 shows that the up flow configuration resulted in correct metal ion concentrations. Elemental analyses for Bi, Fe and Mo by arc plasma on these samples gave satisfactory values. The figure also shows that the down flow preparation of these compounds led to high Bi concentrations. This was likely due to the fact that a water scrubber was used to collect the catalyst particles which selectively removed molybdenum oxide through solubilization.   Results were in agreement with those obtained from very careful preparations of the same materials by co-precipitation, except that the HTAD syntheses take considerably shorter preparation times compared to the co-precipitated materials.   The as prepared HTAD materials demonstrated BET surface areas which were between 1 to 14 m^2/g depending on bismuth concentrations.   The surface areas of the iron modified series started out at low Fe concentrations, 0.03 atom fraction Fe, of 1 m^2/g and steadily increased to 7 m^2/g at an iron atom fraction of 0.40.   All of these samples were calcined to the maximum temperature (450°C) used in the catalytic investigation.   This calcination led to a decrease in the surface areas of all of the catalysts into the league of 1-2 m^2/g, and were nearly the same as that determined on the co-precipitation series of 1-2.5 m^2/g.   SEM analysis showed that the particles had a morphology of 1-10μm spherical particles and were hollow.   Upon calcination, these particles sintered with the above mentioned loss in surface area without changing size or spherical morphology of the parent particle.   The XRD patterns for the material where the iron substitution was x= 0.40 as obtained from the HTAD reactor (top curve) and after calcination in air at 450°C (bottom curve) are shown in Figure 3. These X-Ray diffraction patterns are typical of the entire two series of bismuth molybdates.   An examination of the XRD patterns of the HTAD materials in comparison to the patterns collected on identical compositions by co-precipitation and fusion showed that the phase purities observed for the HTAD materials were equal or better than either the co-precipitated or fusion prepared materials. Over the entire iron substitution series, the XRD reflections broadened as the iron concentration in the synthesis increased.   This suggests a hindrance to crystal particle growth as evidenced by the substantially broadened reflections.   The phase analysis of the HTAD iron substitutional series showed that the mole fraction of the compound containing no bismuth, $Fe_3Mo_3O_{12}$, continuously increased as the iron substitution increased; however, high concentrations of iron led to separate phase MoO_3 and Fe_2O_3. Thus, the synthesis did not provide a homogeneous solid ion substitution by iron over the entire composition range studied.   However, substitution may have occurred into the α-$Bi_2Mo_3O_{12}$ host oxide at low concentrations below 5 atom % (based on total metal ions).

Silver, Platinum and Noble Metals Supported On Alumina

The preparation of precious metal supported catalysts by the HTAD process is illustrated by the synthesis of a wide range of silver on alumina materials, and Pt-, Pt-Ir, Ir-alumina catalysts. It is interesting to note that the aerosol synthesis of alumina without any metal loading results in a material showing only broad reflections by XRD.   When the alumina sample was calcined to 900°C, only reflections for α-alumina were evident.   The low temperature required for calcination to the alpha-phase along with TEM results suggest that this material was formed as nano-phase, α-alumina.   Furthermore, the use of this material for hexane conversions at 450°C indicated that it has an exceptionally low surface acidity as evidenced by the lack of any detectable cracking or isomerization.
   Seven compositions of Ag-alumina were synthesized between 0-65% w/w silver at 900 °C by the HTAD process. Characterization of these materials by XRD

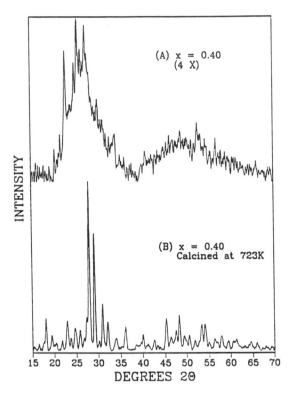

Figure 3. XRD patterns for the iron substituted composition having the empirical formula of $Bi_{(2-2x)}Fe_{2x}Mo_3O_{12}$ where $x = 0.40$. The top powder diffraction pattern is for the aerosol prepared material as directly obtained from the reactor. The bottom diffraction pattern is for the material after calcining in air for 4 hr at 450°C.

showed that only in the case of the 65% material did reflection for metallic silver appear. Phase analysis on all lower compositions of 1, 5, 15, 30, and 50% w/w Ag showed no reflections for any known silver or silver-alumina compound either as the reduced metal or oxide form. When all of the compositions were calcined at 900°C, near the melting point of metallic silver, strong reflections appeared for metallic silver and α-alumina. These data suggest that well dispersed metallic, supported catalysts may be directly synthesized by the aerosol process to very high concentrations.

In support of the conclusion based on silver, series of 0.2, 0.5, 1.0, 2.0, and 5.0 % w/w of platinum, iridium, and Pt-Ir bimetallic catalysts were prepared on alumina by the HTAD process. XRD analysis of these materials showed no reflections for the metals or their oxides. These data suggest that compositions of this type may be generally useful for the preparation of metal supported oxidation catalysts where dispersion and dispersion maintenance is important. That the metal component is accessible for catalysis was demonstrated by the observation that they were all facile dehydrogenation catalysts for methylcyclohexane, without hydrogenolysis. It is speculated that the aerosol technique may permit the direct, general synthesis of bimetallic, alloy catalysts not otherwise possible to synthesize. This is due to the fact that the precursors are ideal solutions and the synthesis time is around 3 seconds in the heated zone.

Phosphorus-Vanadium Oxide Catalysts for Butane to Maleic Anhydride Oxidation

To illustrate the capabilities of the aerosol process for the synthesis of solid state oxidation catalysts not normally obtained from classical synthesis, the preparation of a series of P-V-O catalysts was examined. The original objective was to determine whether the synthesis of the reported (4) active phase, β-$(VO)_2P_2O_7$, of the catalyst responsible for the selective oxidation of butane to maleic anhydride could be synthesized by the aerosol technique.

Although a variety of synthesis, compositions and reactor parameters were studied, the P-V-O catalysts in the temperature series were synthesized in the up flow HTAD reactor using a 0.12 M solution of ammonium vanadate in water which contained the required amount of 85% phosphoric acid to result in a 1.2/1.0 P/V atom ratio. This atom ratio is normally preferred for the most selective oxidation of butane to maleic anhydride. Table I shows that the P/V atom ratios obtained for the analyzed, finished (green colored) catalysts were approximately the same as the feed composition when a series of preparations were studied between 350°C and 800°C. This was typical for all of the catalysts synthesized under a variety of conditions.

Table I Analytical Data on HTAD Synthesized P-V-O Catalysts

Reactor Temperature	P/V Ratio	P/V Ratio
T°C	Feed	Catalyst
350	1.20	1.18
400	1.20	1.13
425	1.20	1.19
450	1.20	1.26
500	1.20	1.21
550	1.20	1.22
650	1.20	1.23
700	1.20	1.19
800	1.20	1.19

One interesting aspect of the synthesized catalysts was that their XRD patterns measured immediately after synthesis resulted in broad reflections as illustrated in Figure 4a for the catalyst synthesized at 600°C using an 8 seconds residence time of the aerosol in the reactor and a feed concentration of 0.8 M in vanadium. When the fresh catalyst was immediately calcined in nitrogen at 450°C for 3 hrs, the XRD shown in Figure 4b still showed only broad bands along with three reflections emerging at 21.3, 29.0, and 35.9 degree 2θ. However, when the uncalcined material was allowed to stand in an air tight container for 14 days at room temperature, the well defined XRD pattern shown in Figure 4c was recorded, demonstrating peaks at 12.1, 24.7, and 29.2 degree 2θ. The optimum process parameters for the synthesis of this type of material was at 700°C, using a vanadium concentration of 0.12 M and a residence time of 8 seconds. and a P/V ratio of 1.2. The XRD of the material synthesized under these conditions is shown in Figure 5 exhibiting only three reflections at 12.3, 25.1, and 29.1 degree 2θ. This P-V-O phase is not the same as the catalytically active species, $\beta-(VO)_2P_2O_7$, obtained from classical synthesis where the reported (4) XRD reflections for the three main peaks were at 22.8, 28.3, and 29.9 degree 2θ. It is also not the same as the reported (5) alcohol solvent prepared precursor. Aerosol preparations at different conditions resulted in similar but not identical reflection positions for all of the compounds after standing in closed containers for more than a week, where the positions of the three bands reported above were either shifted or of different intensities. Our interpretation of the large d-spacing for the principal reflection, its variability with process conditions and crystal development on standing is that the catalyst synthesis afforded disordered lamella compounds which were microcrystalline immediately after synthesis but slowly crystallized upon standing. The surface areas of the compounds prepared between 600 and 800°C were between 6-14 m^2/g, and the measurements on the freshly prepared materials were the same as those after calcining in nitrogen at 450°C for 3 hrs. Another series of synthesis studies showed that the addition of small amounts of oxalic acid to the synthesis solution resulted in an ability to systematically vary the oxidation state of the vanadium ion between 4 and 5.

A wide variety of compositions and materials of slightly different crystal properties were prepared for evaluation as butane oxidation catalysts. These materials were active for butane oxidation, and the activity and selectivity varied according to the process parameters used for synthesis. This chemical performance data will be reported separately.

Perovskite Catalysts for Methane Conversion

Two series of catalysts were synthesized for subsequent evaluation as methane dimerization catalysts. The first series was alkali modified zinc oxide (6) and magnesium oxide catalysts (7), which were reported to be active for methane activation, while the second series was ion modified perovskites described by Machida and Enyo (8). The objective of the present study was to determine whether the aerosol technique could provide a wide range of ion substitutions as homogeneous solid solutions, and to determine whether moderately high surface area catalysts could

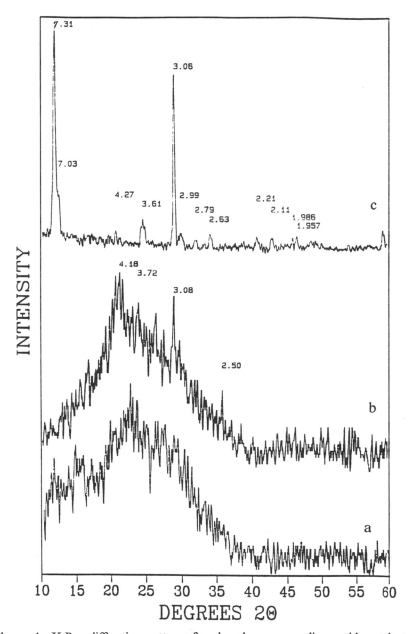

Figure 4. X-Ray diffraction patterns for phosphorous-vanadium oxide catalysts prepared by aerosol technique at 600°C, 8 seconds residence time, and 0.8 M V in feed: a. Catalyst analyzed by XRD immediately after synthesis. b. catalyst calcined at 450°C in nitrogen for 3 hrs immediately after synthesis. c. Catalyst allowed to stand for 14 days in an air tight container at ambient temperature without calcining.

be obtained by the aerosol method in a geometry having low micro-porosity. It was previously shown (1) that catalysts prepared by this technique generally demonstrate moderate to high surface areas depending on synthesis conditions, and the materials so obtained have no micro-porosity. The surface area is generally a result of the external surface of the particles which are formed either as thin walled hollow spheres with one or more holes through the walls or sub-micron fragmented particles. It is expected that this characteristics would be favorable for methane conversion to ethane and ethylene without over oxidation of the products.

The first set of alkali modified materials was $Li_xZn_{(1-x)}O_{(1-x/2)}$. The value of x was 0.00, 0.03, 0.07, 0.10, 0.20, and 0.40. These compounds were synthesized from aqueous solutions of their nitrates using the down flow reactor reported previously (1) but using a porous stainless steel filter and a HTAD process temperature of 1300°C. Processing conditions were adjusted so that a hollow spherical morphology resulted for all of the materials synthesized. Particle sizes of the spheres ranged from 0.5 to 6 microns. Surface areas varied from 4-8 m^2/g. Phase analysis of the entire series by XRD showed that only reflections for hexagonal zinc oxide were observed except in the sample of highest lithium concentration. This material showed a trace of an unidentified separate phase (intensity ratio to the main ZnO reflection was 1.1%) with the only observable reflection at 42.9 degree 2θ. Thus, the aerosol method evidently resulted in a well dispersed form of lithium within the ZnO matrix. Computation of the lattice parameters for ZnO for all of the compounds using 5 seconds scans at a step rate of 0.05 degree per minute between 55 and 65 degree 2θ showed that the lattice parameters varied little. This suggests that either 1.) the compounds in this series did not result in homogeneous solid solutions of lithium in zinc oxide, but the structure is that of exceptionally well dispersed lithium oxide within the ZnO host, or 2.) the lithium ion substituted into the interstices of the host crystal giving little change in the lattice parameters. Synthesis of solid state materials within the substitutional series $M_{0.1}Zn_{0.9}O_{0.95}$ where M was Li, Na, K, Rb, and Cs resulted in well dispersed alkali metals for Li and Na and progressively more intense separate phase reflections for M_2O progressing down the alkali metal series for K, Rb, and Cs. The data suggest that the alkali metal component in the Li-ZnO series and Li and K in this series are stabilized by the ZnO host and not simply physically admixed micro-crystallites of the alkali metal component in ZnO. Preparation of these same materials where a water collector was used in the process resulted in hollow spheres which consisted of pitted, porous walls as demonstrated by SEM analysis. The holes in the walls were most visible for the Cs and Rb compounds at a magnification of 5,000 and were not visible at this magnification for the Li, Na, and K compounds. EDX analysis of the Cs and Rb materials did not detect either element indicating that the pores were make in the walls of the sphere due to the leaching effect of the water scrubber of large crystallites after the synthesis. Materials prepared by the dry collection method exhibited no pores in the walls of the spheres and they demonstrated normal alkali metal analyses.

Methane dimerization catalysts in the perovskite substitutional series $SrCe_{(1-x)}Yb_xO_{(3-x/2)}$ where x was varied from 0.0 to 1.0 were synthesized by the HTAD process. These same catalysts were synthesized by Machida and Enyo (8) by fusion methods and were shown to be active for methane dimerization. The objective of the present studies was to determine 1.) whether the HTAD process provided catalysts which were substantially free from separate phases, and 2.) whether the Yb substitution into the host perovskite could be extended to high levels of substitution. The perovskites were synthesized at 1300°C using total solution molarities in each preparation of 0.30 M and residence times of 2-4 seconds. No further calcination was necessary to obtain the pure perovskite phase.   When x was varied in $SrCe_{(1-x)}Yb_xO_{(3-x/2)}$ from 0.0 to 1.0, and the resulting materials examined for their phase purity and composition, it was found that the materials were composed of either one of two phases which predominated the composition at each end of the compositional series.   When x was between 0.00 and 0.30, the perovskite structure dominated the composition, and when x was 0.40 to 1.00, the fluorite structure dominated the phase composition.   The substitution of the Yb ion in the perovskite portion of the substitutional series (x=0.00 to 0.40) gave rise to a homogeneous solid solution of Yb in $SrCeO_3$ as evident from the linear change of the d-spacing *vs* Yb concentration for the principal reflection from 3.05 at x=0.00 to 2.96 at x=0.40. Furthermore, a plot of d-spacing *vs* Yb concentration for x=0.60 to 1.00 showed that the Ce ion formed a homogeneous solid solution in the fluorite structure, $SrYb_2O_4$, as shown by the linear decrease in a main reflection from 2.054 at x=0.06 to 2.016 at x=1.00. No separate phases for SrO, CeO_2, or Yb_2O_3. were detected in the catalysts by XRD analyses. By adjusting the process parameters, the materials were synthesized as hollow spheres having surface areas between 2-4 m^2/g. The surface areas smoothly increased as the mole fraction of Yb was increased from 0.00 to 1.00. The preparation of this series of compounds by classical means of fusion at 1300°C resulted in surface areas of 0.9 to 1.4 m^2/g. This method generally results in low surface areas in the range of 0.1 to 1 m^2/g. High temperature fusion techniques also usually result in the formation of separate phases of the component metal oxides along with the desired perovskite phase unless the materials are extensively thermally processed at high temperatures for several days. This processing usually leads to very low surface areas near 0.1 m^2/g. The synthesis of these materials having thin walls to aid in the dissipation of heat from the particle coupled with a low particle micro-porosity should be favorable for the selective conversion of methane to ethane and ethylene

The catalytic properties for methane conversion were examined using this series of solid state materials mainly between 710 and 810°C in a quartz reactor with on-line GC analysis using an inlet feed composition of $He/CH_4/O_2/N_2$ =11.5/2.0/1.0/0.5. Figure 6 illustrates the effect on methane conversion of a variation in the concentration of Yb from x=0.00 to 1.00 at 810°C. The figure shows that conversions normalized to unit surface areas could be increased smoothly up to 60% across the substitutional series under these conditions.   On the other hand the selectivity of C-2 products rose smoothly as a result of homogeneous substitution in the perovskite structure (x=0.00 to 0.40) from 34% to 46%. However, during homogeneous substitution in the fluorite material (x=0.60 to 1.00) the selectivity was surprisingly flat between 30 to 33%. Non-selective products were CO and CO_2. Although the maximum C-2 yields under optimum conditions were around 20%, the interesting observation is the fact that the chemical performance could be

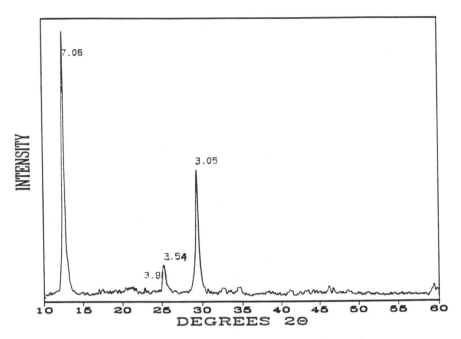

Figure 5. XRD of optimum synthesis of P-V-O catalyst by HTAD process at 700°C, 12 seconds residence time, using feed solution of 0.12 M in V.

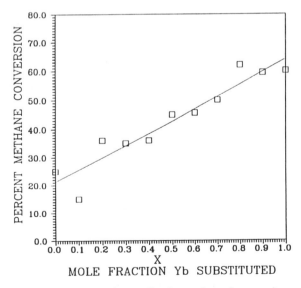

Figure 6. Methane conversions (normalized to unit surface area) as a function of the mole fraction for the substitution (x) of ytterbium in $SrCe_{(1-x)}Yb_xO_{(3-x/2)}$. Conditions for catalytic experiment were: 810°C in a quartz reactor using an inlet feed composition of $He/CH_4/O_2/N_2 = 11.5/2.0/1.0/0.5$

systematically changed as a function of substitution in both the perovskite and fluorite regions of the substitutional series.

Perovskite Catalysts for CO Conversion

A series of perovskites catalysts was synthesized to illustrate the capabilities of the process to produce materials as homogeneous solid solutions and in unusually high phase purities. This type of catalyst was reported by Voorhoeve and co-workers (9) to be active for CO oxidation. Compounds in the substitutional series $La_{(1-x)}M_xCoO_{(3-y)}$ and $La_{(1-x)}M_xFeO_{(3-y)}$ were synthesized by the HTAD process (1) where M was Ca and Sr and x = 0.0 to 1.0. In addition, nine lanthanide based perovskites were synthesized in the series $LnFeO_3$ where Ln was La, Ce, Pr, Nd, Sm, Gd, Dy, Er, and Yb. An examination of the XRD patterns for all of the compounds synthesized demonstrated reflections only for the perovskite. Surface areas recorded for the Ca and Sr substitutional series, $La_{(1-x)}M_xFeO_{(3-y)}$, varied from 10 to 25 m^2/g, and a few experiments using different process conditions to produce fragmented materials rather than hollow spheres resulted in perovskites having 60 m^2/g. All of these compounds required 1200°C HTAD reactor temperatures using residence times between 2-4 seconds. for synthesis of the pure phase perovskite. Figure 7 illustrates a typical XRD pattern for the $La_{0.8}Ca_{0.2}FeO_3$ perovskite. It is seen from the figure that the material was synthesized without phase impurities, and a plot of d-spacings for the main reflection *vs* Ca mole fraction led to a linear decrease in d-spacings as Ca increased. This series of catalysts was not studied for CO oxidation but was extensively examined for $CO-H_2$ conversions to hydrocarbons where systematic changes in chemical performance were observed for both the Ca and Sr substitutional series across the entire substitutional series.

Catalytic Studies using Bismuth Molybdate Compositions

The catalytic reactivity of the HTAD bismuth molybdates using a series of 14 compositions varying the bismuth ion concentration from 0 to 100% and the iron substitutional series of $Bi_{(2-2x)}Fe_{2x}Mo_3O_{12}$ were evaluated for the partial oxidation of propylene between 352 and 452°C. The conditions used were: propylene/O_2/N_2/H_2O = 5.4/14.5/55.7/25.4 at atmospheric pressure and contact times of 10 seconds. Figure 8 shows the unusual reactivity data as a function of iron substitution in the latter series. The reactivity from an iron atom substitution of 5% to 35% iron (based on the total g. at of metal atoms) is similar to data obtained in other laboratories using classically prepared Fe-Bi-Mo-O catalysts. However, the sharp spike in rate in the substitutional range of 0-5% is noteworthy, and may point to an important relationship between structure and reactivity. The XRD of these materials in this low range of iron substitution showed only reflections for the host oxide α-$Bi_2Mo_3O_{12}$ indicating that the iron was well dispersed. The data do not permit the conclusion that the iron, even at this low concentration, entered the lattice as a homogeneous solid solution since the concentration range is only moderately above the detection limits for a separate iron oxide phase by XRD.

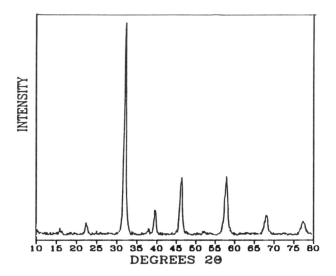

Figure 7   X-Ray diffraction pattern for the aerosol synthesis of the calcium substituted perovskite $La_{0.8}Ca_{0.2}FeO_3$ at 1200°C using a 3 seconds residence time.

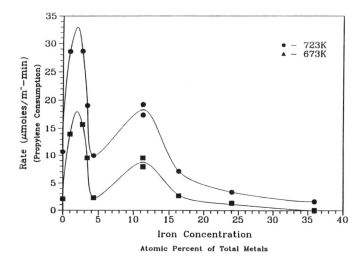

Figure 8. Propylene consumption rate data for aerosol prepared catalysts having the composition of $Bi_{(2-2x)}Fe_{2x}Mo_3O_{12}$ where x = 0.0 - 1.0.   Top curve represents data at a catalytic reactor temperature of 450°C and the bottom curve represents data taken at 400°C.   The catalytic reactor conditions were: propylene/O_2/N_2/H_2O = 5.4/14.5/55.7/25.4 at atmospheric pressure and contact times of 10 seconds.

The selectivity to acrolein as a function of iron substitution is shown in Figure 9. It is interesting to note that the selectivity profile vs iron concentration follows that observed for rates *vs* iron concentration. The unusual aspect of these data is the fact that selectivity increased in the regime where the catalyst was performing at highest conversion.

Finally, Arrhenius treatments of the catalytic data were examined for the HTAD synthesized substitutional series, $Bi_{(2-2x)}Fe_{2x}Mo_3O_{12}$, and the binary bismuth molybdate series where Bi/Mo ratios were varied from pure Mo oxide to pure Bi oxide. The noteworthy aspect of the oxidation results is that in the most reactive regime of x = 0-5% atom fraction Fe, before separate phase $Fe_3Mo_3O_{12}$ begins to dominate the catalyst composition in the iron series, the apparent activation energies were all in the range of 19-20 kcal/mol. Furthermore, the activation energies for the pure Bi-Mo series were between 19-20 kcal/mol while the activities were considerable different. Thus, the chief difference in the reactivities in both series is in the pre-exponential factor, i.e. the number of active sites.

The conclusion that we reach from these data is that the iron substitutional series and the catalytic studies on the unmodified Bi-Mo-O series prepared by the HTAD process suggest a single type of active site for the most active and selective oxidation of propylene. The most active catalysts in both series demonstrated activation energies which were nearly identical, suggesting that the modification by iron in the Fe-Bi-Mo-O series or bismuth concentrations in the Bi-Mo-O series to provide higher rates and selectivities resulted not from an electronic modification but solely from alterations in the number of active sites. In addition, the HTAD, aerosol synthesis results in nano-phase crystallites as contrasted to much larger particles obtained from classical synthesis. The fact that selectivities increased at the same time as conversions is unusual and suggest that the most active sites are also the most selective ones. Recent studies by Volta and Portefaix (10) and Desikan and Oyama (11) suggest that the activity of MoO_3 catalyzed oxidation of propylene increases as does selectivity as the edge site concentration in the catalysts increases. In the present bismuth substitution study, a reasonable possibility to rationalize the superior rates and improved selectivities of the HTAD catalysts over classically formed catalysts is due to the inherently higher edge site concentrations in aerosol generated materials. This is likely due to the fact that the micro-crystallite particle size of the HTAD materials is much smaller than co-precipitation materials. In the case of the low level of iron substitution, it is likely that iron at this level inhibits crystal growth which results in smaller particles with inherently higher ratios of edge to basal planes. Thus, the high activity and selectivity of the low iron modifications was not due to changing the electronics of the catalyst through substitution, but may have been due to increasing the edge to basal plane concentration at low iron concentrations.

Conclusions

Aerosol processes afford materials having an optimum phase purity as contrasted to catalysts prepared by other synthetic methods. In addition, their syntheses are reproducible and metal ion concentrations are predictable based on starting

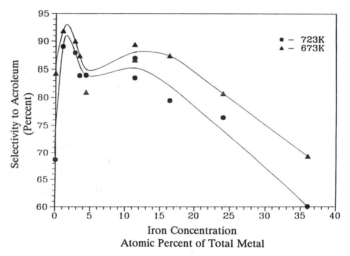

Figure 9. Selectivities to acrolein for the series of aerosol catalysts described in Figure 8.

compositions. The method offers a rapid and efficient method of synthesis of pure phase materials for fundamental or optimization research where reactivity correlations are sought without complications due to separate phase components mixed in with the desired catalyst. The HTAD process offers a means for the synthesis of catalytic materials not obtainable by other means, such as exceptionally high states of dispersions of reduced states of precious metals stabilized by a host metal oxide support. In addition, the technique offers a way to synthesize a wide range of metal alloys in high dispersion. Although the process has demonstrated capabilities to fabricate higher alcohol catalyst from syngas, Fischer-Tropsch catalysts, hydrodesulfurization catalysts, and hydrocarbon dehydrogenation catalysts, their utilization in hydrocarbon partial oxidation seems most promising since the method offers an efficient and rapid method of synthesis of pure phase materials having a controllable surface area and as particles having no micro-porosity.

Acknowledgments

We acknowledge the National Science Foundation (Grant number, CB 85-16935) for support of this research. Part of the studies on the synthesis of maleic anhydride catalysts was supported by NASA in the "Synthesis of Catalysts in Microgravity" program.

Literature Cited

(1) Moser, W.R. and Lennhoff, J.D, Chemical Engineering Communications **1989** *83*, 241.
(2) Ebner, K., U.S. Patent 2,155,119, April 18, 1939 (Assigned to American Lurgi Corporation).
(3) Roy, D.M.; Neurgaonkar, R.R.; O'Holleran, T.P.; and Roy, R. Ceramics Bulletin **1977**, *56*,1023.
(4) Pepera, M.A.; Callahan, J.L.; Desmond, M.J.; Milberger, E.C.; Blum, P.R.; and Bremer, N.J. J. Am. Chem. Soc. **1985**,*107*, 4883.
(5) Brusca, G.; Cavani, F.; Centi, G.; and Trifuro, F. J. Catal. **1986**, *99*, 400.
(6) Matsura, I.; Utsumi, Y.; Nakai, M.; and Doi, T. Chem. Lett. **1986**,*11*,1981.
(7) Ito, T.; and Lundsford, J.E. Nature, *1985*,*314*, 721.
(8) Machida, K.; and Enyo, M. J. Chem Soc., Chem. Commun. **1987**, 1639.
(9) Voorhoeve, R.J.H.;Remeika, J.P.; Freeland, P.E.; and Matthias, B.T. Science **1972**, *177*,353.
(10) Volta, J.C. and Portefaix, J.L., Rev. Appl. Catal, 18, 1 (1985).
(11) Desikan, A.N. and Oyama, S.T., in ACS Symposium Series, 482, 260 (1992); Ed. D. J. Dwyer and F. M. Huffmann.

RECEIVED November 6, 1992

Chapter 19

Synergy Effect of Multicomponent Co, Fe, and Bi Molybdates in Propene Partial Oxidation

H. Ponceblanc[1], J. M. M. Millet, G. Coudurier, and J. C. Védrine

Institut de Recherches sur la Catalyse, Centre National de la Recherche Scientifique, 2 avenue Albert Einstein, 69626 Villeurbanne, France

Mechanical mixtures of $Bi_2Mo_3O_{12}$ and solid solution $Fe_xCo_{1-x}MoO_4$ phases of variable relative composition and variable x value have been studied for propene oxidation to acrolein. A huge synergy effect was observed when Bi, Co and Fe were present and a maximum in selectivity and activity was observed for a mixture of a β phase form of $Fe_xCo_{1-x}MoO_4$ with x=0.67, with 15 wt.% $Bi_2Mo_3O_{12}$. Electrical conductivity measurements have clearly shown that Fe^{3+} cations were also present in the solid solution and strongly increased the conductivity (σ x more than 10^4). EDX-STEM analyses showed that the mixtures are composed of the two phases separated but that under catalytic reaction conditions the solid solution particles deposit on the large bismuth molybdate particles while part of the bismuth molybdate spreads over the solid solution deposited on bismuth molybdate large particles. The intimate contact between bismuth molybdate and solid solution, due to the spreading of the former on the latter deposited on large $Bi_2Mo_3O_{12}$ particles, facilitates electrons and oxygen ions mobility i.e. the redox mechanism involved in the Mars and Van Krevelen mechanism which is known to occur in propene partial oxidation reaction.

The oxidation of propene to acrolein has been one of the most studied selective oxidation reaction. The catalysts used are usually pure bismuth molybdates owing to the fact that these phases are present in industrial catalysts and that they exhibit rather good catalytic properties (1). However the industrial catalysts also contain bivalent cation molybdates like cobalt, iron and nickel molybdates, the presence of which improves both the activity and the selectivity of the catalysts (2,3). This improvement of performances for a mixture of phases with respect to each phase component, designated synergy effect, has recently been attributed to a support effect of the bivalent cation molybdate on the bismuth molybdate (4) or to a synergy effect due to remote control (5) or to more or less strong interaction between phases (6). However, this was proposed only in view of kinetic data obtained on a prepared supported catalyst.

[1]Current address: Rhône–Poulenc Recherche, CRA, 52 rue La Haie Coq, 93508 Aubervilliers, France

0097–6156/93/0523–0262$06.00/0

In a recent work we were able to show that an electronic effect was detected between $Bi_2Mo_3O_{12}$ and a mixed iron and cobalt molybdate with an enhancement of the electrical conductivity of the cobalt molybdate with the substitution of the cobaltous ions by the ferrous ions (7). However this effect alone cannot explain the synergy effect and we have investigated the influence of both the degree of subtitution of the cobalt with the iron cations in the cobalt molybdate and the ratio of the two phases (for a given substituted cobalt molybdate) on the catalytic properties of the mixture. We have tried to characterize by XPS and EDX-STEM the catalysts before and after the catalytic reaction in order to detect a possible transformation of the solid. The results obtained are presented and discussed in this study.

Experimental

$Bi_2Mo_3O_{12}$ was prepared by dissolving H_2MoO_4 and $BiONO_3$ into water and letting the solution boil under stirring for 2 hours (8). Mixed iron and cobalt molybdate were prepared by adding ammonia to an aqueous solution of $(NH_4)_6Mo_7O_{24}.4H_2O$ which was boiled for two hours under argon. A solution of $FeCl_2.4H_2O$ and $Co(NO_3)_2.6H_2O$, in stoichiometric amounts was added to the mixture, which was boiled for one hour (9). The precipitates were filtered, washed with deoxygenated water, (evaporated to dryness under vacuum for the bivalent molybdate) and calcined at 450°C for ten hours under a deoxygenated and dehydrated nitrogen flow. Mechanical mixtures were obtained by mixing the respective powders and hand grinding them for 5 to 10 minutes.

Crystal structures of the bismuth molybdate and of the mixed iron and cobalt solid solution molybdate samples were controlled by X-ray diffraction (10). The chemical compositions of the samples were determined by atomic absorption and their surface areas measured by nitrogen adsorption using the BET method.

EDX-STEM analyses were performed with a Vacuum Generator VG HB 501 electron microscope with field emission gun and a beam area varying from $0.1 \mu m^2$ down to 5 nm². XPS measurements were performed on a Hewlett-Packard HP 5950, at room temperature. Qualitative analysis of the peaks, in terms of elemental ratios, was carried out as described previously and the estimated error in such an analysis is approximately 10% (11).

Selective oxidation of propene to acrolein was carried out in a dynamic differential microreactor containing 40 to 60 mg of catalyst as described previously (12). Reaction conditions were as follows : propene/O_2/N_2 (diluting gas) = 1/1.69/5; total flow rate 7.2 dm³.h⁻¹; total pressure 10^5 Pa; and reaction temperature 380 °C.

Results

Since a polymorphic transition (α/β) of the mixed iron and cobalt molybdate occurs in the temperature range of the catalytic reaction (10,13,14), and since the high temperature form (β) can metastably be maintained at low temperature, the catalysts were tested directly after heating to 380°C ($Fe_{0.67}Co_{0.33}MoO_4$ is in the α form) and after a subsequent heating to 430°C for several hours and return to 380°C ($Fe_{0.67}Co_{0.33}MoO_4$ is in the β form). The oxidation of propene was conducted at 380°C under the conditions given in the experimental section. The products of the reaction were in all cases exclusively acrolein (ACRO) and CO_2.

a) Study of mixtures of $Bi_2Mo_3O_{12}$ and $Fe_xCo_{1-x}MoO_4$ with various iron contents and a fixed phase composition.

Eight mixtures composed of $Bi_2Mo_3O_{12}$ and $Fe_xCo_{1-x}MoO_4$, with a weight content in $Bi_2Mo_3O_{12}$ of 15% (i.e. 4.3 mol.%) were prepared and tested. The iron content of the solid solution varied from 0 to 100%.

The results are given in Table I and the variations of the rates of formation of acrolein and the selectivity to acrolein observed at 380°C on these catalysts as a function of the iron content in the solid solution are presented in Fig. 1 and 2. When the solid solution is in the α form, the activity increased slowly to reach a maximum for an iron content of 70 to 80% whereas it increased sharply to a maximum for an iron content of 67% when it is in the β form. The increase observed was much more important with a β type solid solution than with a α type since the maximum rate obtained in the first case is five times that obtained in the second case. The loss of activity at high iron content could be explained by the decomposition of ferrous molybdate

($6FeMoO_4 + 3/2\ O_2 \rightarrow Fe_2O_3 + 2Fe_2Mo_3O_{12}$) since both ferric oxide and ferric molybdate have been detected by X-rays diffraction and Mössbauer spectroscopy in the sample after the catalytic run (9).

b) Study of mixture of $Bi_2Mo_3O_{12}$ and $Fe_{0.67}Co_{0.33}MoO_4$ with a fixed iron content and various phase compositions

Four mixtures composed of $Bi_2Mo_3O_{12}$ and $Fe_{0.67}Co_{0.33}MoO_4$, with weight contents in $Bi_2Mo_3O_{12}$ of 7,15,25,and 50% (i.e. 1.8, 4.3, 7.4, 19.3 mol.%) were prepared and tested.

The results are presented in Table II and the variations of the rates of formation of acrolein and the selectivity to acrolein observed at 380°C on these catalysts as a function of their bismuth molybdate content are plotted in Fig. 3 and 4. The activity increased slowly to reach a maximum for a mass content of bismuth molybdate of 50% when the solid solution is in the α form, whereas it increased sharply to a maximum for a mass content of bismuth molybdate of 20% when it is in the β form. The increase observed was again much more important with a β type solid solution than with an α type. The maximum rate obtained in the first case is three times that obtained in the second case.

The analysis by X-ray diffraction after catalysis showed only the presence of the α or β phase of the mixed iron and cobalt molybdates depending upon heat treatment, 380 or 430°C respectively. No phase suspected to be present in the conditions of the catalysis reaction have been detected. This was confirmed by IR spectroscopy and EPR which did not detected any new ferric species (9).

XPS has been used to characterize the three mixtures containing respectively 7,25 ,and 50 weight % of $Bi_2Mo_3O_{12}$ (Table II samples J,K and L). These samples have been characterized before and after catalytic reaction (table III). Bi, Mo, Fe, Co and O have been analyzed. The Mo/O ratio remains equal to 0.25 for all the samples, before and after catalysis which confirms that no new phase was formed since the molybdates suspected to have formed, have a much lower Mo/O ratio (0.17 for Bi_2MoO_6 and $Bi_3FeMo_2O_{12}$). Concerning the Bi/(Fe+Co) ratio, it can first be observed that before catalysis this ratio was always lower than that calculated from chemical analysis. This can be explained by the difference between the particles size of the bismuth molybdate and the iron and cobalt molybdates which is in a ratio of more than 30 as calculated from differences in surface area values, 0.3 and 9 to 22 $m^2.g^{-1}$. Secondly the Bi/(Fe+Co) ratio increased systematically after catalysis which could be explained by the decrease in size of the bismuth molybdate particles or by the covering of the iron and cobalt molybdate particles by the bismuth molybdate or by both effects.

The analysis by EDX-STEM has been focused on a $Bi_2Mo_3O_{12}$-$Fe_{0.67}Co_{0.33}MoO_4$ mixture containing 15 weight % of $Bi_2Mo_3O_{12}$ (sample F). Two types of particles have been observed before catalysis. The first particles type with a size of 0.1 to 0.2 μm only contained mixed iron and cobalt molybdate with an Fe/(Fe+Co) atomic ratio calculated from 16 individual analyses equal to 0.64($\pm$0.03). The second particles type

TABLE I. Catalytic data of the mechanical mixtures of $Bi_2Mo_3O_{12}$ and $Fe_xCo_{1-x}MoO_4$ (with a weight content in $Bi_2Mo_3O_{12}$ of 15% (i.e. 4.3 mol.%)) 380°C; S_{BET} : specific surface area; ACRO : acrolein

sample	x %	S_{BET} $m^2.g^{-1}$	rate of formation 10^{-8} mol.s⁻¹.m⁻² ACRO	selectivities % ACRO	CO₂	rate of formation 10^{-8} mol.s⁻¹.m⁻² ACRO	selectivities % ACRO	CO₂
			with $Fe_xCo_{1-x}MoO_4$ under the α form			with $Fe_xCo_{1-x}MoO_4$ under the β form		
A	0	19	0.49	32	54	0.63	54	38
B	12	16	2.7	51	40	5.8	80	12
C	25	19	6.7	54	45	30	91	8
D	36	10	19	60	37	86	93	6
E	53	12	46	80	17	170	94	6
F	67	9	41	75	24	280	96	3
G	86	12	58	80	18	130	97	3
H	100	22	7.5	46	52	4.1	40	58

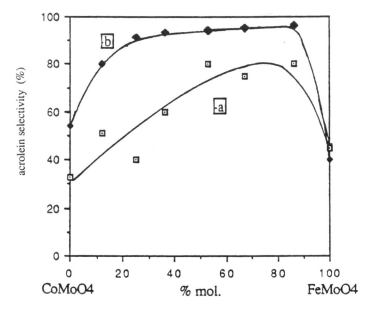

Figure 1. Variations of the selectivity (%) in acrolein on $Bi_2Mo_3O_{12}$-$Fe_xCo_{1-x}MoO_4$ catalysts at 380 °C, versus iron content of the solid solution (x); a with α and b with β $Fe_xCo_{1-x}MoO_4$.

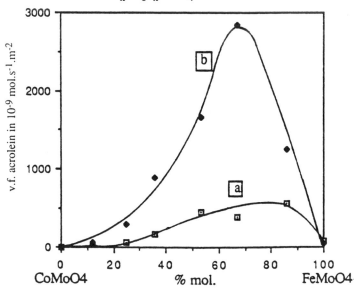

Figure 2. Variations of rate of formation of acrolein on $Bi_2Mo_3O_{12}$-$Fe_xCo_{1-x}MoO_4$ catalysts at 380 °C, versus iron content of the solid solution (x); a with α and b with β $Fe_xCo_{1-x}MoO_4$.

TABLE II. Catalytic data of the mechanical mixtures of $Bi_2Mo_3O_{12}$ and $Fe_{0.67}Co_{0.33}MoO_4$ in the partial oxidation of propene at 380°C; S_{BET} : specific surface area; ACRO : acrolein

sample	weight % of $Bi_2Mo_3O_{12}$	S_{BET} $m^2.g^{-1}$	rate of formation 10^{-8} mol.s^{-1}.m^{-2} ACRO	selectivities % ACRO	selectivities % CO$_2$	rate of formation 10^{-8} mol.s^{-1}.m^{-2} ACRO	selectivities % ACRO	selectivities % CO$_2$
			with $Fe_{0.67}Co_{0.33}MoO_4$ under the α form			with $Fe_{0.67}Co_{0.33}MoO_4$ under the β form		
I	0	17	2.6	22	77	1.8	27	72
J	7	13	37	67	32	200	96	3
F	15	9	41	75	24	280	96	3
K	25	10	49	83	15	280	96	3
L	50	9	100	94	5	150	98	2
M	100	0.3	3.7	94	5	3.7	98	2

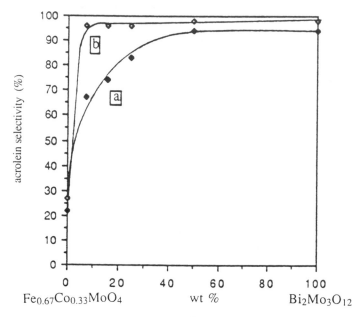

Figure 3. Variations of the selectivity in acrolein on $Bi_2Mo_3O_{12}$-$Fe_{0.67}Co_{0.33}MoO_4$ catalysts at 380 °C, versus $Bi_2Mo_3O_{12}$ weight content; a with α and b with β $Fe_{0.67}Co_{0.33}MoO_4$.

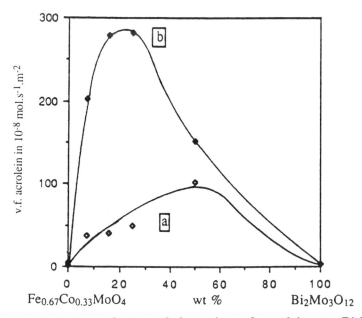

Figure 4. Variations of rate of formation of acrolein on $Bi_2Mo_3O_{12}$-$Fe_{0.67}Co_{0.33}MoO_4$ catalysts at 380 °C, versus $Bi_2Mo_3O_{12}$ weight content; a with α and b with β $Fe_{0.67}Co_{0.33}MoO_4$.

TABLE III. Comparison of the surface elementary ratios of the cations in the mechanical mixtures of $Bi_2Mo_3O_{12}$ and $Fe_{0.67}Co_{0.33}MoO_4$ (B, D, and E) calculated from XPS analysis data before and after catalytic reaction and those calculated from chemical analysis data before catalysis; Σ cat.: sum of all the cations

sample	weight % of $Bi_2Mo_3O_{12}$		Mo/O	(Fe+Co)/Σcat.	Bi/Σcat.	Mo/Σcat.	Bi/(Fe+Co)	Fe/(Fe+Co)
J	7	before catalysis	0.26	0.45	0.0018	0.55	0.0039	0.60
		after catalysis	0.25	0.41	0.036	0.56	0.089	0.63
		chemical analysis	0.25	0.48	0.018	0.51	0.038	0.67
K	25	before catalysis	0.25	0.44	0.012	0.55	0.03	0.57
		after catalysis	0.25	0.36	0.062	0.58	0.17	0.48
		chemical analysis	0.25	0.41	0.067	0.52	0.16	0.67
L	50	before catalysis	0.25	0.44	0.056	0.50	0.14	0.72
		after catalysis	0.26	0.38	0.082	0.54	0.22	0.63
		chemical analysis	0.25	0.31	0.15	0.54	0.49	0.67

with a larger size (2 to 3 μm) contained only the bismuth molybdates. No cobalt and iron was detected in the analyses (table IV and Fig. 5a).

After catalysis the two types of particles were always present. The smaller particles were exactly the same composition and no bismuth was detected. The larger particles had a size reduced by a factor 2 and were not any more composed exclusively of bismuth and molybdenum. The presence of cobalt and iron, in the same proportions was also detected whatever the size of the analyzed surface (from 5 nm^2 to 1μm^2). These results show that the large bismuth molybdates particles are covered with a number of small particles of the solid solution (Table IV and Fig. 5b).

Discussion

The synergy effect observed for the intimate $Bi_2Mo_3O_{12}$ and $Fe_xCo_{1-x}MoO_4$ mixtures may partly be explained by the fact that the presence of iron in the solid solution increases its electric conductivity as it was observed by electrical conductivity measurements (σ multiplied by a factor of 10^4) (7). The electrons stemming from the oxidation of the olefin on the bismuth molybdate are more easily transfered to the solid solution which should play an important role in the redox mechanism. Such synergy effect is possible only if intimate contacts between the two phases exist. It is difficult to imagine that only a mechanical mixture of the two solid phases can give rise to these intimate contacts and a change in the morphology of the sample has to occur. This is what is observed in the case of the study of the $Bi_2Mo_3O_{12}$ and $Fe_{0.67}Co_{0.33}MoO_4$ mixtures. Before catalysis the two phases are separated while after catalysis the large particles of bismuth molybdates are covered with a number of small particles of solid solution as shown by EDX-STEM. These small particles were themselves covered by bismuth molybdate since the Bi/(Fe+Co) ratio as observed by XPS, increased instead of decreasing after catalysis. It can be seen that this ratio becomes for certain catalysts even larger than the chemical ratio (Table III). Such a phenomenon can only be explained by the covering of particles of the solid solution by the bismuth molybdate. A schematic representation of the catalyst particles is shown in Fig. 6. Note that only part of the solid solution particles was deposited on bismuth molybdate particles since some solid solution crystallites were observed without bismuth by EDX-STEM analysis. Such particles have obviously negligeable importance in catalytic activity.

The change of morphology related to the spreading of the bismuth molybdate on the solid solution could be temperature dependent . As a matter of fact the activation of the catalysts was observed but it is difficult to determine whether it is due to the α/β phase transition of the solid solution or to the spreading which occurs in the range of temperature of the polymorphic transition. We may conclude that the observed activation is due to both transformations.

The maximum of activity observed for a relative mass ratio of $Bi_2Mo_3O_{12}$ and $Fe_{0.67}Co_{0.33}MoO_4$ may correspond to the maximum of covering of the bismuth molybdate by the solid solution particles. The difference of the maximum between the α and the β phase could be explained by the fact that the spreading may occur more easily on the β phase. It is not possible to determine if it is the type of polymorph or the heat treatment which is related to this phenomenon.

These results and the comparison between the catalyst particles before and after catalytic run point out the ability for these particles both to exchange electrons and oxygen anions and to change morphology under the conditions of the catalytic reaction with spreading of the oxides one over the other. These two phenomena should be at the basis of the explanation of synergy effect in molybdates based catalysts. The fact that some $Fe_xCo_{1-x}MoO_4$ particles remain free (i.e. not deposited on bismuth molybdate particles) show that even more active and selective catalysts may be obtained in more reliable preparation conditions.

TABLE IV. EDX-STEM analyses of particles of sample F (a $Bi_2Mo_3O_{12}$-$Fe_{0.67}Co_{0.33}MoO_4$ mixture containing 15 weight % of $Bi_2Mo_3O_{12}$) before (**A**) and after catalysis (**B**) shown respectively in Fig. 5a and 5b

analysis	analysed area nm^2	Mo	Fe	Co	Bi
			% mol.		
A before catalysis					
1(total)	1.6 10^7	61.6	4.1	2.6	31.7
2	3.0 10^4	52.4	35.3	12.3	0
3	1.0 10^4	60.5	3.8	2.7	33.0
4	3.0 10^4	54.4	30.2	15.5	0
5	1.0 10^4	54.0	29.6	16.4	0
6	120	59.5	0.7	0.9	39.0
7	120	51.8	30.0	18.2	0
B after catalysis					
1	1.0 10^4	56.7	16.7	9.6	16.0
2	120	54.1	28.5	17.4	0
3	48	51.8	31.3	16.9	0
4	48	54.6	27.6	17.8	0
5	12	58.7	24.3	17.0	0
6	300	61.5	10.2	5.7	22.5
7	300	62.0	9.3	5.2	23.5
8	300	57.0	16.6	9.5	16.9
9	300	54.9	21.0	12.3	11.8
10	300	56.9	20.0	11.3	11.8

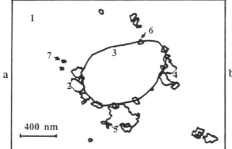
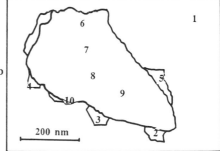

Figure 5. Electron micrograph schematic representation of a particle of sample F (a $Bi_2Mo_3O_{12}$-$Fe_{0.67}Co_{0.33}MoO_4$ mixture containing 15 weight % of $Bi_2Mo_3O_{12}$). The numbers correspond to the EDX-STEM analyses which results are given in Table IV: a) sample before and b) after catalytic reaction.

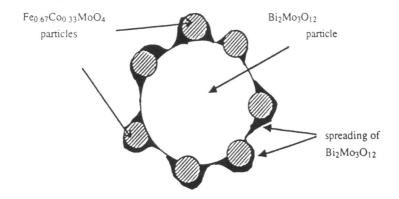

Figure 6. Schematic representation of the catalysts particles for explaining synergy effect for mixtures of $Bi_2Mo_3O_{12}$ and $Fe_{0.67}Co_{0.33}MoO_4$.

Acknowledgments

Financial support for this work by RHONE POULENC company is gratefully acknowledged.

Literature Cited

(1)   M. El Jamal, M. Forissier, G. Coudurier and J.C. Vedrine, in Proceed. of the 9th Intern. Congress on Catalysis, M.J. Phillips and M. Ternan (Ed.), Chem. Soc. of Canada, Ottawa, 4, 1617 (1988).
(2)   J.L. Callahan, R.W. Foreman and F. Veatch, US Patent 2,941,007 (1960).
(3)   J.C. Daumas, J.Y. Derrien and F. Van den Bussche, Fr. Patent 2,364,061 (1976).
(4)   Y. Moro-oka, D.H. He and W. Ueda, in "Symposium on Structure-Activity Relationships in Heterogeneous Catalysis", R.K. Grasselli and A.W. Sleight (Ed.) Stud. in Surf. Sci. and Catal., Elsevier, Amsterdam, 67, 57 (1991).
(5)   L.T. Weng and B. Delmon, Appl. Catal. A, 81, 141 (1992).
(6)   O. Legendre, Ph. Jaeger and J.P. Brunelle, in "New Developments in Selective Oxidation by Heterogeneous Catalyis", P. Ruiz and B. Delmon (Ed.), Stud. in Surf. Sci. and Catal., Elsevier, Amsterdam, 72, 387 (1992).
(7)   H. Ponceblanc, J.M. M. Millet, G. Coudurier, J.M. Hermann and J.C. Védrine, submitted to J. Catal. august 1992.
(8)   P.A.Batist, J. Chem. Tech. Biotechn., 29 451 (1979).
(9)   H. Ponceblanc, Thesis Lyon n° 259-90 (1990).
(10) H. Ponceblanc, J.M. M. Millet, G. Coudurier, O. Legendre and J.C.Védrine, J. Phys. Chem. in press november 1992.
(11) J.C. Vedrine and Y. Jugnet, in "Les techniques physiques d'étude des catalyseurs", B. Imelik and J.C. Vedrine (Ed.), Technip Paris, 365 (1988) Chap.10.
(12) M. Forissier, A. Larchier, L. De Mourgues, M. Perrin and J.L. Portefaix, Rev Phys. Appl., 11 639 (1976).
(13) A.W. Sleight and B.L. Chamberland, Inorg. Chem., 7, 1672 (1968).
(14) H.Ponceblanc, J.M.M.Millet, G.Thomas, J.M.Herrmann and J.C. Védrine, J. Phys. Chem. in press november 1992.

RECEIVED December 2, 1992

Chapter 20

Selective Oxidation of Alkanes, Alkenes, and Phenol with Aqueous H₂O₂ on Titanium Silicate Molecular Sieves

C. B. Khouw, H. X. Li, C. B. Dartt, and M. E. Davis

Department of Chemical Engineering, California Institute of Technology, Pasadena, CA 91125

Titanium containing pure-silica ZSM-5 (TS-1) materials are synthesized using different methods. The activity of the titanium containing catalysts for the oxidation of alkanes, alkenes and phenol at temperatures below 100 °C using aqueous H_2O_2 as oxidant is reported. The relationships between the physicochemical and catalytic properties of these titanium silicates are discussed. The effects of added aluminum and sodium on the catalytic activity of TS-1 are described. The addition of sodium during the synthesis of TS-1 is detrimental to the catalytic activity while sodium incorporation into preformed TS-1 is not. The framework substitution of aluminum for silicon appears to decrease the amount of framework titanium.

The discovery of titanium substituted ZSM-5 (TS-1) and ZSM-11 (TS-2) have led to remarkable progress in oxidation catalysis [1,2]. These materials catalyze the oxidation of various organic substrates using aqueous hydrogen peroxide as oxidant. For example, TS-1 is now used commercially for the hydroxylation of phenol to hydroquinone and catechol [1]. Additionally, TS-1 has also shown activity for the oxidation of alkanes at temperatures below 100 °C [3,4].

Several preparation methods have been reported for the synthesis of TS-1. In this work, we have investigated the physicochemical properties of TS-1 samples synthesized by different preparation methods and tested these materials as catalysts for the oxidation of n-octane, 1-hexene and phenol using aqueous hydrogen peroxide (30 wt%) as oxidant at temperatures below 100 °C. For comparison, TiO_2 (anatase) and the octahedral titanium-containing silicate molecular sieve (ETS-10) [5] have been studied. The effect of the presence of aluminum and/or sodium on the catalytic activity of TS-1 is also discussed.

Experimental

Samples. TS-1 samples were synthesized by modifications of the preparation methods reported in the patent literatures [6,7]. As shown in Table I, samples TS-1(A) and TS-1(B) were crystallized from clear solutions prepared by mixing titanium

0097–6156/93/0523–0273$06.00/0

butoxide (TNBT), tetraethylorthosilicate (TEOS), tetrapropylammonium hydroxide (TPAOH, 1 M) and double distillated water (6). TNBT, TEOS and TPAOH were purchased from Aldrich. The reaction mixture for the crystallization of TS-1(A) was prepared in an ice-bath, while the one for TS-1(B) was mixed at room temperature. TS-1(C) was synthesized by wetness impregnation of a TiO_2-SiO_2 co-precipitate (Type III no.2, obtained from W.R. Grace) with the TPAOH solution (7). Aluminum and/or sodium containing titanium silicate (TAS-1, NaTS-1) were prepared by adding $Al(NO_3)_3$ and/or $NaNO_3$ (from Aldrich) into the TiO_2-SiO_2 co-precipitate/TPAOH mixture. All samples were calcined at 500-550 °C for 10 hours prior to physicochemical characterization and catalytic studies.

The TiO_2 used here was made by hydrolyzing TNBT in distillated H_2O with subsequent calcination at 500 °C. ETS-10, which is a titanium silicate molecular sieve with titanium in octahedral coordination, was provided by Engelhard, Co. For comparison, pure-silica ZSM-5 was also synthesized in the absence of alkali metal cations. Its synthesis involves the use of tetrapropylammonium bromide (TPABr) and piperazine.

Table I. Sample Preparations

| Sample | Composition | | | Crystallization | |
	Si/Al	Si/Na	Si/Ti	Temp.(°C)	Time (d)
TS-1(A)[a]	-	-	30	175	10
TS-1(B)[a]	-	-	30	175	4
TS-1(C)[b]	-	-	56	150	10
TAS-1(D)[b]	50	-	56	175	7
TAS-1(E)[b]	200	-	56	175	7
NaTS-1[b]	-	10	56	175	7
NaTAS-1[b]	50	10	56	175	7

[a.] synthesized from a solution containing TEOS, TNBT, TPAOH and water.
[b.] synthesized from a mixture containing TiO_2-SiO_2 (Grace), TPAOH, water and $NaNO_3$ and/or $Al(NO_3)_3$.

Characterization. X-ray powder diffraction (XRD) patterns were collected on a Scintag XDS-2000 diffractometer that is equipped with a liquid-nitrogen-cooled Germanium solid-state detector using Cu-Kα radiation. Fourrier Transform Infrared (FTIR) spectra were recorded on a Nicolet System 800 Spectrometer using KBr pellets that contain 2 wt% of sample. Raman spectra were obtained on a Nicolet Raman accessory. Nitrogen adsorption isotherms were collected at liquid nitrogen temperature (77 K) on an Omnisorp 100 analyzer.

Catalytic Reactions. Phenol hydroxylation was carried out in a batch reactor using 30% aqueous H_2O_2 in acetone at reflux conditions at a temperature of ~80 °C. Hydrogen peroxide was introduced slowly via a syringe pump. The products were analyzed on a HP 5890 Series II Gas Chromatography (GC) equipped with a 50 m long HP-1 (non-polar) capillary column.

The oxidation of n-octane and the epoxidation of 1-hexene were performed in a 25 ml Parr reactor using 30% aqueous H_2O_2 as an oxidant and acetone as solvent at 100 °C and 80 °C, respectively and stirred at 500 RPM. Prior to product analysis, the product mixtures were diluted with acetone in order to obtain a single liquid-phase. The products were analyzed on a HP 5890 Series II GC equipped with a 25 m long HP-FFAP (polar) capillary column.

Results and Discussion

Titanium Silicates. XRD data show that all the TS-1 samples are very crystalline and have the MFI structure. The TiO_2 obtained after calcination has the anatase structure.

Figure 1 shows the IR spectra of TS-1(A), TS-1(B), TS-1(C), TiO_2-SiO_2 co-precipitate and pure-silica ZSM-5. The absorption band at 960 cm^{-1} is characteristic of TS-1 *(8)*. All of TS-1 samples used in this study show this band. The band is not present in TiO_2 and ETS-10 (not shown in the figure). However, this band is present in TiO_2-SiO_2 which is the precursor to TS-1(C). The relative intensities of the peak at 960 cm^{-1} are listed in Table II.

Table II. Catalytic hydroxylation of n-octane, 1-hexene and phenol with aqueous H_2O_2

Catalyst	Si/Ti[a]	Conversion[b] (%)			IR peak ratio[f]
		n-octane[c]	1-hexene[d]	phenol[e]	
TS-1(A)	30	19	4.3	8	1.11
TS-1(B)	30	15	4.0	3	0.91
TS-1(C)	56	12	4.7	4	1.05
TiO_2-SiO_2	56	<1 (0)[g]	0 (3.1)[g]	0	0.47
TiO_2	0	<1	0	0	-
ETS-10	5	0 (0)[g]	0 (0)[g]	0	0
ZSM-5	-	<1	0	0	0

[a]. For TS-1, the values reported are Si/Ti in the solution.
[b]. Conversion is based on the substrates.
[c]. Reaction conditions: 100 mg catalyst, 30 mmol of n-octane, 24 mmol of H_2O_2 (30% in H_2O), 6 ml acetone, 100 °C, stirred at 500 RPM for 3 hr.
[d]. Reaction conditions: 100 mg catalyst, 40 mmol of 1-hexene, 29 mmol of H_2O_2 (30% in H_2O), 6 ml acetone, 80 °C, stirred at 500 RPM for 1.5 hr.
[e]. Reaction conditions: 100 mg catalyst, 25 mmol of phenol, 5 mmol of H_2O_2 (30% in H_2O, injected at a rate of 1 ml/hr), 5 ml acetone, reflux temp., stirred for 2 hr.
[f]. Ratio of peaks areas of the band at 960 cm^{-1} to the one at 800 cm^{-1}.
[g]. tert-Butyl hydroperoxide (3 M in 2,2,4 trimetylpentane) is used as oxidant.

The catalytic activity of the materials used in this study are shown in Table II. The hydroxylation of phenol produces a mixture of catechol and hydroquinone. The oxidation of n-octane yields a mixture of 2-, 3- and 4- octanols and octanones, however no terminal alcohol is observed. The epoxidation of 1-hexene gives 1,2-epoxy hexane without any hexanediol observed.

The activity data confirm that an IR absorption band at 960 cm^{-1} is a necessary condition for titanium silicates to be active for the selective oxidation of hydrocarbons with aqueous H_2O_2 as suggested by Huybrechts et al. (9). However, this band is not a sufficient condition for predicting the activity of the TS-1 catalyst. Although TS-1(B) and TS-1(C) show intensities for the 960 cm^{-1} band similar to TS-1(A), their activities are different. First of all, the reaction data reveal that TS-1(A) is much more active than TS-1(B) for phenol hydroxylation, while both samples show similar activity for n-octane oxidation and 1-hexene epoxidation. Therefore, the presence of the IR band at 960 cm^{-1} in TS-1 catalysts may correlate with the activities for the oxidation of n-octane and the epoxidation of 1-hexene but not for phenol hydroxylation. However, note that the amorphous TiO_2-SiO_2 also has an IR absorption band at 960 cm^{-1} and it does not activate either substrate.

Raman spectra of ETS-10, TS-1(A), TS-1(B) and anatase are shown in Figure 2. The spectrum for TS-1(A) (compared to the one for TiO_2 anatase) reveals that TS-1(A) contains anatase. No anatase is detected in TS-1(B), TS-1(C) (not shown in the figure) and ETS-10. The amount of TiO_2 in TS-1(A) is small and probably nanophase because 1) it is not observable by XRD; and 2) the nitrogen adsorption data from TS-1(A) and TS-1(B) indicate that they have pore volumes that are similar to that obtained from pure-silica ZSM-5. Figure 3 illustrates the UV-VIS diffuse reflectance spectrum of TS-1(A). The spectrum shows two bands at ~220 nm and ~270 nm, which can be assigned to framework (10) and extra-framework nanophase (11) titanium species, respectively. The other TS-1 samples do not show the band for extra-framework titanium. From the catalytic activity data, it appears that the presence of TiO_2 anatase in a titanium silicate (TS-1(A)) does not inhibit its catalytic activity for selective oxidation. Although TiO_2 anatase itself is not active, TS-1(A), which contains TiO_2 anatase, shows higher activity for phenol hydroxylation compared to the other TS-1 catalysts. To study this phenomena, TNBT was hydrolyzed to TiO_2 anatase in the presence of TS-1(C) and pure-silica ZSM-5. Table III shows the catalytic activity of these materials

Table III. Influence of TiO_2 on the activity of TS-1(C) and pure-silica ZSM-5 for phenol hydroxylation

Catalyst	Phenol Conversion (%)[a]
TS-1(C) + TiO_2[b]	7
TS-1(C) + TiO_2[c]	4
ZSM-5 + TiO_2[b]	0

a. Reaction conditions: 200 mg catalyst, 25 mmol of phenol, 5 mmol of H_2O_2 (30% in H_2O, injected at a rate of 1 ml/hr), 5 ml acetone, reflux temp., stirred for 2 hr.

b. TiO_2 (50%wt) was impregnated by hydrolyzing Ti-butoxide in the presence of uncalcined TS-1(C)/pure-silica ZSM-5, then calcined at 550 °C.

c. TiO_2 (50%wt) was added as a physical mixture.

for phenol hydroxylation. The conversion for phenol hydroxylation was increased by the impregnation of TiO_2 on TS-1(C). However, TiO_2 impregnated pure-silica ZSM-5 does not show any activity. When TiO_2 anatase was physically mixed with TS-1(C),

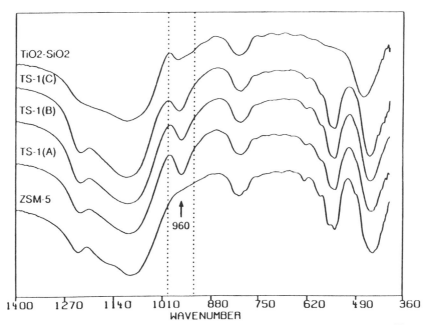

Figure 1. Infrared spectra of titanium containing materials and pure-silica ZSM-5.

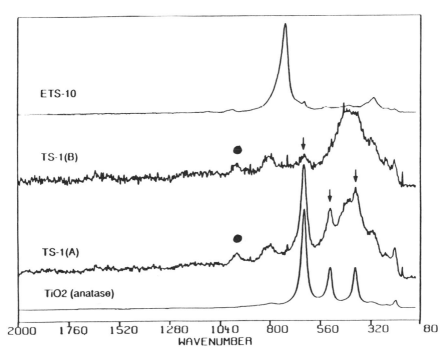

Figure 2. Raman spectra of titanium containing materials. Band marked by ● is at 960 cm⁻¹.

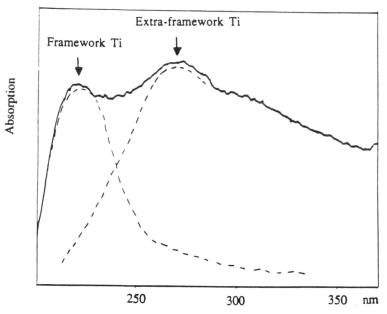

Figure 3.   UV-visible diffuse reflectance spectra of TS-1(A).

the activity of this physical mixture is the same as pure TS-1(C). These results suggest that TiO_2 works synergistically with the titanium in the framework of TS-1.

In contrast to the high activity of TS-1(C), its amorphous precursor, TiO_2-SiO_2 shows no catalytic activity for alkane, alkene and phenol hydroxylations using aqueous H_2O_2 as oxidant (see Table II). However, TiO_2-SiO_2 is catalytically active for epoxidation using a non-aqueous alkyl hydroperoxide as oxidant. Moreover it has also been reported that silica supported titanium is used commercially for epoxidation of alkenes using a non-aqueous alkyl hydroperoxide as oxidant (1). It seems that the mechanism for epoxidation is different than for the other reactions shown here. Although the state of Ti in TiO_2-SiO_2 co-precipitate may be the same as the one in TS-1(C) (IR absorption at 960 cm^{-1}), the environmental conditions are different. This may suggest that the presence of titanium in a hydrophobic environment (e.g. inside the ZSM-5 micropores) is necessary for epoxidation of alkenes using aqueous H_2O_2 as oxidant. For alkane and phenol hydroxylations, clearly the Ti environment is unique amongst the materials investigated in this study.

ETS-10 shows no catalytic activity. This observation indicates that titanium in octahedral coordination in ETS-10 is not active for selective oxidation and epoxidation using either aqueous hydrogen peroxide or non-aqueous alkyl hydroperoxide as the oxidant.

Effect of sodium and aluminum on TS-1. The catalytic activities of aluminum and/or sodium containing TS-1 are depicted in Table IV. The data show that the addition of aluminum during the synthesis of TS-1 yields a material (TAS-1(D)) that has a lower conversion of n-octane oxidation and a smaller IR peak ratio. The existence of the acid sites due to the incorporation of aluminum into the framework of TS-1 may accelerate the decomposition of H_2O_2 to water and oxygen during the reaction. However, reducing the number of acid sites by exchanging with sodium ions only increases the conversion by 1% (Na/TAS-1(D)). Therefore, the addition of aluminum into the synthesis mixture most likely reduce the amount of titanium present in the sample.

Table IV. n-Octane oxidation on aluminum and/or sodium containing TS-1

Catalyst	n-Octane Conversion (mol %)	IR peak ratio
TAS-1(D)	7	0.36
TAS-1(E)	12	0.96
NaTS-1	0	0
NaTAS-1	0	0
Na/TS-1(C)[a]	11	1.05
Na/TAS-1(D)[b]	8	0.36

[a.] TS-1(C) impregnated with $NaNO_3$ (10%wt).
[b.] TAS-1(D) exchanged with 1 M solution of $NaNO_3$.

The addition of sodium during the synthesis of TS-1 completely eliminates the activity for n-octane oxidation and also the IR band at 960 cm^{-1} (this IR band is present in the amorphous precursor, TiO_2-SiO_2). It has been shown (12) that the presence of sodium in the synthesis gel prevents the incorporation of titanium into the zeolite framework. However, the addition of sodium after the zeolite crystallizes does not have any significant effect on the catalytic activity, as shown by the catalytic data in Table IV.

Conclusions

Several preparation methods have been used to synthesize TS-1. TS-1 is catalytically active for the oxidation of alkanes, alkenes and phenol using aqueous H_2O_2 as oxidant at temperatures below 100 °C. Although some of the physicochemical properties of these materials are similar, significant differences are found amongst the catalysts prepared in different ways. Framework titanium in TS-1 appears to be necessary for alkane and phenol hydroxylation. For epoxidation, the presence of titanium in a hydrophobic environment is necessary for catalytic activity if aqueous H_2O_2 is used as oxidant. Octahedral titanium in ETS-10 and titanium species in the TiO_2-SiO_2 co-precipitate are not catalytically active for the oxidation of alkanes, alkenes and phenol using aqueous H_2O_2 as oxidant.

The addition of aluminum during the synthesis of TS-1 reduces its activity for n-octane oxidation. The presence of sodium in the synthesis gel of TS-1 completely eliminates the catalytic activity for alkane oxidation. However, the presence of sodium in preformed TS-1 does not have a significant effect on its catalytic activity.

Acknowledgement

Support of this work was provided by the Amoco Oil Co. We thank Dr. S. M. Kuznicki for providing a sample of ETS-10 and Dr. R. J. Davis for recording the UV-VIS diffuse reflectance spectrum of TS-1 samples.

Literature Cited

(1) Notari, B., *Stud. Surf. Sci. Catal.* 1988, *37*, 413.
(2) Reddy, J. S., Kumar, R. , Ratnasamy, P., *Appl. Catal.* 1990, *58*, L1.
(3) Huybrechts, D. R. C., DeBruyker, L., Jacobs, P. A., *Nature*, 1990, *345*, 240.
(4) Tatsumi, T., Nakamura, M., Negishi, S., Tominaga, H., *J. C. S. Chem.Commun.*, 1990, 476.
(5) Kuznicki, S. M., U.S. Patent 4,853,202, 1989.
(6) Taramasso, M., Perego, G., Notari, B., U.S. Patent 4,410,501, 1983.
(7) Padovan, M., Leofanti, G., Roffia, P., Eur. Patent 0 311 983, 1989.
(8) Kraushaar, B., Van Hooff, J. H. C., *Catal. Lett.*, 1988, *1*, 81.
(9) Huybrechts, D. R. C., Vaesen, I., Li, H. X., Jacobs, P. A., *Catal. Lett.*, 1991, *8*, 237.
(10) Reddy, J. S., Kumar, R., *J. Catal.*, 1991, *130*, 440.
(11) Kim, Y. L., Riley, R. L., Huq, M. J., Salim, S., Le, A. E., Mallouk, T. E., *Mat. Res. Soc. Symp. Proc.*1991, *233*, 145.
(12) Bellussi, G., Fattore, V., *Stud. Surf. Sci. Catal.* 1991, *69*, 79.

RECEIVED November 6, 1992

Chapter 21

Nature of Vanadium Species in Vanadium-Containing Silicalite and Their Behavior in Oxidative Dehydrogenation of Propane

G. Bellussi[1], G. Centi, S. Perathoner, and F. Trifirò

Department of Industrial Chemistry and Materials, University of Bologna, Via le Risorgimento 4, 40136 Bologna, Italy

V-containing silicalite samples prepared hydrothermally were characterized using a combination of physicochemical techniques and the activity of these silicalites in the oxidative dehydrogenation of propane using O_2 and N_2O was studied. FT-IR data on the mechanism of propane transformation in the presence of gaseous oxygen on these samples were also reported. The results indicate the presence of various types of vanadium species, and in particular, the formation of small amounts of a tetrahedral V^{5+} species stabilized in its configuration by interaction with the framework. This species probably forms at defect sites and is the more active and selective species in the oxidative dehydrogenation ot propane. Activated oxygen species generated by the interaction of O_2 or N_2O with reduced vanadium sites are suggested to be responsible for the selective transformation of propane to propylene. However, the good selectivity of this catalyst in propylene formation from propane is also connected to the type of chemisorption of the intermediate propylene on these tetrahedral V^{5+} sites and to the relative inertness towards its further transformation due to the specific coordination environment of vanadium. In particular, the data indicate that propylene is coordinatively adsorbed as a π-complex with the CH_3 group interacting with a nearlying weakly basic silanol. Infrared darta suggest that the rate of formation of propylene from propane in the presence of O_2 is higher than the rate of its consecutive transformation to surface oligomers or acrolein and acrylic acid, probably intermediates to aromatics and COx, respectively. The model of the coordination environment and the peculiar characteristics of these tetrahedral V^{5+} sites in the silicalite in relation to their selective behavior in propane oxidative dehydrogenation are discussed.

Transition-metal substituted or modified zeolites are currently receiving increasing attention as gas-phase heterogeneous partial oxidation catalysts, because they offer the

[1]Current address: Eniricerche, Via Maritano 26, San Donato Milanese (MI), Italy

opportunity for well defined and isolated active centers in an ordered oxide matrix. Consequently, they can provide useful information about the structure-activity relationship in selective oxidation reactions. Furthermore, the inclusion of the transition metal in the zeolite framework or the stabilization effect due to the coordination at defined sites of the zeolite framework can stabilize the transition metal in an unusual coordination with a consequent change in its reactivity behavior. It is thus possible to expect a change in the specific catalytic behavior of the transition metal, in addition to possible effects of shape-selectivity.

V-containing silicalite, for example, has been shown to have different catalytic properties than vanadium supported on silica in the conversion of methanol to hydrocarbons, NO_x reduction with ammonia and ammoxidation of substituted aromatics, butadiene oxidation to furan and propane ammoxidation to acrylonitrile (*1 and references therein*). However, limited information is available about the characteristics of vanadium species in V-containing silicalite samples and especially regarding correlations with the catalytic behavior (*1- 6*).

Recently, interesting preliminary results have also been obtained in propane oxidative dehydrogenation to propylene (*2*). In this reaction, several advantages can be considered in having well defined vanadium atoms stabilized at specific coordination sites of the zeolite framework: *i)* atomically dispersed vanadium atoms, *ii)* a coordination of vanadium which limits alkene readsorption and activation through an allylic-type mechanism and *iii)* a coordination environment where the strength of bridging V-O-S (S = support) oxygen is sufficiently high to avoid O insertion on activated alkenes. In fact, in alkane oxidative dehydrogenation the more critical problem is not the selective activation of the alkane, but rather the inhibition of the consecutive transformation of the alkene formed. The factors cited favour this inhibition.

Experimental

V-containing silicalite (Al- and Na-free) samples were prepared hydrothermally and then treated with an ammonium acetate solution at room temperature in order to remove extralattice vanadium. Three samples with SiO_2/V_2O_3 ratios of 117, 237 and 545, respectively, were prepared. Hereinafter these samples will be referred to as follows: *V-Sil117*, *V-Sil237* and *V-Sil545*. Details on the preparation procedure, and characterization of the samples have been reported previously (*1,2*).

Catalytic tests were performed in an isothermal flow quartz reactor apparatus under atmospheric pressure, provided with on-line gas chromatographic (GC) analysis of the reagent and products by two GC instrument equipped with flame ionization and thermoconducibility detectors. The activity data reported refers to the behavior after at least two hours of time on stream, but generally the catalytic behavior was found to be rather constant in a time scale of around 20 hours.

Fourier-transform infrared (FT-IR) spectra (resolution 2 cm^{-1}) were recorded with a Perkin-Elmer 1750 instrument in a cell connected to grease-free evacuation and gas manipulation lines. The self-supporting disk technique was used. The usual pretreatment of the samples was evacuation at 500°C.

Results

Characterization Data. A complete characterization of V- containing silicalite samples has been reported in a previous paper (*1*), but the main relevant aspects useful for a better understanding of the nature of vanadium species in V- silicalite samples and for the correlation with the catalytic behavior will be briefly summarized.

The XRD powder patterns of V-containing silicalite samples indicate in all cases the presence of only a pentasyl-type framework structure with monoclinic lattice symmetry, characteristic of silicalite-1; no evidence was found for the presence of vanadium oxide crystallites. The analysis of cell parameters of *VSil545* does not indicate significant modifications with respect to those found for pure silicalite-1. This is in agreement with that expected on the basis of the small amount of V atoms present in V-containing silicalite.

The wide line ^{51}V-NMR spectra of *VSil117*, *VSil237* and *VSil545* samples are characterized by a symmetrical sharp line centered at -480 ppm. However, in *VSil117* and possibly also in *VSil237*, the narrow symmetrical sharp line overlaps a broad signal in the 0-(-1000) ppm range which may suggest the presence in these samples of an additional V^{5+}- species in an undefined environment. The line-shape observed for *VSil* samples can be interpreted as being due to the presence of V^{5+} sites in a nearly symmetrical tetrahedral environment with relatively short (about 0.160- 0.165 nm) V-O bonds, which, however, differ from those present in an orthovanadate such as Na_3VO_3.

The line-shape of this V^{5+} tetrahedral species does not change after evacuation of the sample or exposure to moisture, whereas upon evacuation the submonolayer tetrahedral V^{5+} species detected on SiO_2 (*7*) or other oxides easily coordinate water molecules after exposure of the samples to ambient conditions and thus reform the more stable octahedral coordination. This indicates that vanadium is stabilized in nearly tetrahedral coordination by a direct specific interaction with the silicalite framework and the interaction is stronger than that observed for vanadium supported on silica.

ESR analysis of *VSil117* indicates the presence of isolated nearly octahedral vanadyl species and of vanadium-oxide polynuclear species containing pairs of V^{3+} ions and of VO^{2+} ions. Compared to *VSil117*, *VSil237* and *VSil545* are less heterogeneous and no polynuclear V-oxide species are present, but rather only isolated nearly octahedral VO^{2+} sites. The ESR parameters are very close to those found for VO^{2+} sites in samples obtained by solid-state reaction in air of V_2O_5 with H- ZSM5 (*8*) and can be attributed to isolated V^{4+} ions in the zeolite interior near to a charge- compensating (OH) site. The migration of V inside the zeolite and the spontaneous reduction of V^{5+} to V^{4+} occurring during calcination derives from the interaction of V with the strong Brønsted groups associated with Al sites in the zeolite framework. The correspondence of ESR parameters of isolated VO^{2+} species in V-containing silicalite samples with those observed in V-ZSM5 samples suggests that the vanadyl groups are probably inside the zeolitic channels of the silicalite and near to -OH groups.

In *VSil237* and *VSil545* samples a new signal with hyperfine structure appears after reduction with H_2 at 773 K. The signal can be attributed to V^{4+} ions in a distorted tetrahedral environment. A new signal with superhyperfine structure appears in the spectrum recorded at 77 K after admission at 293 K of small amounts of gaseous oxygen on the prereduced *VSil545* sample and subsequent evacuation. The new signal overlaps the signal attributed to V^{4+} in a tetrahedral environment and indicates the formation of an O_2^- species by electron transfer from the tetrahedral V^{4+} species to O_2.

The UV-visible diffuse reflectance spectra of *VSil545* shows a well defined band centred at 26000 cm^{-1} and two further broad bands at about 38000 and 43000 cm^{-1}. *VSil117* shows these bands, as well as additional bands at about 30000 and 34000 cm^{-1}. The impregnation of pure silicalite or SiO_2 with ammonium vanadate and subsequent calcination led, on the contrary, to quite different spectra. The presence of an intense absorption band at 26000 cm^{-1} is thus characteristic of V-containing silicalite prepared hydrothermally, and tentatively of V sites specifically interacting with the zeolite framework. The position of this band suggests the presence of a short (double) vanadium-oxygen bond. The second CT band, however, does not agree with that expected for a distorted octahedral or square-pyramidal environment, but rather is indicative of a nearly tetrahedral field. The interpretation of the UV-Visible DR spectra of *VSil545* thus suggests the presence of a V^{5+}-species characterized by a short vanadium-oxygen bond (about 0.16 nm) and three slightly longer (about 0.165-0.170 nm) V-O bonds, according to the correlation observed between $(VO_4)^{3-}$ charge-transfer transitions and V-O distance.

UV-Visible diffuse reflectance spectra also show that vanadium is mainly present as V^{5+} in V-containing silicalite samples. TPR and XPS results are in agreement with this conclusion. In addition, XPS data indicate that V is homogeneously dispersed in *VSil545*, whereas in *VSil117* part of the vanadium is segregated on the external surface of the silicalite samples.

The interaction of vanadium with the silicalite framework induces modifications in the surface acidity properties that were characterized by IR spectroscopy using suitable probe molecules. The analysis of the IR hydroxyl stretching region does not provide evidence of specific differences between the V-containing silicalite (*VSil545*) and pure silicalite (*Sil*). In order to characterize further differences in the Brønsted acidity, the adsorption of ammonia and pyridine on *VSil545* and pure silicalite (*Sil*) was carried out. Additional tests were also performed using CD_3CN and *t*-butyl cyanide as probe molecules to differentiate between the presence of Lewis sites in internal or external positions of the zeolite crystals and surface heterogeneities. On the basis of (i) the higher steric hindrance of *t*-butyl cyanide which does not allow its reaction with Lewis acid sites inside the zeolite crystals, and (ii) the presence of stronger additional sites in the V-containing silicalite as compared to pure silicalite (as shown by deuterated acetonitrile adsorption), it can be concluded that very weak Lewis acid sites are present on the external surface of both *VSil545* and *Sil*, but additional stronger Lewis acid sites are present inside the zeolite channels in V-containing silicalite, and are reasonably related to vanadium sites. Ammonia temperature programmed desorption (NH_3-TPD) results are in agreement with these conclusions.

Propane Oxidative Dehydrogenation on V-containing Silicalite. Reported in

Figure 1 is the catalytic behavior of *VSil545* in propane oxidative dehydrogenation to propylene. Selectivities to propylene in the range of 60-80% are obtained up to propane conversions of about 20-25% and reaction temperatures up to around 450- 500°C. For higher reaction temperatures and conversions the selectivity decreases due both to the formation of carbon oxides and of aromatics. As compared to pure silicalite, a significant increase in both the selectivity to propylene and the activity in propane conversion is observed.

As the amount of vanadium in the silicalite increases (*VSil117* vs. *VSil545*), the selectivity to propylene decreases with a parallel increase in the formation of CO_x (Figure 2). The global activity of the catalyst is higher in the samples containing higher amounts of vanadium (*VSil117*), but the specific rate of propylene formation (moles of propylene formed per second and per V atom) is higher for *VSil545* in which only tetrahedral V^{5+} species are present. This is further evidences of the role this V species plays in the selective transformation of propane to propylene. (*See also Figure 3.*)

In the absence of gaseous oxygen, the activity of the catalyst decreases considerably indicating the specific role of gaseous oxygen in the mechanism of propane activation. When N_2O is used as the oxidizing agent instead of O_2, the activity increases considerably (Fig. 4) (similar conversions are obtained for reaction temperatures around 100-150°C lower using N_2O rather than O_2) and the selectivity to propylene increases up to values in the 90-95% range. This is in agreement with the formation of the more active O^- species by reaction of N_2O with the reduced vanadium sites. The results in terms of selectivity are much better as compared to other Me-silicalites, confirming the peculiar role of vanadium in the formation of selective sites of propane oxidative dehydrogenation (2). The activity both in the presence of N_2O or O_2 is higher than without oxidizing agents or in the absence of the catalyst (Fig. 4).

Infrared Characterization of the Conversion of Propane on V- containing Silicalite. In order to obtain more information about the mechanism of propane transformation on the V-containing silicalite in the presence of gaseous oxygen, the nature of the adsorbed species formed after contact with a mixture of propane and oxygen was characterized by infrared spectroscopy. The study was carried out both at room temperature and at 200°C, but a better identification of the evolution of the adsorbed species was possible at the lower temperature. The study was carried out on *VSil545* after activation by outgassing at 500°C into the IR cell. This pretreatment probably induces a partial reduction of the vanadium, analogous that observed by ESR spectroscopy after similar treatment (1).

In the absence of gaseous oxygen, propane is only physisorbed on *VSil545* at room temperature, as shown by the presence of v_{CH} bands in the 3000-2800 cm^{-1} region and δ_{CH} bands in the 1300-1600 cm^{-1} region (Fig. 5b). Bands in the 2000-1600 cm^{-1} region are due to the overtones of fundamentals of skeletal vibrations of the silicalite framework and indicate that the crystalline structure is preserved after the evacuation procedure. In comparison to gaseous propane, there is an increase in the relative intensities of the v_sCH_3 (2875 cm^{-1}) and $v_{as}CH_2$ (2936 cm^{-1}) and δ_sCH_3 (1370 cm^{-1}) bands with respect to the main band at 2964 cm^{-1} ($v_{as}CH_3$). This modification is due to the weak interaction with the surface, but no shift in the frequency of the bands is observed. A slight perturbation in the v_{OH} region (Fig. 5) is also observed indicating

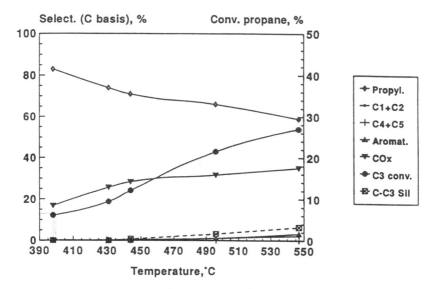

Figure 1. Propane oxidative dehydrogenation to propylene on VSil545. *Exp. conditions*: flow reactor tests with 2.8% C₃, 8.4% O₂ in helium. 4.2 g of catalyst with a total flow rate of 3.1 L/h (STP conditions).

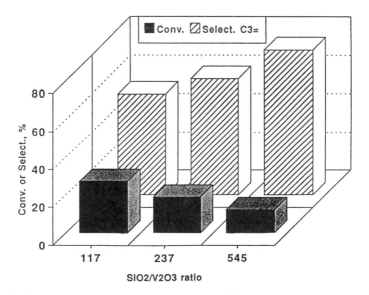

Figure 2. Comparison of the catalytic behavior of *VSil* samples in propane oxidative dehydrogenation to propylene. Conversion of propane and selectivity to propylene at 470°C. Exp. conditions as in Fig. 1.

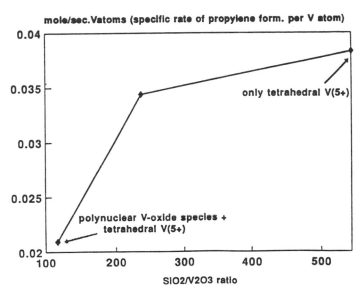

Figure 3. Specific rate of propylene formation at 500°C per vanadium atom in *VSil* samples as a function of the SiO_2/V_2O_3 ratio. Exp. conditions as in Fig. 1

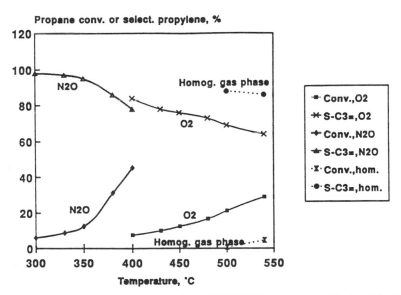

Figure 4. Comparison of the behavior of VSil545 in propane oxidative dehydrogenation using N_2O or O_2 as oxidizing agents. Exp. conditions as in Fig. 1. The dotted lines represent the propane conversion and propylene selectivity observed in the absence of the catalyst (homogeneous gas phase). The activity of the catalyst in the absence of O_2 or N_2O is similar to that observed in the homogeneous gas phase, but the selectivity to propylene (around 50-60%) is lower.

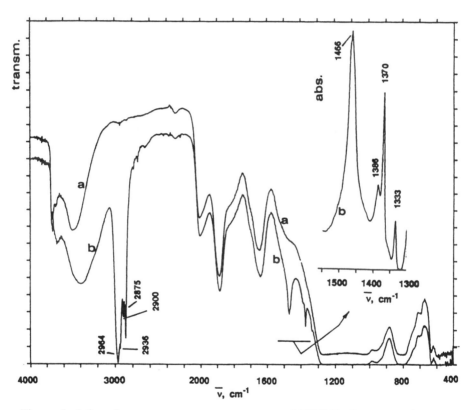

Figure 5. Infrared spectrum at room temperature of *VSil545* after evacuation at 500°C (**a**) and after contact at room temperature with 50 torr of propane (**b**). Reported in the inset is the expansion in the 1300-1550 cm^{-1} region of spectrum b after subtraction of spectrum a. The background spectrum due to gaseous species has been subtracted.

a weak interaction of physisorbed propane with free silanols. Physisorbed propane is easily removed by evacuation at room temperature.

In contact with both propane and oxygen, physisorbed propane is the main adsorbed species, but a new adsorbed species characterized by a band centred at 1623 cm^{-1} may be observed (Fig. 6B). Simultaneously a stronger perturbation in the ν_{OH} region is noted (Fig. 6A). The band at 3728 cm^{-1}, attributed to free silanols (1), disappears leaving a less intense band at higher frequency (3742 cm^{-1}). At the same time a broader band centred at about 3664 cm^{-1} appears. By evacuation at room temperature, all the adsorbed species disappear (only a weak band centred at 1623 cm^{-1} remains) and the original spectrum in the ν_{OH} region is restored (Fig. 6, spectrum c).

The shift in the surface hydroxyl absorption band (about 65 cm^{-1}) and the simultaneous presence of an adsorption band at 1623 cm^{-1}, reasonably attributed to $\nu_{C=C}$ shifted by about 30 cm^{-1} in respect to $\nu_{C=C}$ in the gaseous propylene due to the coordination to an unsaturated cation, indicates the formation of propylene coordinated to the surface in the form of a π-complex with the CH$_3$ group bonded to a hydroxyl group (9):

The spectrum is in agreement with that observed for room temperature adsorption on γ-Al$_2$O$_3$ (9). It should be noted that acidic hydroxyls on the oxide surface usually give rise to stable hydrogen-bonded complexes with the involvement of the alkene double bond (9,10). When the acid strength of the hydroxyl is high as in ZSM zeolites, the complete transfer of the proton to the alkene to give a carbonium ion also occurs at room temperature. In the presence of transition metal oxides containing ions in high oxidation states (V^{5+}, Mo^{6+}) alkene interacts forming allyl-type species or alcoholate species (9,10). All these species give rise to different IR bands than those observed for *VSil545* (Fig. 6).

Therefore, the formation of a π-complex of propylene bonded to a vanadium site and to a nearlying hydroxyl group after contact of a propane and oxygen mixture with activated *VSil545* indicates some interesting aspects of the surface reactivity of V-containing silicalite:

i) The interaction of gaseous oxygen with the activated *VSil545* leads to the formation of dehydrogenation sites able to form propylene from propane even at room temperature. Reasonably, these oxidative dehydrogenating sites can be attributed to activated oxygen species such as (O$_2$)$^-$ (11) formed by interaction of O$_2$ with reduced V sites generated in the stage of evacuation of the zeolite before IR studies.

ii) The formation of a π-bonded propylene instead of an alcoholate or allylic-type species as observed for example for room temperature propylene interaction with V-TiO$_2$ (10), indicates the different reactivity of V sites in the silicalite with respect to V supported on TiO$_2$ towards the reaction of H-abstraction (allyl formation) or O-insertion (alcoholate formation).

iii) The formation of a propylene π-complex with the CH$_3$ group interacting with a vicinal OH group (see Scheme 1) indicates the presence of silanol groups lying near to the vanadium sites, in agreement with the data on the characterization of V-containing silicalite (1). In addition, the formation of this adsorbed species indicates the presence of a weak basic hydroxyl group, reasonably due to the inductive effect connected to the vicinal V sites interacting with the silicalite framework.

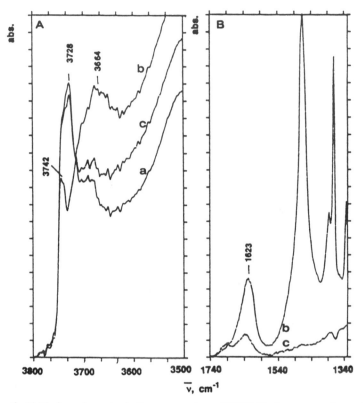

Figure 6. (A) Infrared spectra in the v_{OH} region of *VSil545* after evacuation at 500°C (a), after contact at r.t. with 50 torr of propane and 1 torr of oxygen (b) and after subsequent r.t. evacuation (c). (B) IR spectra of the adsorbed species in the 1340-1740 cm^{-1} region after treatments as for Fig. 6A; the contributions due to evacuated sample and gaseous species have been subtracted from these spectra.

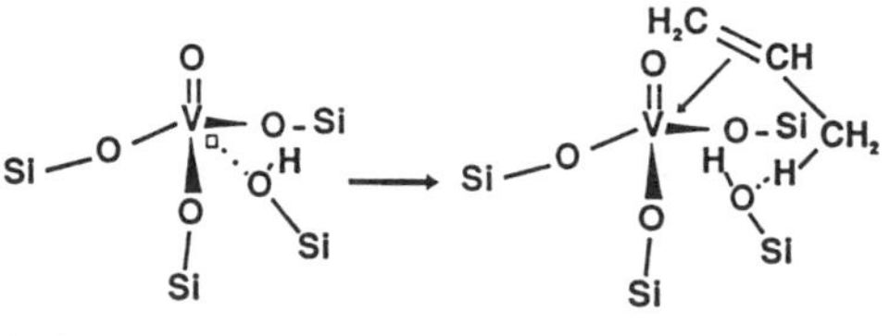

Scheme 1. Model of the room temperature chemisorption of intermediate propylene on *VSil545*.

After 48 hours of contact of *VSil545* with the propane and oxygen mixture, two poorly resolved bands appeared at 1684 and 1705 cm^{-1} as well as a shoulder centred at 1423 cm^{-1} (Fig. 7a). The same spectrum is obtained after 1 hour of contact at 200°C of the propane and oxygen mixture with *VSil545* (Fig. 7b).

The two bands at around 1700 cm^{-1} may be reasonably attributed to $v_{C=O}$ in two different adsorbed species, probably acrolein and acrylic acid. In this compound, in fact, the $v_{C=O}$ band is found at a frequency about 20 cm^{-1} higher than in acrolein (9). In these adsorbed compounds [for example, on V-Mo oxides (9)], the $v_{C=C}$ band is expected at a nearly the same value as in π-bonded propylene (around 1625 cm^{-1}), whereas other IR active bands are covered by the stronger bands due to physisorbed propane. A more clear identification of the above species, therefore, is not possible. The shoulder at about 1425 cm^{-1} may be attributed to v_sCOO in adsorbed acrylate, but the $v_{as}COO$ band expected at around 1550 cm^{-1} is absent. A more reasonable interpretation is the formation of alkene oligomers. In fact, propene adsorbed on HNaY gives rise to the formation of a main band at about 1460 cm^{-1} (9), apart from v_{CH} and δ_{CH} bands that, in our case, are covered by the band of physisorbed propane. However, all adsorbed species are removed by evacuation, indicating their weak interaction with the surface.

Discussion

Nature of Vanadium Species. The data on the characterization of V- containing silicalite indicate the presence of various types of vanadium species: (i) a polynuclear V-oxide containing V in various valence states (V^{3+}, V^{4+} and V^{5+}), (ii) octahedral VO^{2+} sites, preferentially interacting with OH groups localized inside the pore structure of the zeolite crystals, (iii) nearly symmetrical tetrahedral V^{5+} species, attributed to a V species directly interacting with the zeolite framework, and (iv) after reduction, V^{4+} species in a nearly tetrahedral environment.

The first species present mainly in *VSil117* can be removed by extraction at room temperature with an ammonium acetate solution. After extraction, evidence is found for at least two distinct V-species, (i) octahedral VO^{2+} sites shown by ESR and (ii) tetrahedral V^{5+} sites shown by ^{51}V-NMR and UV-Visible DR spectra. The tetrahedral V^{5+} species is the dominating species in *VSil545* (sample after extraction), whereas the octahedral VO^{2+} species is probably present only in smaller amounts. Both species are apparently not removed by the extraction procedure, suggesting their strong stabilization by direct interaction with the zeolite. Furthermore, for the tetrahedral V^{5+} species, the coordination does not change upon dehydration, contrary to that observed for vanadium species on a silica surface. The situation of the tetrahedral V^{5+} species in V-containing silicalite is thus different as compared to that found for vanadium on SiO_2. In agreement, ^{51}V- NMR and UV-Vis diffuse reflectance data indicate the peculiar spectroscopic characteristics of this species.

The following aspects characterize these tetrahedral V^{5+} species: *i)* stable tetrahedral environment, *ii)* presence of a V=O double bond, *iii)* possibility of reversible reduction to a tetrahedral V^{4+} species which is oxidized by O_2 through the

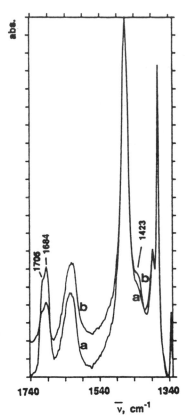

Figure 7. IR spectra of the adsorbed species in the 1340–1740 cm $^{-1}$ region after contact with 50 torr of propane and 1 torr of oxygen for 48 hours at room temperature (a) and after treatments for 1 hour at 200 °C (b); the contributions due to evacuated sample and gaseous species have been subtracted from these spectra.

formation of a O_2^- species, *iv)* presence of relatively strong Lewis acid sites after evacuation, and *v)* influence on the acid- base characteristics of nearlying Brønsted silanol sites. A tentative model for the local coordination environment of V^{5+} sites in the silicalite structure that takes into account these indications is given in Fig. 8. The model is consistent with the presence of Brønsted sites lying near to the vanadium, the localization of this species in the zeolite pore structure, the relative stability of this species against both extraction and changes in coordination by adsorption of water molecules, the monodispersion and immobility of the vanadium species and the reversible redox behavior with formation of V^{4+} tetrahedral species.

Present data do not justify the attribution of this V species to a real substitutional V site in the zeolite framework, because the amount of these V sites is very low and at present the degree of incorporation of these sites in the zeolite cannot be extended. It is therefore reasonable to assume that these V sites form at defect sites, possibly hydroxyl nests, the formation of which may be enhanced by the presence of V during hydrothermal synthesis, in agreement with Rigutto and van Bekkum (*3*).

Relationship between Nature of V Species and Catalytic Activity. ESR spectroscopy indicates the possibility of the formation of O_2^- species upon adsorption of oxygen at low temperature on the tetrahedral V(IV) species formed by reduction of tetrahedral V^{5+}. It is known (*11a and reference therein*) that these activated oxygen species are very reactive towards stoichiometric hydrocarbon selective oxidation, even though there is evidence to indicate that lattice (nucleophilic) oxygen ions rather than adsorbed (electrophilic) oxygen species are involved in the selective oxidation pathway. Adsorbed oxygen species are usually considered responsible for the unselective pathway to carbon oxides and only in few cases is their role in the selective reaction suggested. However, in selective n-butane oxidation, their role in some steps of the mechanism from n-butane to maleic anhydride has been proposed (*12*). It is also known that N_2O interacts with one-electron redox centres forming N_2 and O^-; the latter species is known to be more active in alkane activation (*11b*).

The data obtained on propane oxidative dehydrogenation on V-containing silicalite are consistent with the formation of adsorbed oxygen species by a one-electron redox reaction of reduced V^{4+} species with O_2 or N_2O. In particular, the higher activity and higher selectivity to propylene using N_2O instead of O_2 is indicative of the role of activated oxygen species in the mechanism of alkane oxidation. The detection of adsorbed propylene after contact of a mixture of propane and oxygen with the activated *VSil545*, but not after contact with only propane, is in agreement with this suggestion. However, direct infrared evidence of the presence of O^- or $(O_2)^-$-type species after contact with oxygen was not possible due both to the cut-off of transmission (Fig. 5) and to the low concentration of these species. On the other hand, the role of surface adsorbed oxygen species in the selective oxidative dehydrogenation of ethane using N_2O on V_2O_5-SiO_2 (*13,14*) or MoO_3-SiO_2 (*15*) has already been reported.

V-containing silicalite shows an interesting selective behavior in propane oxidative dehydrogenation to propylene using gaseous oxygen or N_2O. Good selectivities to propylene can be obtained also at relatively high propane conversion. In addition, it has been observed that as compared to alkane oxidative dehydrogenation on vanadium

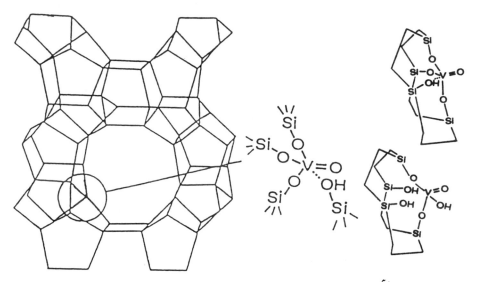

Figure 8. Tentative model of the local coordination environment of V^{5+} sites in the silicalite structure.

on silica, using V-containing silicalite better selectivity to the alkene in the absence of oxygen-containing products (besides carbon oxides) may be obtained (*13- 16*). In comparison to propane oxidative dehydrogenation on V-Mg-O catalysts (*17,18*) similar or slightly better selectivities for similar propane conversion are obtained, even though the productivity with the V-containing silicalite is much lower.

The good catalytic behavior of V-containing silicalite may be associated with the presence of the tetrahedral V^{5+} species stabilized by the interaction with the zeolite framework as regards both redox and coordination changes. In fact, ESR and TPR data indicate the lower rate of reduction of this species as compared to that of supported vanadium-oxide, and ^{51}V-NMR data indicate the stability against changes in the coordination environment. Catalytic data (Fig.s 2 and 3) indicate the better catalytic performances of this species in propane oxidative dehydrogenation as compared to supported polynuclear vanadium-oxide which can be removed by treatment with an ammonium acetate solution.

In addition, the infrared examination of the mechanism of propane and oxygen interaction with the sample (Fig. 6) indicates the different mechanism of interaction of the intermediate propylene as compared to other supported vanadium catalysts such as V-TiO$_2$ (*10*). In particular, the formation of a π-bonded complex stabilized by a nearlying silanol with weak basic character due to the inductive effect of vicinal vanadium is shown. This indicates the relative inertness of the V sites in the silicalite towards O-insertion or allylic H-abstraction on the adsorbed propylene. It is evident that the reduced reactivity of V sites in these reactions limits the consecutive reactions of intermediate propylene, thus enhancing the selectivity in the formation of this product.

Infrared data (Fig. 7) indicate also the presence of two main pathways of further transformation of adsorbed propylene: *i)* the formation of oligomers, due reasonably to the presence of Brønsted acid sites in agreement with the infrared characterization of the surface acidity of this sample (*1*) and *ii)* the consecutive oxidation of propylene with formation of acrolein and acrylic acid. In the catalytic tests, these latter species were not detected and therefore can reasonably be assumed to be intermediates to carbon oxides formation during the catalytic reaction. The formation of small amounts of products with a higher carbon atom number than the starting propane, on the contrary, was detected especially for reaction temperatures higher than 500°C. However, it is interesting to note that the rate of these consecutive reactions on the intermediate propylene, as indicated by IR data (Fig.s 6 and 7), is much lower than the rate of generation of propylene from propane in the presence of O$_2$. This is in agreement with the good selectivity to propylene shown by the V-containing silicalite.

As mentioned above, it is reasonable to assume that this tetrahedral V^{5+} species forms at defect sites (hydroxyl nests) in the zeolite framework, but is stabilized by this interaction in a well defined environment through V-O-Si bonds. As indicated by the characterization data, the local coordination of vanadium must be different from that found for well dispersed vanadium sites on silica. This stabilization probably limits the unselective metal-bonded propane or propylene adsorption, in agreement with the role of adsorbate bonding on the selection of partial and total oxidation pathways of ethane on vanadium supported on silica (*16*) and in agreement with IR evidence (Fig.

6). The presence of stable V-O-Si bonds also limits the possibility of insertion of O in the activated organic molecule and the stable coordination limits the possibility of alkene activation through an allylic mechanism. Both these factors are important in controlling the consecutive transformation of propylene.

The role of adsorbed oxygen species in the mechanism of alkane transformation, on the contrary, is more questionable. The effect induced by the substitution of O_2 with N_2O and IR indications are in agreement with this interpretation, but, on the other hand, activated electrophilic oxygen species form on reduced V^{4+} sites, preferably in tetrahedral coordination (19). The partial reduction of tetrahedral V^{5+}=O with formation of tetrahedral V^{4+} after propane oxidative dehydrogenation can be observed using UV-Visible diffuse reflectance, ESR and ^{51}V-NMR spectroscopies. It is thus not possible to assign unequivocally the active species in propane selective activation to a tetrahedral V^{5+}=O species or to V^{5+}-O$^-$ or V^{5+}-O-O$^-$ species formed in the mechanism of reoxidation of a reduced vanadium site. However, it can be said that charge localization and one-electron redox chemistry is certainly favoured by the presence of isolated tetrahedral V^{5+} species on the zeolite matrix as compared with supported polynuclear vanadium- oxide and thus the role of activated oxygen species in the selective mechanism of propane oxidative dehydrogenation on V- silicalite is reasonable.

Further studies are necessary to clarify these aspects and the details of the mechanism of propane oxidative dehydrogenation on V-containing silicalite. However, it should be noted that the amount of V^{5+} sites which appears to be stabilized in defect site positions is rather low, which certainly affects the activity of the V-silicalite and its possible applications. Alternative methods may enhance their amount and the catalytic performance of this system which appears to be rather interesting, especially for the possibility of having isolated stabilized tetrahedral V^{5+} sites with a redox character.

Literature Cited

(1) Centi, G., Perathoner, S., Trifirò, F., Aboukais, A., Aïssi, C.F., Guelton, M. *J. Phys. Chem.*, **1992**, *96*, 2617.

(2) Zatorki, L.W., Centi, G., Lopez Nieto, J., Trifirò, F., Bellussi, G., Fattore, V. In *Zeolites: Facts, Figures, Future*, Jacobs, P.A., van Santen R.A. Eds.; Elsevier Science Pub.: Amsterdam 1989; p. 1243.

(3) Rigutto, M.S., van Bekkum, H. *Appl. Catal.*, **1991**, *68*, L1.

(4) Bellussi, G., Maddinelli, G., Carati, A., Gervasini, A., Millini, R. In *Proceedings, 9th Int. Zeolite Conference, Montreal (Canada), July 1992, in press.*

(5) Whittington, B.I., Anderson, J.R. *J. Phys. Chem.*, **1991**, *95*, 3306.

(6) Miyamoto, A., Medhanavyn, D., Inui, T. *Appl. Catal.*, **1986**, *28*, 89.

(7) Schrami-Marth, M., Wokaum, A., Pohl, M., Krauss, H.-L. *J. Chem. Soc. Faraday Trans.*, **1991**, *87*, 2635.

(8) Sass, C.E., Chen, X., Kevan, L. *J. Chem. Soc. Faraday Trans.*, **1990**, *86*, 189.

(9) Davydov, A.A. *Infrared Spectroscopy of Adsorbed Species on the Surface of Transition Metal Oxides*, Rochester, C.H. Eds., J. Wiley & Sons Pub.: New York 1990, p. 145-153.

(10) (a) Escribano, V.S., Busca, G., Lorenzelli, V. *J. Phys. Chem.*, **1990**, *94*, 8939. (b) Busca, G., Lorenzelli, V., Ramis, G., Escribano, V.S. *Materials Chem. & Phys.*, **1991**, *29*, 175.

(*11*) (a) Che, M., Tench, A.J., *Adv. Catal.*, **1983**, *32*, 1; (b) *Ibidem*, **1982**, *31*, 77.

(*12*) Ebner, J.R., Gleaves, J.T. In *Oxygen Complexes and Oxygen Activation by Transition Metals*, Martell, A.E., Sawyer D.T. Eds., Plenum Press Pub: New York 1988; p. 273.

(*13*) Iwamatsu, E., Aika, K.-I., Onishi, T. *Bull. Chem Soc. Jpn.*, **1986**, *59*, 1665.

(*14*) Erdöhelvi, A., Solymosi, F. *J. Catal.*, **1990**, *123*, 31.

(*15*) Mendelovici, L., Lunsford, J.H. *J. Catal.*, **1985**, *94*, 37.

(*16*) Oyama, S.T. *J. Catal.*, **1991**, *128*, 210.

(*17*) Chaar, M.A., Patel, D., Kung, H.H. *J. Catal.*, **1988**, *109*, 463.

(*18*) Siew Hew Sam, D., Soenen, V., Volta, J.C. *J. Catal.*, **1990**, *123*, 417.

(*19*) Shvets, V.A., Sarichev, M.E., Kasansky, V.B. *J. Catal.*, **1968**, *11*, 378.

RECEIVED November 6, 1992

Chapter 22

Oxidation of Highly Dispersed Platinum on Graphite Catalysts in Gaseous and Aqueous Media

J. A. A. van den Tillaart, B. F. M. Kuster, and G. B. Marin

Laboratorium voor Chemische Technologie, Eindhoven University of Technology, P.O. Box 513, 5600 MB Eindhoven, Netherlands

To investigate the interaction of oxygen with platinum in gaseous and aqueous media, in-situ X-ray Absorption Spectra (XAS) of a highly dispersed Pt/graphite catalyst were measured at the Pt L_{III} and L_{II} edge for samples exposed to hydrogen or to oxygen. White-line surface areas were determined to evaluate the relative degree of oxidation of the platinum. The Extended X-ray Absorption Fine Structure (EXAFS) was evaluated to determine the local environment of the platinum. Both techniques show a corrosive oxidation for the samples contacted with oxygen gas in contrast to chemisorption of most probably a hydroxide species for the samples contacted with oxygen in water. The reduction of highly dispersed oxidized platinum is easily accomplished by hydrogen in both media.

The selective catalytic oxidation over highly dispersed noble metal catalysts in aqueous media is gaining interest (1-5). One of the main problems consists of the deactivation of the catalyst. The role of oxygen is recognized as being crucial in this matter (6,7).

Bulk platinum is the thermodynamically stable form of platinum in air at ambient conditions as the standard Gibbs energy of formation of platinumoxide, PtO_2, amounts to 167 kJ mol^{-1} (8). Surface platinum atoms and platinum atoms in very small particles however are less noble than platinum atoms in bulk metal because of their incomplete coordination. The standard Gibbs energy of formation of a single platinum atom amounts to 520 kJ mol^{-1} at 298 K (8). Hence, there exists a critical coordination below which platinum is no longer stable in ambient air.

0097–6156/93/0523–0298$06.00/0
© 1993 American Chemical Society

The literature on the oxidation of highly dispersed platinum in gaseous media agrees on the relative ease of oxidation at even ambient conditions. It was shown, using techniques such as EXAFS, Small and Wide Angle X-ray Scattering, chemisorption and Temperature Programmed Reduction, that the oxidation of platinum is passivating and limited to about two platinum layers (*9-13*). For highly dispersed catalysts this leads to an almost complete oxidation of the platinum as almost all platinum atoms are at the surface.

The thermodynamically stable form of bulk platinum in oxygen saturated water at ambient conditions is the completely hydrated platinum(IV)oxide, $PtO_2 \cdot 4H_2O$, also referred to as platinic acid $H_2Pt(OH)_6$ (*14,15*) with a standard Gibbs energy of formation of -84 kJ mol^{-1}. The formation of this compound will be even more favoured in the case of incompletely coordinated platinum.

The literature on the oxidation of platinum in aqueous media is essentially limited to the electrochemical oxidation of bulk platinum. Angerstein-Kozlowska et al. (*16*) investigated the initial stages of electrochemical oxidation in 0.5 M H_2SO_4 at room temperature. It was concluded by deconvolution of the fine structure in the broad anodic oxidation peak of a cyclic voltammogram that the oxidation proceeds by formation of different surface lattices of hydroxide species. Part of the adsorbed hydroxide rearranges, according the authors, to a form of subsurface hydroxide. Ross (*17*) however attributed the observed fine structure to sulphate anion adsorption as it is not present with HF as electrolyte. Peuckert and Bonzel (*18*) concluded from XPS measurements that an electrochemically grown oxide film on platinum (111) in acid medium consists of $Pt(OH)_4$ ($PtO_2 \cdot 2H_2O$), which was confirmed by Wagner and Ross (*19*) for Pt (100) using AES, XPS and Temperature Programmed Desorption. In alkaline media a compound with stoichiometry $PtO(OH)_2$ was proposed on Pt (111) (*18*). Burke and Lyons (*14*) explained anodic charging curves, measured on polycrystalline platinum in different aqueous buffers, by initial formation of a charged anionic species such as $[Pt_2O \cdot OH]^-$.

It is clear that the oxidation of platinum proceeds by different routes and leads to different stable compounds when comparing the oxidation of highly dispersed platinum by oxygen gas and the aqueous electrochemical oxidation of bulk platinum. The present work investigates the extent to which the effects of exposure of highly dispersed platinum to oxygen depend on the medium. For this purpose in-situ characterization has to be applied as highly dispersed platinum is not stable in ambient air. X-ray Absorption Spectroscopy meets this requirement and provides information on the degree of oxidation and on the local environment of the platinum.

Experimental

Catalyst. A highly dispersed platinum on graphite catalyst was prepared following a method described by Richard and Gallezot (*20*).

High surface area graphite (Johnson Matthey, CH10213) with a BET area of 312 m^{2}g^{-1} was activated by partial combustion in flowing air at 773 K for 5 hours. Under these conditions 25 wt% of the graphite was burned off. Subsequently the graphite was suspended at ambient temperature for 24 hours in a solution of concentrated sodium hypochlorite (Janssen p.a.). After this wet activation step the graphite was separated from the solution by filtration on a Millipore filter (HV 0.45μ), carefully rinsed with distilled water and dried in a vacuum oven at 373 K.

Platinum was introduced on the activated support by a competitive cation exchange technique. An amount of 100 g of a 8 wt% Pt solution of platinumtetrammine hydroxide (Johnson Matthey) was added dropwise to a suspension of 40 g graphite in 800 ml 1 M ammonia (Merck p.a.) and stirred at ambient temperature for 24 hours. The catalyst was subsequently separated by filtration on a Millipore filter (HV 0.45μ), washed with distilled water and dried in a vacuum oven at 373 K. The dried catalyst was reduced in flowing hydrogen at 573 K for 2 hours and stored under air before use.

The platinum content of the prepared catalyst amounts to 4.9 wt% as determined by UV/VIS spectrophotometry of a formed Sn-Pt complex (21-23). The volume-surface mean platinum diameter as determined by computer aided TEM micrograph analysis amounts to 1.85 ± 0.04 nm. This is in good agreement with a fraction of exposed platinum atoms of 0.75 as determined from CO chemisorption, assuming a 1:1 stoichiometry (24).

Catalyst samples were pressed as a wafer in the stainless steel or perspex sample holder. It was calculated that, due to diffusion limitations, the aqueous samples needed at least 6 hours pretreatment in order to attain monolayer coverage of the platinum with oxygen or hydrogen throughout the sample. A pretreatment time of 20 hours was chosen.

- The basic sample, referred to as O2G, received no further treatment as it was exposed to ambient air.
- A sample, referred to as H2G, was treated in-situ with flowing hydrogen gas at atmospheric pressure and room temperature.

The aqueous samples, H2L and O2L, were both exposed ex-situ to hydrogen saturated water at atmospheric pressure and 363 K for 20 hours prior to further treatment. The pH of the water after contacting with the catalyst sample amounted to 5 due to the presence of acidic groups on the graphite support.

- The sample, H2L, was subsequently contacted in-situ with hydrogen saturated destilled water at atmospheric pressure and ambient temperature.
- The sample, O2L, was exposed ex-situ to oxygen saturated distilled water at atmospheric pressure and 323 K for 20 hours and subsequently contacted in-situ with oxygen saturated distilled water at atmospheric pressure and ambient temperature. By using a perspex sample holder for this sample absence of cathodic protection was ascertained.

X-ray Absorption Spectra (XAS).   X-ray absorption measurements were performed at station 9.2 of the SRS at Daresbury (UK) with an electron beam energy of 2 GeV and a stored current varying between 290 and 160 mA. The wiggler was operational at 5.0 Tesla. Data were collected in the transmission mode from 11.37 keV to 13.43 keV (Pt L$_{III}$-edge: 11.564 keV, Pt L$_{II}$-edge: 13.273 keV) with a Si (220) monochromator detuned to 50 % of the maximum intensity for harmonic rejection. The ion chambers were filled with gas mixtures optimised to suit the measuring conditions. Energy calibration was monitored using a gold foil and a third ion chamber, and was set at 11.919 keV at the Au L_{III} edge. The spot size was reduced with the entrance slits to give a spot of 1 mm height and 10 mm width. Energy resolution was estimated to be 2.2 eV. Samples were scanned 8 times to determine the error in the data and to improve the data quality by averaging.

The amount of catalyst was determined to give a total absorption of 2.5, leading to samples of about 80 mg. All spectra were recorded at a temperature of 323 K.

White-line Analysis.   There is no generally accepted method for the determination of the white-line surface area of platinum. The methods proposed by Gallezot et al. (*25*) and Mansour et al. (*26*) are sensitive to the local environment of the absorber. Horsley (*27*), however, used a deconvolution based on absorption theory which is quite insensitive to the local structure of the central absorber. This method is therefore more appropriate to compare the white-line surface areas of platinum samples with a different local structure and will be followed in this work.

Prior to averaging, the data were visually examined to determine if the sample had not changed during measurement or if any other disturbance was present. The white-line surface area was determined using the averaged XAS data. The Pt L_{III} and L_{II} edge region (-100 to 100 eV relative to edge) were isolated, the energy rescaled relative to the edge, the pre-edge background was subtracted by using a Victoreen approximation and the data were finally normalised by the edge jump.

The data were subsequently deconvoluted to the sum of an arctangent and a Lorentzian function as described by Horsley (*27*):

$$\mu' = \frac{1}{2} + \frac{1}{\pi}\arctan(p_1(E-p_2)) + \frac{p_3}{1+p_4(E-p_5)^2} \tag{1}$$

in which μ' is the calculated normalised absorption coefficient, E is the energy relative to the edge and p_1 to p_4 are the adjustable regression parameters. Maximum likelihood parameter estimates were obtained by applying a Marquardt algorithm to minimize the sum of squared residuals between the calculated and observed normalised absorption coefficient.

The white-line surface area was obtained as the integrated Lorentzian part of the regression function:

$$A = \int_{-\infty}^{\infty} \frac{p_3}{1+p_4(E-p_5)^2}dE = \frac{p_3\pi}{\sqrt{p_4}} \tag{2}$$

The L_{II} surface area was corrected to the same scale as the L_{III} surface area by multiplication with the quotient of the specific edge jumps.

The L_{III} and L_{II} edge white-line surface area of 5d metals is generally accepted to be a measure for the d-band occupancy, or better the d-band vacancy (28). Small reduced platinum and iridium clusters were reported to be electron deficient compared to the bulk metal (28). For a given particle size the white-line surface area provides information on the relative degree of oxidation of the metal; the higher the white-line surface area the higher the oxidation state of the metal.

Extended X-ray Absorption Fine Structure (EXAFS) Analysis. Data analysis was performed with software developed at the Laboratory of Inorganic Chemistry and Catalysis of the Eindhoven University of Technology (Vaarkamp M., Linders H.L. and Koningsberger D.C., to be published) and follows mainly the procedure described by Sayers and Bunker (29). After averaging, the pre-edge background was subtracted by using a Victoreen approximation. Glitches and jumps were removed by using a spline approximation through the neighbouring data. To isolate the EXAFS oscillations, the background absorption was subtracted using a flexible spline approximation. Chi-data were produced after normalisation of the data by the edge jump. These chi-data were used to estimate the adjustable parameters in the EXAFS model equation. Using phase shift and amplitude functions extracted from reference compounds allows to simplify the EXAFS model equation for an unoriented sample with small or Gaussian disorder to (29):

$$\chi'(k) = \sum_j \frac{N_j}{kR_j^2} F_j(k)\, e^{-2k^2(\Delta\sigma^2)_j} \sin[2kR_j + \delta_j(k)] \tag{3}$$

in which $\chi'(k)$ is the calculated EXAFS oscillation, N_j is the number of atoms in the jth shell, R_j is the mean distance between the central absorber and the jth shell, $F_j(k)$ and $\delta_j(k)$ are respectively the amplitude function and the phase shift function for the atom in the jth shell and $(\Delta\sigma^2)_j$ is the relative mean square displacement of an atom in the jth shell. The photoelectron wave number, k, is defined for each shell as:

$$k = \sqrt{\frac{8m\pi^2}{h^2}(E - E_0 - E_{c,j})} \tag{4}$$

in which m is the electron mass, h is Planck's constant, E is the photon energy, E_0 is the edge energy and $E_{c,j}$ is the energy correction for the jth shell.

The best regression was determined using a minimization routine incorporated into the program which minimized the sum of squared residuals between the calculated and observed EXAFS oscillation, χ' and χ. The result was visually checked comparing the k^1 and k^3 weighted Fourier transforms of the regression and contributions of each regressed shell to the acquired data.

Results and discussion

The results from the white-line regression can be found in Figure 1 and in Table I.

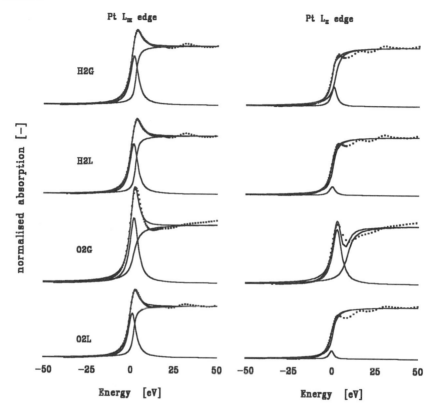

Figure 1. Normalised absorption at the Pt L edges; boxes: experimental data; full lines: calculated Lorentzian contribution, arctangent contribution and complete equation (1).

According to Figure 1 it is obvious that the sample exposed to oxygen gas is completely different from the other samples. The white-line is much more intense at both edges and the first EXAFS oscillations are clearly different. The same conclusion can be drawn from the white-line surface areas given in Table I. The white-line surface area for the sample exposed to oxygen gas is significantly higher. These observations indicate a larger degree of oxidation of the platinum contacted with air in the gas phase compared to the catalyst sample contacted with oxygen in water.

The results of the EXAFS analysis are illustrated in Figure 2 and Table II. The EXAFS regressions of the samples contacted with hydrogen are not depicted because they almost coincide to those of the sample exposed to oxygen in water. It is again clear from Figure 2 that the sample contacted with oxygen gas is quite different from the sample contacted with oxygen in water.

Table I. White-line surface areas from deconvolution of the platinum L edges

sample	L_{III} edge [eV]	L_{II} edge [eV]
H2G	7.38	0.90
H2L	7.62	0.45
O2G	10.24	4.17
O2L	7.43	0.37

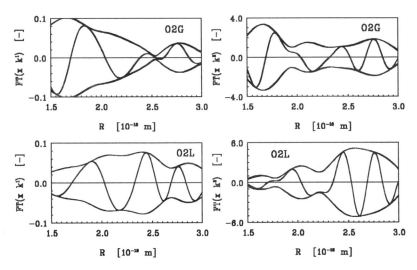

Figure 2. Fourier transformed EXAFS (modulus and imaginary part) of the oxygen exposed samples; full lines: observed; dotted lines: regressed with equation (3). Note that the dotted line is mostly covered by the full line.

As can be seen from Table II, both hydrogen contacted samples have a Pt-Pt coordination of 8.4 at a distance of $2.76 \cdot 10^{-10}$ m. This coordination number is in good agreement with calculations for fcc Pt crystallites with a fraction exposed Pt atoms of 0.75 (*30*). The first shell Pt-Pt distance of $2.76 \cdot 10^{-10}$ m is also in good agreement with the first shell distance in bulk fcc platinum of $2.77 \cdot 10^{-10}$ m. Furthermore the EXAFS of the hydrogen contacted samples contains a Pt-C contribution, attributed to the metal support interaction, with an average coordination of 1.6 at $2.6 \cdot 10^{-10}$ m. This relatively high coordination number indicates an intimate contact of the platinum crystallite with the graphite support.

Table II. Results of EXAFS analysis

sample	shell	N [-]	R [10^{-10} m]	$\Delta\sigma^2$ [10^{-25} m^2]	E_c [eV]
H2G	C	1.65	2.612	637	3.56
	Pt	8.37	2.760	411	4.45
H2L	C	1.57	2.609	222	2.62
	Pt	8.50	2.759	416	4.41
O2G	O	2.95	2.089	262	-0.92
	C	1.66	2.274	150	2.91
	Pt	2.66	2.759	532	-0.65
	Pt	0.40	3.145	330	-4.56
O2L	O	1.10	2.050	1457	11.09
	C	1.77	2.593	687	3.21
	Pt	7.40	2.754	469	3.28

The results for the sample contacted with oxygen gas show a coordination of 2.7 Pt-Pt at a distance of 2.76 10^{-10} m and a coordination of 3.0 Pt-O at a distance of 2.1 10^{-10} m and 0.4 Pt-Pt at a distance of 3.1 10^{-10} m, both typical for PtO$_2$. The metal support interaction in this sample has changed to a coordination of 1.7 Pt-C at a decreased distance of 2.3 10^{-10} m.

The results for the sample contacted with oxygen in water are comparable to the results for the hydrogen contacted samples. The Pt-Pt coordination is somewhat decreased to 7.4 at a distance of 2.75 10^{-10} m. The Pt-C coordination has not significantly changed with 1.8 at a distance of 2.6 10^{-10} m. Furthermore a small contribution from a 1.1 Pt-O coordination at a distance of 2.1 10^{-10} m can be recognized.

When the sample contacted with oxygen gas is compared with the sample contacted with oxygen in water it is clear from both the white-line surface area and the EXAFS analysis that the oxidation of highly dispersed platinum by gaseous oxygen is extremely corrosive. The strongly decreased coordination number of the metallic Pt-Pt shell at 2.76 10^{-10} m and the coexistence of the Pt-O and Pt-Pt shell of PtO$_2$ at respectively 2.1 10^{-10} m and 3.1 10^{-10} m can be explained by corrosive oxidation of the platinum particles. The white-line surface area indicates a high degree of oxidation. These observations are consistent with oxidized platinum particles made up of a core of metallic platinum covered by a shell of platinum oxide, as was discussed in the introduction (*9-13*). The decrease in the coordination number for the first shell Pt-Pt contribution is not attributed to a decrease in particle size. The TEM micrographs used for particle size analysis were obtained on with air contacted freshly prepared catalyst, i.e. the O2G sample, and show clearly an average particle size of 1.85 nm.

The observations on the samples contacted with hydrogen show the ease of reduction of the oxidized sample. Exposure to hydrogen gas at room temperature or to hydrogen saturated water at 363 K is sufficient to reduce the platinum particles completely.

The EXAFS of the sample contacted with oxygen in water contains a Pt-O coordination of 1.1 at a typical distance for PtO_2. The white-line analysis indicates however no significant degree of oxidation. Therefore this oxygen contribution results most probably from a chemisorbed hydroxide species at a bonding distance equal to that of Pt-O in platinum oxide. This is in agreement with the literature on the electrochemical oxidation of bulk platinum in aqueous media (14,16,18,19) and with the conclusions of Schuurman et al. (Schuurman Y., Kuster B.F.M, van der Wiele K. and Marin G.B., *Appl.Catal.A: Gen.*, **1992**, in press) regarding the deactivation of the highly dispersed Pt/graphite catalyst during the aqueous phase oxidation of methyl α-D-glucoside. The measured increasing oxygen surface coverage during reaction did not lead to formation of a platinum oxide but was attributed to chemisorbed oxygen.

The somewhat decreased first shell Pt-Pt coordination, when compared to the reduced samples, could then result from restructuring of the platinum surface into a rearranged subsurface hydroxide (16).

Conclusions

This work indicates a large difference between gaseous and aqueous oxidation of highly dispersed platinum. This is not in contradiction with the literature on gaseous oxidation of highly dispersed platinum and aqueous electrochemical oxidation of bulk platinum. The oxidation of highly dispersed platinum by gaseous oxygen is highly corrosive. The aqueous contacting of highly dispersed platinum with oxygen leads most probably to the formation of surface and subsurface hydroxide species as was reported for the anodic oxidation of bulk platinum. The reduction of highly dispersed oxidized platinum is easily accomplished by molecular hydrogen independently of the used medium.

Acknowledgements

This research has been made possible by a grant from the Netherlands Organization for Scientific Research (NWO). We would like to thank prof. D.C. Koningsberger and M. Vaarkamp for valuable comments and for the use of EXAFS analysis software.

Literature Cited

(1) Gallezot P., de Mésanstourne R., Christides Y., Mattioda G. and Schouteeten A., *J.Catal.*, **1992**, *133*, 479
(2) Mallat T. and Baiker A., *Appl.Catal.A: Gen.*, **1991**, *79*, 41

(3) Mallat T. and Baiker A., *Appl.Catal.A: Gen.*, **1991**, *79*, 59

(4) Schuurman Y., Kuster B.F.M, van der Wiele K. and Marin G.B., in *New Developments in Selective Oxidation by Heterogeneous Catalysis*, Ruiz P. and Delmon B., Eds., Elsevier Science Publishers, Amsterdam, **1992**, *72*, 43

(5) Vinke P., van Dam H.E. and van Bekkum H., in *New Developments in Selective Oxidation*, Centi G. and Trifiró F., Eds, Elsevier Science Publishers, Amsterdam, **1990**, 147

(6) Dijkgraaf P.J.M., Rijk M.J.M., Meuldijk J. and van der Wiele K., *J.Catal.*, **1988**, *112*, 329

(7) Dijkgraaf P.J.M., Duisters H.A.M., Kuster B.F.M. and van der Wiele K., *J.Catal.*, **1988**, *112*, 337

(8) Weast R.C., *Handbook of Chemistry and Physics*, CRC Press, **1988**

(9) Bassi I.W., Lyttle F.W. and Parravano G. *J.Catal.*, **1976**, *42*, 139

(10) Joyner R.W., *J.C.S.Faraday I.*, **1980**, *76*, 357

(11) McCabe R.W., Wong C. and Woo H.S., *J.Catal.*, **1988**, *114*, 354

(12) Nandi R.K., Molinaro F., Tang C., Cohen J.B., Butt J.B. and Burwell R.L. jr., *J.Catal.*, **1982**, *78*, 289

(13) Ratnasamy P., Leonard A.J., Rodrique L. and Fripiat J.J., *J.Catal.*, **1973**, *29*, 374

(14) Burke L.D. and Lyons M.E.G., in *Modern Aspects of Electrochemistry*, White R.E., Bockris J.O'M. and Conway B.E., Eds., Plenum Press, NY, **1986**, *18*, 169

(15) Pourbaix M., *Atlas of Electrochemical Equilibria in Aqueous Solutions*, Pergamon Press, Oxford, **1966**

(16) Angerstein-Kozlowska H., Conway B.E. and Sharp W.B.A., *Electroanal. Chem. & Interface Electrochem.*, **1973**, *43*, 9

(17) Ross P.N., *J. Electroanal.Chem.*, **1977**, *76*, 139

(18) Peuckert M. and Bonzel H.P., *Surf.Sci.*, **1984**, *145*, 239

(19) Wagner F.T. and Ross P.N. jr., *Appl.Surf.Sci.*, **1985**, *24*, 87

(20) Richard D. and Gallezot P., in *Preparation of Catalysts IV*, Delmon B., Grange P., Jacobs P.A. and Poncelet G., Eds., Elsevier, Amsterdam, **1987**, 71

(21) Ayres G.H. and Meyer A.S., *Anal.Chem.*, **1951**, *23*, 299

(22) Ayres G.H. and Meyer A.S., *J.Am.Chem.Soc.*, **1955**, *77*, 2671

(23) Charlot G., *Les méthodes de la chimie analytique*, Masson, Paris, **1961**

(24) Scholten J.J.F., Pijpers A.P. and Hustings A.M.L., *Cat.Rev.-Sci.Eng.*, **1985**, *27*, 151

(25) Gallezot P., Weber R., Dalla Betta R.D. and Boudart M., *Z.Naturforsch.*, **1979**, *34a*, 40

(26) Mansour A.N., Cook J.W.Jr. and Sayers D.E., *J.Phys.Chem.*, **1984**, *88*, 2330

(27) Horsley J.A., *J.Chem.Phys.*, **1982**, *76*, 1451

(28) Lyttle F.W., Greegor R.B. and Marques E.C., in *Proceedings 9th International Congress on Catalysis*, Phillips M.J. and Ternan M., Eds., Chemical Institute of Canada, Ottawa, **1988**, *5*, 54

(29) Sayers D.E. and Bunker B.A., in *X-ray Absorption: Principles, Applications, Techniques of EXAFS, SEXAFS and XANES*, Koningsberger D.C. and Prins R., Eds., John Wiley & Sons, New York, **1988**, 211

(30) Kip B.J., Duivenvoorden F.B.M., Koningsberger D.C. and Prins R., *J.Catal.*, **1987**, *105*, 26

RECEIVED November 6, 1992

Chapter 23

Promotion and Deactivation of Platinum Catalysts in Liquid-Phase Oxidation of Secondary Alcohols

T. Mallat, Z. Bodnar, and A. Baiker

Department of Chemical Engineering and Industrial Chemistry, Swiss Federal Institute of Technology, ETH–Zentrum, Zürich CH–8092, Switzerland

Promotion and deactivation of unsupported and alumina-supported platinum catalysts were studied in the selective oxidation of 1-phenyl-ethanol to acetophenone, as a model reaction. The oxidation was performed with atmospheric air in an aqueous alkaline solution. The oxidation state of the catalyst was followed by measuring the open circuit potential of the slurry during reaction. It is proposed that the primary reason for deactivation is the destructive adsorption of alcohol substrate on the platinum surface at the very beginning of the reaction, leading to irreversibly adsorbed species. Over-oxidation of Pt^0 active sites occurs after a substantial reduction in the number of free sites. Deactivation could be efficiently suppressed by partial blocking of surface platinum atoms with a submonolayer of bismuth promoter. At optimum Bi/Pt_s ratio the yield increased from 18 to 99 %.

The transformation of alcohols to the corresponding carbonyl compounds or carboxylic acids is one of the few examples in which a heterogeneous (solid) catalyst is used in a selective, liquid phase oxidation (*1,2*). The process, which is usually carried out in an aqueous slurry, with supported platinum or palladium catalysts and with dioxygen as oxidant, has limited industrial application due to deactivation problems.

There are numerous indications in the literature on catalyst deactivation attributed to over-oxidation of the catalyst (*3-5*). In the oxidative dehydrogenation of alcohols the surface M^0 sites are active and the rate of oxygen supply from the gas phase to the catalyst surface should be adjusted to that of the surface chemical reaction to avoid "oxygen poisoning". The other important reason for deactivation is the by-products formation and their strong adsorption on active sites. This type of

0097–6156/93/0523–0308$06.00/0

"chemical" deactivation has been ascribed to the formation of acidic or polymeric compounds during the reaction (*6-8*).

It has been discovered that the performances of platinum and palladium catalysts may be improved by promotion with heavy metal salts. However, there is little information available about the role and chemical state of the promoter (*8,9*). We have recently found that a geometric blocking of active sites on a palladium-on-activated carbon catalyst, by lead or bismuth, suppresses the by-product formation in the oxidation of 1-methoxy-2-propanol to methoxy-acetone (*10*).

In this paper we report the application of bimetallic catalysts which were prepared by consecutive reduction of a submonolayer of bismuth promoter onto the surface of platinum. The technique of modifying metal surfaces at controlled electrode potential with a monolayer or sub-monolayer of foreign metal ("underpotential" deposition) is widely used in electrocatalysis (*11,12*). Here we apply the theory of underpotential metal deposition without the use of a potentiostat. The catalyst potential during promotion was controlled by proper selection of the reducing agent (hydrogen), pH and metal ion concentration.

The air-oxidation of 1-phenylethanol to acetophenone in an aqueous alkaline solution has been chosen as a model reaction. The catalytic experiments were completed with the application of an in-situ electrochemical method for studying catalyst deactivation and the role of promoters. The potential of the catalyst, which was considered as a slurry electrode, was measured during the oxidation reaction. More details of the method can be found elsewhere (*13,14*).

Materials and Methods

Distilled water (after ion exchange) and purum or puriss grade reagents were used for the experiments. For the preparation of an unsupported Pt powder, 12 mmol H_2PtCl_6 in 100 cm^3 water was dropped into 300 cm^3 0.4 M aqueous $NaHCO_3$ solution at 95 °C. After refluxing it for 3 h the slurry was cooled to 30 °C and treated with hydrogen for 3 h. The catalyst was filtered off, washed to neutral with water and dried in air. The metal dispersion was 0.052 determined from the hydrogen region of a cyclic voltammogram (*15*).

The 5 wt% Pt-on-alumina was a commercial catalyst (Engelhard 4462). The metal dispersion (D = 0.30) was determined from TEM pictures. Different fields were examined and about 1000 particles were counted and their size determined. The degree of dispersion was calculated from the surface average diameter (*16*).

Before bismuth-promotion the Pt-on-alumina catalyst was pre-reduced in water with hydrogen. The pH was decreased to 3 with acetic acid and the appropriate amount of bismuth nitrate dissolved in water (10^{-3} - 10^{-4} M) was added into the mixed slurry in 15-20 min, in a hydrogen atmosphere. Promotion of unsupported Pt was carried out similarly. The metal composition of the bimetallic catalysts was determined by atomic absorption spectroscopy.

The oxidation reactions were performed in a 200 cm^3 glass reactor, equipped with gas distributer, condenser, thermometer, measuring and reference electrodes. The mixing frequency of the magnetic stirrer was 1500 min^{-1}. 75 mg Pt or 450 mg Pt-on-alumina catalyst was prereduced in nitrogen atmosphere at 60 °C with 3.67 g or 3.00 g 1-phenylethanol, respectively. The solvent composition was 35 cm^3 water +

0.32 g Na_2CO_3 + 0.37 g dodecylbenzenesulfonic acid Na salt + 5 cm^3 dioxane (for unsupported Pt). After 30 minutes the alcohol was oxidized with air (7.5 cm^{3}min^{-1}) at 60 °C. Conversion and selectivity were determined by GC analysis. Only alterations from this procedure will be indicated in the text.

All the potentials in the paper are referred to a Ag/AgCl/KCl$_{sat}$ electrode (E=197 mV). The electrochemical cell and polarization method used for cyclic voltammetric measurements have been described previously (9). 2 mg catalyst powder on a carbon paste electrode was polarized with 1 mVs^{-1} scan rate in a 0.085 M aqueous Na_2CO_3 solution at 25 °C.

For the measurement of the open circuit potential of the catalyst during the oxidation reaction a Pt rod measuring electrode and a Ag/AgCl/KCl$_{sat}$ reference electrode were applied (13,14).

Promotion of Platinum-on-Alumina

We have found that platinum catalysts quickly loose their activity in the selective oxidation of aliphatic and aromatic secondary alcohols to the corresponding ketones in aqueous solutions. Acceptable yields could be achieved with an extremely high catalyst loading or with long reaction times. Catalyst deactivation could be suppressed by depositing a promoter metal submonolayer onto the surface of the platinum particles. For example, total conversion with more than 99 % selectivity could be reached in 6 hours in the oxidation of 1-phenylethanol to acetophenone with Bi-Pt/alumina catalysts. It is shown in Table I that even a moderate coverage of platinum by bismuth had a substantial influence on the final conversion. Lead-promotion gave similar results, but gold, tin or ruthenium were less efficient. The general sequence of promoter efficiency in suppressing catalyst deactivation in the partial oxidation of secondary alcohols was:
Bi $\succ$ Pb $\succ$ Sn $\approx$ Au $\approx$ Ru.

Table I. Bi-promotion of a Pt-on-alumina catalyst

No	Bi/Pt$_s$ at/at	Conv. %	Sel. %
1	0	18.4	99.1
2	0.1	96.1	-
3	0.15	99.5	-
4	0.2	99.9	99.5

Role of Bismuth Promoter

In general, the influence of promotion may be explained by a geometric (blocking) effect or by the formation of new active sites. We suggest, that in our case the suppression of by-product formation by blocking a fraction of Pt^0 active sites is of decisive importance. Unfortunately, it is difficult to prove or exclude the existence of new active centers after promotion as reaction rates are always distorted due to oxygen transport limitation (to avoid oxygen poisoning). We found that our reactor worked in a "mixed" regime, as the measured reaction rates were influenced by both the rate of oxygen supply and that of the surface chemical reaction.

The correlation between the coverage of surface platinum atoms by bismuth adatoms (θ_{Bi}) and the measured rate of 1-phenylethanol oxidation was studied on unsupported platinum catalysts. An electrochemical method (cyclic voltammetry) was applied to determine θ_{Bi} and a good electric conductivity of the sample was necessary for the measurements. The usual chemisorption measurements have the disadvantage of possible surface restructuring of the bimetallic system at the pretreatment temperature. Another advantage of the electrochemical polarization method is that the same aqueous alkaline solution may be applied for the study of the surface structure of the catalyst and for the liquid phase oxidation of the alcohol substrate.

The anodic polarization curves of a Pt powder modified by adsorbed Bi are shown in Figure 1. The ionization of adsorbed hydrogen on unmodified Pt (curve a) ranges between -0.8 and -0.4 V:

$$H_{ad} + OH^- \rightarrow H_2O + e^-$$

Above this value a further oxidation of the surface by OH adsorption occurs, which becomes considerable above -0.2 V. A simplified reaction route of the step-by-step surface oxidation of Pt by OH⁻ is (*17*):

$$Pt_s \rightarrow Pt_sOH \rightarrow Pt_sO$$

Bismuth promotion suppresses the hydrogen sorption on platinum (curves b-d). The peak at -0.05 V indicates the oxidation of adsorbed bismuth, which overlaps the OH adsorption on uncovered platinum surface sites (*18*). Bismuth adatoms are discharged in the low potential region (Bi^0) and occupy three platinum sites at low surface coverages. The structure of the oxygen-containing species are $(BiOH)_{ad}$, $(BiO)_{ad}$ and $[Bi(OH)_2]_{ad}$ (*19*).

Bi does not adsorb hydrogen, thus a Bi/Pt coverage can be calculated from the hydrogen chemisorption data. It is seen in Figure 2 that there is an excellent correlation between the Bi-coverage of Pt and the rate of 1-phenylethanol oxidation. It seems that the hydrogen chemisorption ability of Pt or the size of active sites ensembles has to be minimized to avoid deactivation. There are indications in the literature that the suppression of hydrogen sorption on a Pt electrode can eliminate the poison formation (*20*).

Oxidation State of the Catalyst during Reaction

The measurement of open circuit potential of the catalyst during the liquid phase oxidation of alcohols provides a unique insight into the redox processes taking place on the catalyst surface. A Pt catalyst stored in air contains surface oxides and in an aqueous Na_2CO_3 solution it behaves as an oxygen electrode. Its potential is 250-280 mV when referred to a Ag/AgCl/KCl$_{(sat)}$ electrode (Figure 3). When the catalyst is

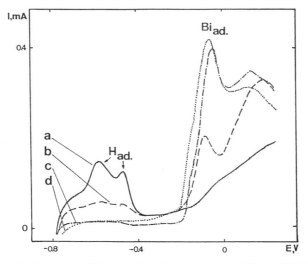

Figure 1. Anodic branches of the cyclic voltammograms of Pt and Bi-Pt catalysts; a - Pt powder, b - Bi/Pt_s=0.39, c - Bi/Pt_s=1.15, d - Bi/Pt_s=1.92.

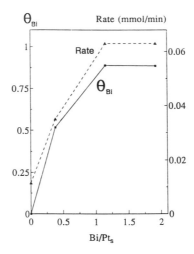

Figure 2. Bi coverage of unsupported Pt and the rate of 1-phenylethanol oxidation as a function of the overall Bi/Pt_s ratio.

pre-reduced before the oxidation reaction to obtain Pt^0 active surface sites, its potential is shifted to the negative direction by more than 1 V. 1-phenylethanol is a good reducing agent and decreases the catalyst potential in nitrogen atmosphere close to the value measured after hydrogen reduction (1 bar). After the addition of alcohol, the measured value is a mixed potential determined by two (or more) electrode processes (*21,22*). The two main components are the hydrogen electrode process and the alcohol/ketone reaction.

The oxidation state of the catalyst during the air-oxidation of 1-phenylethanol is shown in Figure 4. The anodic polarization curve of a bismuth-promoted platinum catalyst (Bi/Pt_s = 0.39) is taken from Figure 1b, as a reference. The lower part of Figure 4 represents the conversion of alcohol as a function of the potential of an unsupported and an alumina supported Bi-Pt catalyst. When air is introduced to the reactor, the open circuit potential of the Bi-Pt/alumina catalyst slightly increases to the anodic direction. The catalyst potential reaches the region of the oxidation of Bi_{ad} only at the end of the reaction. Almost up to total conversion the potential of the active Bi-Pt/alumina catalyst remains in the "hydrogen region". This behaviour indicates that both platinum and bismuth are in a reduced (discharged) state and platinum is partially covered with hydrogen during the oxidation reaction. This is in a good agreement with the dehydrogenation mechanism of alcohol oxidation, according to which only the Pt^0 sites are active (*23,24*). A similar situation was found in the oxidation of several other types of secondary alcohols like diphenyl carbinol or α-tetralol.

Different results were obtained when the reaction was catalyzed with unsupported Bi-Pt catalysts. A few minutes after introduction of air into the reactor the catalyst potential was around -400 mV (Figure 4b). It is clear from the anodic polarization curve above that at this potential the hydrogen coverage is close to zero. It is interesting that the oxidation of 1-phenylethanol occurs on alumina-supported Bi-Pt catalysts covered by hydrogen up to total conversion, while the similarly prepared but unsupported bimetallic catalysts are practically free of hydrogen at above 2-4 % conversion. This behaviour is attributed to some by-product formation and strong adsorption which shifts the mixed potential of the unsupported catalyst by 400 mV to the positive direction. The presence of high boiling point by-products in the latter case was evidenced by GC analysis.

Chemical Deactivation and Oxygen Poisoning

In Figure 5 the conversion of 1-phenylethanol and the open circuit potential of alumina-supported catalysts are plotted as a function of reaction time. There is a striking difference between the curves of unpromoted (a, a') and bismuth-promoted (c, c') catalysts. When air is introduced to the reactor, the potential of the platinum-on-alumina catalyst quickly increases to the anodic direction and after one minute the catalyst potential is above -300 mV. One may conclude that there is practically no hydrogen on the platinum surface and after a short period an increasing fraction of platinum is covered by OH. The influence of bismuth promotion is a higher reaction rate (final conversion) and lower catalyst potential during reaction.

It seems that the reason of deactivation of the unpromoted Pt-on-alumina catalyst is the over-oxidation of the Pt^0 active sites, due to the high rate of oxygen

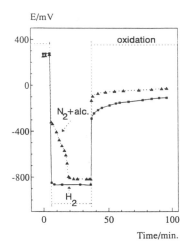

Figure 3. The variation of the potential of the Pt powder catalyst during pretreatment and oxidation.

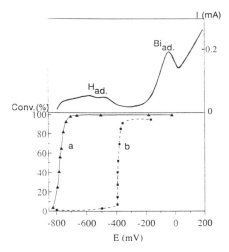

Figure 4. Anodic polarization curve of a Bi-Pt catalyst ($Bi/Pt_s=0.39$) and the conversion - catalyst potential relationship in the oxidation of 1-phenylethanol, in an aqueous Na_2CO_3 solution; a - Bi-Pt/alumina, $Bi/Pt_s=0.20$, b - unsupported Bi-Pt, $Bi/Pt_s=0.39$.

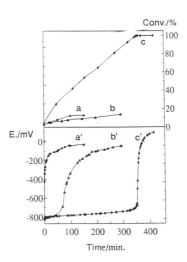

Figure 5. The influence of Bi-promotion on the deactivation of a 5 wt% Pt/alumina catalyst in the oxidation of 1-phenylethanol;   a,a'- Pt/alumina,   21 vol% O_2/N_2, 1500 min^{-1},   b,b' - Pt/alumina,   5 vol% O_2/N_2,   750 min^{-1},   c,c' - Bi-Pt/alumina, Bi/Pt_s=0.20, 21 vol% O_2/N_2, 1500 min^{-1}.

supply to the catalyst surface related to the rate of its consumption in the surface chemical reaction. However, a decrease in the rate of oxygen supply by decreasing both the oxygen partial pressure and mixing speed did not solve the problem. There is only a short period during which the catalyst is partially covered by hydrogen and the conversion is still below 20 % after a twofold reaction time (Figure 5b and 5b').

When the air flow was temporarily substituted by a nitrogen flow for 15-20 minutes in the reaction represented by Figure 5a, the rate of alcohol oxidation did not increase. These experiments also prove that the reason of catalyst deactivation is not the over-oxidation of Pt^0 active sites, but a partial coverage of active sites by impurities (chemical deactivation).

The best promoters of the partial oxidation of secondary alcohols are Bi and Pb. Neither of them adsorb hydrogen and a partial coverage of platinum by any of them decreases the hydrogen sorption on the bimetallic system. Nevertheless, there is no hydrogen on the platinum-on-alumina catalyst during the oxidation reaction, while a partial hydrogen coverage and a reduced state of the promoted catalysts were observed up to almost total conversion. The explanation of this apparent contradiction is the partial coverage of unpromoted platinum by impurities. A decrease in the number of Pt^0 active sites by a factor of 3-10 due to Bi or Pb promotion may still be advantageous compared to an almost complete coverage of them by irreversibly adsorbed species.

We believe that the primary reason of deactivation is the formation of irreversibly adsorbed species and oxygen poisoning occurs when the blocking of active sites reaches a critical level. By-products can be formed during the oxidation reaction. A frequently observed side-reaction is the aldol-dimerization of the carbonyl compound and a further oxidation of the product. The process is catalyzed by bases, including the basic functional groups of a carbon support (25).

Another important source of by-product formation is the dissociative adsorption of alcohol on the platinum surface. It was found that methanol adsorption on platinum in aqueous solutions is a step-by-step dehydrogenation process resulting in triply bound *C-OH surface species (26). The formation of adsorbed CO, COH and HCOH species and C-C bond cleavage of higher alcohols leading to the formation of alkanes were also confirmed by classical analytical and in-situ spectroscopic methods (27-29). As there is a positive potential shift of unpromoted platinum from the beginning of the reaction (Figure 5a) we propose that the initial adsorption of the alcohol substrate is the real reason of deactivation, and the side reactions during the oxidation reaction are of secondary importance.

The product composition of the irreversible adsorption of alcohols on platinum depends on whether the alcohol molecule comes into contact with a "free" metal surface or with hydrogen or oxygen covered sites (29). This effect can be seen in Figure 3: the catalyst potential during the oxidation reaction is higher by about 110-150 mV, when the platinum oxide has been pre-reduced by the alcohol reactant itself, compared to the potential corresponding to the reaction after pre-hydrogenation. The potential difference is an indication of the different composition of adsorbed species on platinum or the different surface coverage of active sites by impurities.

Acknowledgment

Financial support of this work by the Swiss National Foundation (Support Program "Eastern Europe") is kindly acknowledged.

Literature cited

(1)   Heyns, K.; Blazejewitz, L. *Tetrahedron*, **1960**, *9*, 67.
(2)   van Bekkum, H., in *Carbohydrates as Organic Raw Materials*, Lichtenthaler, F. W., Ed., VCH: Weinheim, Germany, 1990, pp. 289-310.
(3)   Dijkgraaf, P. J. M.; Duisters, H. A. M.; Kuster, B. F. M.; van der Wiele, K. *J. Catal.*, **1990**, *112*, 337.
(4)   Dirkx, J. M. H.; van der Baan, H. S. *J. Catal.*, **1981**, *67*, 1.
(5)   Dirkx, J. M. H.; van der Baan, H. S. *J. Catal.*, **1981**, *67*, 14.
(6)   Nicoletti, J. W.; Whitesides, G. M. *J. Phys. Chem.*, **1989**, *93*, 759.
(7)   Smits, P. C. C.; Kuster, B. F. M.; van der Wiele, K.; van der Baan, H. S. *Appl. Catal.*, **1987**, *33*, 83.
(8)   Mallat, T.; Baiker, A. *Appl. Catal. A: Gen.*, **1991**, *79*, 41.
(9)   Mallat, T.; Allmendinger, T.; Baiker, A. *Appl. Surf. Sci.*, **1991**, *52*, 189.
(10) Mallat, T.; Baiker, A.; Patscheider, J. *Appl. Catal. A: Gen.*, **1991**, *79*, 59.
(11) Kolb, D. M. *Adv. in Electrochem. and Electrochem. Eng.*, **1978**, *11*, 125.
(12) Szabo, S. *Int. Rev. Phys. Chem.*, **1991**, *10*, 207.
(13) Baria, D. N.; Hulburt, H. M. *J. Electrochem. Soc.*, **1973**, *120*, 1333.
(14) van der Plas, J. F.; Barendrecht, E.; Zeilmaker, H. *Electrochim. Acta*, **1980**, *25*, 1471.
(15) Commission on Electrochemistry, *Pure Appl. Chem.*, **1991**, *63*, 711.
(16) Mallat, T.; Petro, J. *React. Kinet. Catal. Lett.*, **1979**, *11*, 307.
(17) Angerstein-Kozlowska, H.; Conway, B. E.; Barnett, B.; Mozota, J. *J. Electroanal. Chem.*, **1979**, *100*, 417.
(18) Kadirgan, F.; Beden, B.; Lamy, C. *J. Electroanal. Chem.*, **1983**, *143*, 135.
(19) Clavilier, J.; Feliu, J. M.; Aldaz, A. *J. Electroanal. Chem.*, **1988**, *243*, 419.
(20) Watanabe, M.; Horiuchi, M.; Motoo, S. *J. Electroanal. Chem.*, **1988**, *250*, 117.
(21) Wagner, C.; Traud, W. *Z. Electrochem.*, **1938**, *44*, 391.
(22) Koryta, J.; Dvorak, J. *Principles of Electrochemistry*, Wiley & Sons, Chichester, England, 1987, pp. 367-369.
(23) DiCosimo, R.; Whitesides, G. M. *J. Phys. Chem.*, **1989**, *93*, 768.
(24) Horànyi, G.; Vèrtes, G.; König, P. *Acta Chim. Hung.*, **1972**, *72*, 179.
(25) Mallat, T.; Baiker, A.; Botz, L. *Appl. Catal. A: Gen.*, **1992**, *86*, 147.
(26) Bagotzky, V. S.; Vassiliev, Yu. B.; Khazova, O. A. *J. Electroanal. Chem.*, **1977**, *81*, 229.
(27) Lopes, M. I. S.; Beden, B.; Hahn, F.; Léger, J. M.; Lamy, C. *J. Electroanal. Chem.*, **1991**, *313*, 323.
(28) Iwasita, T.; Nart, F. C. *J. Electroanal. Chem.*, **1991**, *317*, 291.
(29) Damaskin, B. B.; Petrii, O. A.; Batrakov, V. V., *Adsorption of Organic Compounds on Electrodes*, Plenum Press, New York, U. S., 1971, pp. 321-349.

RECEIVED November 6, 1992

Chapter 24

Toward Catalysis of Selective Epoxidation and Its Applicability to Multistage Organic Synthesis

Pierre Laszlo, Michel Levart, Ezzeddine Bouhlel,
Marie-Thérèse Montaufier, and Girij Pal Singh

Laboratoire de Chimie Fine aux Interfaces, Ecole Polytechnique,
91128 Palaiseau, France

By analogy between the oxo forms of vanadium(V) and iron(IV), the latter being the active species in oxidations by cytochrome P-450, the system constituted by vanadium oxide as the catalyst, and t-butylhydroperoxide, as the oxidant, gives good results in the conversion of olefins to the corresponding epoxides. With the supported "clayniac" catalyst, in the presence of i-butyraldehyde as a sacrificial reducer, olefins are epoxidized in good yields by compressed air at room temperature, in a convenient procedure.

The synthetic organic community yearns for a simple and practical new epoxidation routine. This urge has several determinants: the standard reagent used for this purpose, m-chloroperbenzoic acid *(1-2)* (acronymed as MCPBA), is too dangerous for handling and for storage; accordingly, it is on its way out, because of the newer and more stringent safety requirements. The biomimetic studies seeking to better understand and to emulate the enzymatic cytochrome P-450 system *(3)* have come up with quite a few porphyrin catalysts for epoxidation *(4-5-6-7)*. The Sharpless asymmetric epoxidation of allylic alcohols *(8)* has achieved stardom, and it has led to numerous attemps to emulate its success, and to extend its applicability to other olefins *(9-10-11-12-13-14)*.

The search for a new epoxidation method that would be appropriate for organic synthesis should also, preferably, opt for a catalytic process. Industry has shown the way. It resorts to catalysis for epoxidations of olefins into key intermediates, such as ethylene oxide and propylene oxide. The former is prepared from ethylene and dioxygen with silver oxide supported on alumina as the catalyst, at 270°C *(15-16)*. The latter is prepared from propylene and an alkyl hydroperoxide, with homogeneous catalysis by molybdenum complexes*(17)* or better (with respect both to conversion and to selectivity) with an heterogeneous Ti(IV) catalyst *(18)*. Mixtures of ethylene and propylene can be epoxidized too *(19)* by tert-butylhydroperoxide *(20)* (hereafter referred to as TBHP).

0097–6156/93/0523–0318$06.00/0

Catalysis by Vanadium Oxide of Epoxidation by TBHP

Our initial interest in epoxidation stemmed from a perceived analogy. The cytochrome P-450 enzymatic system oxidizes hydrocarbons to make them more hydrophilic and water-soluble, hence more easily excreted. It also converts olefins into epoxides *(21)*. The active species is an oxo iron O=Fe^{IV} entity, with the geometry of a square pyramid; the corners of the square are the porphyrin nitrogens *(22)*. After oxygen atom transfer, e.g. to a hydrocarbon RH, the enzyme reverts to its resting state, from which the stepwise: reduction to Fe(II) / binding of dioxygen / cleavage of one oxygen into water / and oxidation regenerates the Fe^{IV} = O oxo form *(22)*. Vanadium oxide V$_2$O$_5$, alone, supported, or in combination, is a potent oxidation catalyst, that effects alkane activation *inter alia (23-24-25-26-27-28-29)*. The active centers on the surface of vanadium oxide are V = O double bonds *(30)*. These vanadium oxo active sites belong to a square pyramid, the four corners of the square are also oxygen atoms *(31)*. Such catalysts as vanadyl pyrophosphate often operate by oxo transfer, followed by regeneration (effected by an oxidant) of the V = O double bond *(32)*. Thus, the analogy *(33)* is firm and compelling. We could base ourselves on it for devising a novel and efficient procedure for the epoxidation of olefins.

Typical procedure The mole ratio of alkene:t-BuOOH:catalyst was 10:10:0.25 and 40 mmol of the olefin serving as its own solvent. Thus, 10 mmol of TBHP (80% in di-tert-butyl peroxide) and 0.25 mmol of vanadium pentoxide were added to 50 mmol of the olefin and the reaction mixture was stirred at 60°C under a dinitrogen atmosphere. The products formed were analyzed by GC by comparison of their retention time with those of authentic samples. Good yields of epoxides were obtained only with an excess of olefin to TBHP of 5:1. That the olefin doubles up as the solvent makes for a more practical procedure. Typical results *(34)* are shown in Table 1:

Table I. Epoxidation of alkenes by TBHP catalyzed by V$_2$O$_5$

Olefin	Temp,°C	Time,h	TBHP consumed,%	Epoxide yield,%	Selectivity,%
Cyclopentene	45	3	100	45	80
Cyclohexene	60	6	100	60	87
Cycloheptene	60	8	96	40	60
Cyclooctene	60	4	100	70	82
Hex-1-ene	60	16	100	39	88
Norbornene	60	5	100	80	88

These results compare well with those from other methodologies *(35-36-37-38-39)*, including some using porphyrin-derived catalysts, that had been conceived also on the analogy with the cytochrome P-450 system.

One should mention another, concurrent and near-contemporary methodology, also based on catalysis by an oxovanadium species. Mukaiyama and coworkers used oxovanadium (IV) complexes with the general formula OVL$_2$ to perform epoxidations *(40)* A catalytic amount (4 mol%) of the complex is used. Molecular oxygen is the oxidant, under 3 atm of pressure. An alcohol in stoichiometric amount (1.5 equivalent) is introduced as a sacrificial reducer. The efficiency of the catalyst increases with a decrease in the oxidation potential *(40)*. For instance, in the epoxidation of styrene, whereas L=acac does not work, L=mac (3-methyl-2,4-pentanedionato) does.

Furthermore, the best ligands L are β-diketones with an electron-donating substituent at position 2 *(40)*. As for the sacrificial reducer, only primary or secondary alcohols are effective; i-propanol does a good job, and was used in the standard procedure that was developed.

Catalysis by Nickel Complexes of Aerobic Epoxidation.

The same Japanese group went on to devise a general procedure for epoxidation using homogeneous catalysis by nickel complexes *(41)*. In a very similar way as with the vanadium(IV) species, a nickel(II) complex such as Ni(mac)$_2$ or Ni(dmp)$_2$ (dmp=1,3-propanedionato) in catalytic amount (4 mol%) allows the compressed oxygen(4 atm of O$_2$) epoxidation of numerous olefins, provided that a stoichiometric amount (2 equivalents) of a primary alcohol, such as 1-butanol, -octanol, -undecanol, or -tetradecanol is present as a co-reactant and as a sacrificial reducer. In this manner, trisubstituted and exo-terminal olefins or norbornene analogs are smoothly oxygenated into the corresponding epoxides in high-to-quantitative yields. Olefins bearing ester groups are epoxidized with no harm to these functional groups. Aromatic olefins in the styrene family substituted with electron-withdrawing groups can also be epoxidized in high yields. The choice of nickel(II) as a catalyst for an oxidation reaction has few precedents *(42)*, apart from the nickel bromide-catalyzed oxidation of primary and secondary alcohols to carbonyls by benzoyl peroxide *(43)* and, more generally, oxidation by "nickel peroxide" of organic compounds *(44)*.

We have tested successfully a variant of this Mukaiyama procedure, that has been described in more detail subsequently *(45-46)*. Our prior experience with nickel(II) catalysts impregnated on clays and nicknamed *(47)* "claynick" *(48)* and the convenience of a supported catalyst *(49-50)* made us opt for clay-impregnated nickel acetylacetonate as the catalyst.

"Clayniac": a solution of nickel acetylacetonate (2g in 2mL of acetone, ca. 0.4M) is stirred under moderate heating (50°C). The K10 montmorillonite (Süd-Chemie; 2g) is then added to the suspension. The solvent is evaporated under reduced pressure, on a rotary evaporator. After careful washing by methylene chloride, to remove non-impregnated nickel, the residue is dried in an atmospheric oven at 100°C overnight.

Standard procedure: a mixture of olefin (5mM), the catalyst (50mg), and i-butyraldehyde (1mL:15mmol) in methylene chloride (10 mL) was stirred at room temperature in an autoclave under 10 bars of compressed air. After completion or interruption of the reaction, the solid catalyst was filtered through a short plug of silicagel. The products were analyzed by GC, by comparison with authentic samples (Aldrich; Janssen) and a n-tetradecane standard was used to determine the yield. The results are shown in Table 2.

This methodology, while commendable for its simplicity, has some drawbacks: yields are not quantitative; one has to invest i-butyraldehyde as a sacrificial reducer; the resulting i-butyric acid has to be separated from the epoxide product, which detracts from the usefulness of the methodology. Nevertheless, as also shown by others recently *(51)*, the Mukaiyama procedure can be adapted to good use.

A simple one-electron oxidation process appears to be excluded by the lack of a correlation between the observed reactivities and the half-wave oxidation potentials of the olefins from the literature. What is the role of the sacrificial aldehyde? The answer to this question is a good entry point into the mechanism of the transformation. We

Table II. Aerobic epoxidation of olefins catalyzed by Ni(acac)$_2$

Olefin	Conversion, %	Yield, %	Selectivity, %
2,3-dimethyl-2-butene	90	50	95
1-hexene	61	53	85
2-methyl-1-pentene	98	67	94
2-methyl-2-pentene	97	76	95
3-methyl-2-pentene	100	73	93
2,3,3-trimethyl-1-butene	100	92	92
1-octene	57	29	84
2-methyl-1-heptene	100	67	91
2,4,4-trimethyl-1-pentene	95	81	94
2,4,4-trimethyl-2-pentene	100	76	76
1,3,5-trimethyl-1-cyclohexene	97	67	93
2-methyl-1-undecene	100	83	97
cyclopentene	89	34	82
cyclohexene	97	71	92
cycloheptene	100	89	94
1-methylcyclohexene	100	87	95
cis cyclooctene	100	100	95
cyclododecene	94	76	97
1,5,9-cyclododecatriene (*trans-trans-cis*)	99	43	89
4-vinylcyclohexene	90	85	95
4-methylstyrene	100	48*	74
1,1-diphenylethylene	98	68	88
trans stilbene	70	55	84
norbornene	100	74	94
(+)-limonene	92	75 (1.8:1)	95
1-chloro-3-methyl-2-butene	70	36	93
1,4-dibromo-2-butene	95	51	95

* p-Me-C$_6$H$_4$-CH$_2$-CHO is the product.

find experimentally that aromatic aldehydes ArCHO basically are inefficient. Crotonaldehyde has intermediate capability, 14% that of i-butyraldehyde. Aliphatic aldehydes, such as 3-cyclohexylpropionaldehyde, cyclohexane carboxaldehyde, pivalaldehyde, and i-butyraldehyde work best. These findings are consistent with the proposal *(52)* of a radical chain reaction for co-oxidation by molecular dioxygen of an olefin and an aldehyde, viz.:

initiation

$$RCHO + \frac{1}{2}O_2 \rightarrow RC{\bullet}O + \frac{1}{2}H_2O_2$$

propagation

$$RC{\bullet}O + O_2 \rightarrow RCO_3{}^{\bullet}$$

$$RCO_3{}^{\bullet} + RCHO \rightarrow RCO_3H + RC{\bullet}O$$

radical epoxidation

$$C{=}C + RCO_3{}^{\bullet} \rightarrow \overset{O}{\overset{\triangle}{C{-}C}} + RCOO^{\bullet}$$

peracid epoxidation

$$C{=}C + RCO_3H \rightarrow \overset{O}{\overset{\triangle}{C{-}C}} + RCO_2H$$

non radical oxidation

$$RCO_3H + RCHO \rightarrow 2\, RCO_2H$$

An obvious test of such a mechanism is use of standard free-radical traps (p-benzoquinone, TEMPO): indeed they block the reaction; which, to us, also indicates that the reaction is not exclusively interfacial, but occurs also in the solution. We believe that the role of the nickel (II) centers is joint coordination of the sacrificial aldehyde and of dioxygen in the initiation step.

Activation of dioxygen by nickel (II) in the presence of a sacrificial aldehyde auxiliary makes a very efficient oxidizer; of sufficient potency to oxidize also saturated hydrocarbons, as a Japanese group has shown recently *(53)*.

Acknowledgements

This work was supported by the award of postdoctoral fellowships to SLR by Orkhem, to EB by L'Oréal and to GPS by the Ecole Polytechnique. We have also benefited from support by a grant no. 89T0296 from Ministère de la Recherche et de la Technologie.

References

1   Camps, F.; Coll, J.; Messeguer, A.; Pericas, M.A. *Tetrahedron Lett.* **1981**, *22*, 3895 .
2   Paquette, L.A.; Barrett, J.H. *Org. Synth. Collective Volume* **1973**, *5*, 467.
3   Guengrich, F.P.; Macdonald, T.L. *Accounts Chem. Res.* **1984**, *17*, 9.
4   Tabushi, I.; Kodera, M. *J. Am. Chem. Soc.* **1984**, *108*, 1101.
5   Mansuy, D.; Fontecave, M.; Bartoli, J-F. *J. Chem.Soc.Chem.Comm.* **1981**, 874.
6   Chin, .D.H.; LaMar, G.N.; Balch, A.L. *J. Am. Chem. Soc.* **1980**, *102*, 1446-4344.
7   Groves, J.T.; Quinn, R. *J. Am. Chem. Soc.* **1985**, *107*, 5790 .
8   Katsuki, T.; Sharpless, K.B. *J. Am. Chem. Soc.* **1980**, *102*, 5974.

9   Halterman, R.L.; Jan, S-T. *J. Org. Chem.* **1991**, *56*, 5253.
10   Groves, J.T.; Viski, P. *J. Org. Chem.* **1990**, *55*, 3628.
11   O'Malley, S.; Kodadek, T. *J. Am. Chem. Soc.* **1989**, *111*, 9116.
12   Groves, J.T.; Myers, R.S. *J. Am. Chem. Soc.* **1983**, *105*, 5791.
13   Jacobsen, E.N.; Zhang, W.; Güler, M. L. *J. Am. Chem. Soc.* **1991**, *113*, 6703.
14   O'Connor, K.J.; Wey, S-J.; Burrows, C. *J. Tetrahedron Lett.* **1992**, *33*, 1001.
15   Heider, R.L. U.S. Patent 2 554 459, 1951; *Chem. Abstr.* **1951**, *45*, 9076e.
16   Kilty, P.A.; Sachtler, W.M.H. *Catal. Rev.* **1974**, *10*, 1.
17   Landau, R.; Sullivan, G.A.; Brown, D. *Chemtech* **1979**, *9*, 602.
18   Sheldon, R.A.; *J. Mol. Catal.* **1980**, 7, 107.
19   Farberov, M.I.; Bobylev, B.N.; Epstein, D.I. *Proc. Acad. Sci. USSR* **1976**, *226*, 28.
20   Simpkins, N.S. (Ed.) 100 Modern Reagents, The Royal Society of Chemistry and Moss Publishing, Nottingham, England, p.35.
21   Ortiz de Montellano, P.R. (Ed.) Cytochrome P-450, Structure, Mechanism and Biochemistry, Plenum Press, New York, 1986.
22   McMurry, T.J.; Groves, J.T. in the above reference, pp.1-28; Ortiz de Montellano,P.R. in the above reference, pp. 217-272.
23   Vanhove, D.; Blanchard, M. Bull. Soc. Chim. France, 1971, 3291 ; *J. Catal.* **1975**, *36*, 6.
24   Bond, G.C.; Sarkany, J.; Parfitt, G.D. *J. Catal.* **1979**, *57*, 476.
25   Ai, M. *J. Catal.* **1986**, *101*, 389.
26   Cavani, F.; Centi, G.; Riva, A.; Trifiro, F. *Catal. Today* **1987**, *1*, 17.
27   Centi, G.; Trifiro, F.; Etner, J.R.; Franchetti, V.M. *Chem. Rev.* **1988**, *88*, 55-80.
28   Komatsu, T.; Urugami, Y.; Otsuka, K. *Chem. Lett.* **1988**,1903.
29   Bosch, H.; Janssen, F. *Catal. Today* **1988**, *2*, 369.
30   Iwamoto, M.; Furukawa, H.; Matsukami, K.; Takenaga, T.; Kagawa, S. *J. Am. Chem. Soc.* **1983**, *105*, 3719.
31   Kobayashi, H.; Yamaguchi, M.; Tanaka, T.; Nishimura, Y.; Kawakami, H.; Yoshida, S. *J. Phys. Chem.* **1988**, *92*, 2516.
32   Busca, G.; Centi, G.; Trifiro, F. *Appl. Catal.* **1986**, *68*, 74.
33   Leatherdale, W.H. The Role of Analogy, Model and Metaphor, North-Holland, Amsterdam, 1974.
34   Laszlo, P.; Levart, M.; Singh, G.P. *Tetrahedron Lett.* **1991**, *32*, 3167.
35   Lindsay-Smith, J.R.; Sleath, P.R. J.Chem.Soc. , *Perkin Trans. II* **1982**, 1009-1015.
36   Groves, J.T.; Nemo, T.E. *J. Am. Chem. Soc.* **1983**, *105*, 5786.
37   Battioni, P.; Bartoli, J.F.; Leduc, P.; Fontecave, F.; Mansuy, D. *J. Chem. Soc. Chem. Comm.* **1987**, 791.
38   Nishiki, M.; Satoh, T.; Sakurai, H. *J. Mol. Catal.* **1990**, *62*, 79.
39   Kamiyama, T.; Inoue, M.; Kasiwagi, H.; Enomoto, S. *Bull. Chem. Soc. Japan* **1990**, *63*, 1559.
40   Takai, T.; Yamada, T.; Mukaiyama,T. *Chem. Lett.* **1990**, 1657-1660.
41   Mukaiyama,T.; Takai, T.; Yamada, ,T.; Rhode, O. *Chem. Lett.* **1990**, 1661-1664.
42   Haines, A.H. Methods for the Oxidation of Oreganic Compounds, Alkanes, Alkenes, Alkynes, and Arenes, Academic Press, London, 1985.
43   Doyle, M.P.; Patrie, W.J.; Williams, S.B. J.Org. Chem. 1979, *44*, 2955.
44   Mijs, W.J.; DeJonge,C.R.H.I. (Eds.) Organic Syntheses by Oxidation with Metal Compounds, Plenum Press, New York, 1986, ch.6, pp.373-422.
45   Yamada,T.; Takai, T.; Rhode,O.; Mukaiyama, T. *Chem. Lett.* **1991**, 1-4.
46   Yamada,T.; Takai, T.; Rhode,O.; Mukaiyama, T. *Bull. Chem. Soc. Jpn.* **1991**, *64*, 2109.

47 We apologize to the reader for this atrocious pun.
48 In conjunction with ferric chloride, it catalyzes the Michael reaction: Laszlo, P.; Montaufier, M-T.; Randriamahefa, S.L. *Tetrahedron Lett.* **1990**, *31*, 4867.
49 Little or no contamination of organic products or solvents by inorganic materials. After repeated washing of the freshly prepared catalyst by methylene chloride, followed by the epoxidation procedure, only traces of nickel are found in the solvent. Product isolation involves only decantation or filtration, followed by solvent evaporation.
50 Laszlo, P.,(Ed.) Preparative Chemistry Using Supported Reagents, Academic Press, San Diego, CA, USA 1987, 2 vols.
51 Tsuchiya, F. and Ikawa, T. *Can. J. Chem.* **1969**, *47*, 3191-3197.
    Vreugdenyhil, A.D. and Reit, H. *Rec. Trav. Chim. PB* **1972**, *91*, 237-245.
    Haber, J. ; Mlodnicka, T. and Poltowicz J., *J. Mol. Catal.* **1989**, *54*, 451-461.
    Iwanejka, R. ; Leduc, P. ; Mlodnicka, T. and Poltowicz, J. in *Dioxygen Activation and Homogeneous Catalytic Oxidation*, L.I. Simandi, Ed., Elsevier, Amsterdam, 1991.
52 Irie, R.; Ito, Y.; Katsuki,T. *Tetrahedron Lett.* **1991**, *32*, 6891.
53 Murahashi, Shun-Ichi; Oda Y.; Naota T. *J. Am. Chem. Soc.* **1992**, *114*, 7913-7914.

RECEIVED November 12, 1992

Activation and Selective Oxidation of C_1–C_4 Alkanes

Chapter 25

Oxidative Coupling of Methane over Praseodymium Oxide in the Presence and Absence of Tetrachloromethane

Yasuyuki Matsumura, Shigeru Sugiyama[1], and John B. Moffat

Department of Chemistry and Guelph—Waterloo Centre for Graduate Work in Chemistry, University of Waterloo, Waterloo, Ontario N2L 3G1, Canada

The catalytic oxidative coupling of methane to ethane and ethylene has been investigated on praseodymium compounds. Although praseodymium oxide produces predominantly carbon oxides from methane, the addition of small quantities of tetrachloromethane (TCM) to the reactant stream significantly improves its catalytic activity. The X-ray diffraction pattern for the catalyst after the reaction with TCM shows the presence of praseodymium oxychloride. Further, the oxychloride, generated from praseodymium chloride by heating under oxygen at 750°C is found to catalyze the reaction effectively. The catalytic activity of praseodymium oxychloride is stabilized with addition of TCM to the feedstream presumably because TCM hinders the formation of praseodymium oxide on the surface during the reaction.

The oxidative coupling of methane to ethane and ethylene is an intriguing process for the utilization of natural gas, which is predominantly methane. Although the search for effective catalysts has been vigorous during the last decade the conversions of methane and selectivities to C_2 hydrocarbons remain less than desirable for an economically practical process (1). The presence of chlorine in the reaction system has been found to enhance the conversion of methane and the selectivity to C_{2+} compounds. Work in this laboratory has examined the effect of the continuous addition of small quantities of chlorine compounds into the catalyst or the feedstream (2) and work in other laboratories has focused on chlorine-promoted catalysts (3). The addition of a small quantity of tetrachloromethane (TCM) to the reactant stream often results in high conversion of methane and high selectivity to C_{2+} compounds (4-7). Although participation of TCM in the gas phase reaction cannot be excluded, experimental observations show that TCM interacts with and alters the surface of the

[1]Current address: Department of Chemical Science and Technology, University of Tokushima, Minamijosanjima, Tokushima 770, Japan

catalysts (7). However, the source of the effect of the introduction of chlorine is not yet clear.

In this work, we will show that the addition of TCM to the feedstream in the methane conversion process results in the enhancement of the conversion of methane and the selectivity to C_2 hydrocarbons on praseodymium oxide primarily as a result of the formation of praseodymium oxychloride, in contrast with the production of carbon oxides on praseodymium oxide in the absence of TCM (8-10). The surface properties of these catalysts are characterized by application of adsorption experiments and X-ray photoelectron spectroscopy (XPS).

Experimental

Praseodymium oxide (Pr_6O_{11}) was obtained from Aldrich and used without further purification. Praseodymium chloride ($PrCl_3$) was prepared from praseodymium chloride hexahydrate (Aldrich 99.9%) by heating at ca. 150°C in air.

The catalytic experiments were conducted in a fixed-bed continuous flow reactor operated under atmospheric pressure. The reactor was designed to minimize the free volume in the hottest zone to reduce the contribution of the noncatalytic homogeneous reaction. The sample was placed in a quartz tube (7-mm i.d. and 35 mm in length sealed at each end to a 4-mm i.d. tube) and sandwiched with quartz wool plugs. Pretreatment of the sample was carried out immediately prior to the reaction under a helium stream (0.90 dm^3 h^{-1}) or an oxygen stream (0.75 dm^3 h^{-1}) at the desired temperature. Tetrachloromethane (TCM) was admitted to the main flow of reactants (CH_4, O_2, and diluent He) by passing a separate stream of helium through a gas dispersion tube in a glass saturator containing the liquid at ice-water temperature. Appropriate adjustments were made to ensure that the residence time was unchanged by the addition of TCM. The total flow rate of the feedstream was 0.90 dm^3 h^{-1}.

The reaction gas was analyzed with an on-stream gas chromatograph equipped with a TC detector and integrator. Porapak T (5.40 m) and Molecular sieve 5A (1.25 m) were used as separation columns.

Blank experiments conducted with methane absent from the feed (O_2 + He + TCM) indicated that TCM undergoes oxidation producing carbon monoxide and/or carbon dioxide. The data reported were corrected by running duplicate experiments with methane absent under otherwise identical sets of values of the process variables.

The surface area of the catalysts was measured by the conventional B.E.T. nitrogen adsorption method.

Powder X-ray diffraction (XRD) patterns were recorded with a Siemens Model D500 diffractometer. Patterns were recorded over the range $2\theta = 5$-$70°$.

The adsorption of carbon dioxide or oxygen on praseodymium samples was measured by a constant-volume method using a calibrated Pirani vacuum gauge. Praseodymium oxide was heated in oxygen (4 kPa) at 775°C for 1 h, then evacuated at 750°C for 0.5 h just before the measurement. The sample of praseodymium oxychloride was prepared from praseodymium chloride by heating under oxygen flow at 750°C for 1 h. This sample and praseodymium chloride were preheated *in vacuo* at 750°C for 0.5 h and adsorption experiments were carried out at room temperature. The amounts of the adsorbates reversibly adsorbed on the samples were estimated as follows. In the case of carbon dioxide, the adsorbate in the gas phase was removed with a liquid nitrogen trap to 1 Pa and the number of molecules reversibly adsorbed on the sample was determined. For oxygen, two successive adsorption experiments with intervening evacuation at room temperature for 0.5 h provided the number of oxygen molecules reversibly adsorbed. The number of the molecules irreversibly adsorbed on the samples was calculated by subtraction of the amount of the reversible adsorption from that of the total adsorption.

Surface analyses by XPS were carried out using a Phi (Perkin Elmer) ESCA 5500 MT spectrometer. The samples were mounted with indium foil (0.1 mm-thick) in air and set into the spectrometer. After measurement xenon-ion etching of the sample was carried out (3 kV, 0.5 min), and the spectra were measured again after etching.

Results

Methane Conversion. The results for the conversion of methane on praseodymium oxide are shown in Figure 1 and Table I. The major products were carbon monoxide, carbon dioxide, ethylene, and ethane both in the presence and absence of TCM in the feedstream while small amounts of formaldehyde and C_3 compounds were detected. Water and hydrogen were also produced. The catalyst produced low methane conversion (ca. 6%) and selectivity to C_{2+} compounds (ca. 30%) in the absence of TCM in the feedstream. On addition of TCM the conversion of methane after 0.5 h on-stream was increased by almost two-fold (11.9%) and increased still further to 17.2% after 6 h on-stream. The selectivity to C_{2+} also increased with time on-stream to 43.3% after 6 h on-stream. It is noteworthy that over the 6 h on-stream with TCM present the C_2H_4/C_2H_6 ratio increased from 1.0 to 2.1. No methyl chloride was detected in the product stream. After the reaction in the presence of TCM, the colour of the catalyst in the inlet portion of the bed was found to have been converted from the original black to white-green. The quantity of the white-green portion recovered was 0.21 g with main XRD peaks at 25.6, 31.2 and 34.4° in 2θ which were identical with those of praseodymium oxychloride (*11*). The BET surface area of the white-green compound was 3.7 m^2 g^{-1} while that for praseodymium oxide measured after the reaction in the absence of TCM was 7.9 m^2 g^{-1}.

Praseodymium chloride pretreated in a helium flow at 750°C for 1 h produced a low conversion of methane and selectivity to C_{2+} compounds after 0.5 h on-stream both in the absence and presence of TCM (Figure 2 and Table I). When TCM was present, the conversion and selectivity increased to 17.1 and 46.4% after 1.8 h on-stream, respectively, and then the values remained almost constant. In the absence of TCM, the conversion and selectivity also increased to 16.0 and 54.5%, respectively, after 1.8 h on-stream while in the latter case the values decreased gradually to 11.7 and 37.2% over 6 h on-stream. Although no TCM was added to the feedstream, methyl chloride was formed in the reaction. After 0.5 h on-stream, the selectivity to methyl chloride was 2.6% but decreased to 0.1% over 6 h on-stream. The XRD pattern of the catalyst after the reaction with TCM present in the

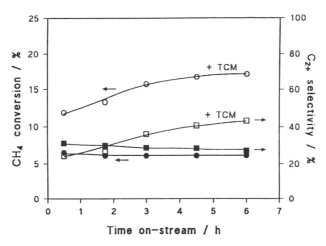

Figure 1   Conversion and C_{2+} selectivity for oxidative coupling of methane over praseodymium oxide in the presence and absence of TCM.

Table I. Oxidative Methane Coupling over Praseodymium Compounds

Catalyst	Time on-stream/h	CH$_4$	O$_2$	CO	CO$_2$	C$_2$H$_4$	C$_2$H$_6$	C$_3$
Pr$_6$O$_{11}$ heated in O$_2$ at 775°C for 1 h	0.5	6.5	98	4.4	62.5	9.7	22.1	1.2
	6.0	6.1	100	5.6	67.2	7.4	18.6	1.2
Pr$_6$O$_{11}$ heating in O$_2$ at 775°C for 1 h (TCM)	0.5	11.9	99	10.5	65.3	11.4	11.8	1.1
	6.0	17.2	100	25.0	31.8	28.1	13.2	2.0
PrCl$_3$ heated in He at 750°C for 1 h	0.5	9.6	100	39.9	48.5	7.6	0.5	0.8
	6.0	11.7	100	7.0	55.7	15.7	20.4	1.1
PrCl$_3$ heated in He at 750°C for 1 h (TCM)	0.5	11.0	94	32.2	47.9	14.4	1.2	0.6
	6.0	17.9	100	31.7	15.9	41.6	6.9	1.8
PrCl$_3$ heated in O$_2$ at 750°C for 1 h	0.5	16.0	99	22.8	27.8	27.9	20.1	2.5
	6.0	11.4	99	5.0	60.2	11.0	22.7	1.2
PrCl$_3$ heated in O$_2$ at 750°C for 1 h (TCM)	0.5	21.1	94	31.8	11.3	46.8	8.4	1.0
	6.0	20.1	94	34.1	13.4	41.7	7.3	2.2
PrCl$_3$ heated in O$_2$ at 750°C for 15 h	0.5	18.6	91	26.2	15.0	42.4	13.7	2.0
	6.0	17.0	90	28.7	17.3	33.2	18.1	2.6
PrCl$_3$ heated in O$_2$ at 750°C for 15 h (TCM)	0.5	19.7	94	25.7	15.1	43.0	11.7	3.6
	6.0	18.6	92	25.5	15.9	42.2	13.7	2.3

[a] Reaction conditions: catalyst, 0.7 g; CH$_4$, 29 kPa; O$_2$, 4 kPa; TCM, 0.2 kPa (when present); reaction temperature, 750°C.

feed was also identical with that of praseodymium oxychloride. In the case of the reaction in the absence of TCM, the inlet portion of the catalyst was dark white-green while the outlet portion was white-green although XRD patterns for both portions of the catalyst were identical with that of praseodymium oxychloride. The BET surface areas for the white-green compounds produced by the reaction in the presence and absence of TCM were 2.8 and 1.1 $m^2 g^{-1}$, respectively.

The catalytic activity of praseodymium chloride pretreated in an oxygen flow at 750 K for 1 h was high at the initial stage (Figure 3) while the XRD pattern of the sample just after the pretreatment was identical with that of praseodymium oxychloride. In the presence of TCM, the conversion of methane and selectivity to C_{2+} hydrocarbons were 21.1 and 56.4%, respectively, after 0.5 h on-stream and the activity was nearly constant throughout the reaction. After the reaction, the colour of the catalyst from the reactor was white-green. When TCM was not fed to the reaction stream, the conversion of methane and selectivity to C_{2+} compounds were 16.0 and 50.5%, respectively, after 0.5 h on-stream but decreased to 11.4 and 34.9%, respectively, over 6 h on-stream. The colour of the inlet portion of the catalyst after the reaction was dark white-green and the outlet portion was white-green as observed with praseodymium chloride pretreated under a helium stream followed by the reaction in absence of TCM. The XRD patterns of the samples just before and after the reactions were identical with that of praseodymium chloride. The BET surface areas of the catalysts measured after the reactions in the presence and absence of TCM were 4.7 and 2.3 $m^2 g^{-1}$, respectively.

Stable catalytic activity was observed with praseodymium chloride pretreated under an oxygen stream at 750°C for 15 h (Figure 4). The longer pretreatment produced high selectivity to C_{2+} compounds, that is, 58.8% in the presence of TCM and 58.5% in the absence of TCM after 0.5 h on-stream while methane conversions were 19.7 and 18.6%, respectively. Without TCM the conversion and the selectivity decreased gradually to 17.0 and 54.0% after 6 h on-stream, respectively. The catalytic activity was partially restored by addition of TCM to the reaction stream for 1 h. The conversion was 17.3% and the selectivity was 55.0% after 0.5 h on-stream following the period of the TCM feeding. The BET surface areas of the catalysts measured after the reactions in the presence and absence of TCM were both 1.5 $m^2 g^{-1}$. The XRD patterns of the catalysts used for the reactions were identical with that of praseodymium oxychloride. A small amount of methyl chloride was detected when praseodymium chloride was used as a precursor of the catalyst both in the presence and in the absence of TCM.

Adsorption of Carbon Dioxide and Oxygen on Praseodymium Samples. Adsorption of carbon dioxide or oxygen on the praseodymium samples was carried out in the pressure range of 1-40 Pa to evaluate the number of chemisorption sites on the samples. Praseodymium oxide irreversibly adsorbed 9.5 x 10^{-6} mol g^{-1} of carbon dioxide. The amount of oxygen irreversibly adsorbed on the sample was 15.2 x 10^{-6} mol g^{-1}. Carbon dioxide or oxygen was not adsorbed on the samples containing chlorine, i.e., praseodymium chloride and praseodymium oxychloride prepared from the chloride by heating under oxygen flow at 750°C for 1 h.

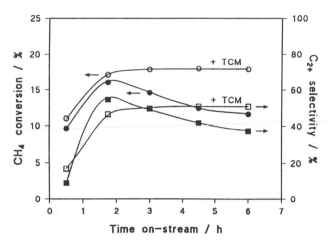

Figure 2   Conversion and C_{2+} selectivity for oxidative coupling of methane in the presence and absence of TCM over praseodymium chloride preheated in helium.

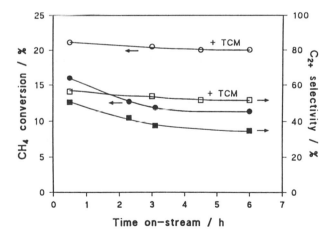

Figure 3   Conversion and C_{2+} selectivity for oxidative coupling of methane in the presence and absence of TCM over praseodymium chloride preheated in oxygen at 750°C for 1 h.

Analyses by XPS. In order to characterize the surface of the catalysts, XPS analyses for praseodymium oxide, praseodymium chloride, and praseodymium oxychloride, the latter formed in the conversion of methane, were carried out. There were two major peaks in the spectrum of C 1s for praseodymium oxide (Figure 5a). After xenon ion-sputtering for 0.5 min, the peak at higher binding energy was removed (Figure 5b). The peak of C 1s is usually observable in XPS measurements because carbon is present as a contaminant of the sample and cannot be removed thoroughly after surface sputtering. The binding energy is generally evaluated as 284.6 eV (12). Hence, the peak at lower binding energy can be attributed to a contaminant of the sample. In the present work this peak was used as a standard peak at 284.6 eV. The binding energy of the peak removed by the sputtering was determined at 287.4 eV. In the spectrum of O 1s for praseodymium oxide, peaks at 531.2 eV and 533.2 eV were observed (Figure 5c). After the xenon ion-sputtering the peak at 533.2 eV disappeared and a new peak appeared at 527.8 eV while the main peak at 531.2 eV shifted to 530.0 eV (Figure 5d). The main peak is attributed to lattice oxygen in the oxide because the composition of oxygen atoms on the surface etched with xenon ions roughly corresponds to its chemical formula (Table II). The composition was calculated from the peak intensities of O 1s, Pr 3d, and Cl 2p (if present) using sensitivity factors of 7.63, 0.71 and 0.89, respectively (13). The peak for Pr 3d was observed at 933.8 eV for praseodymium oxide (Figure 5e). After the sputtering a peak appeared at 927.5 eV while the main peak shifted to 932.0 eV (Figure 5f). In Figure 5, peak intensities are normalized by using the sensitivity factors.

Although oxygen is absent in the chemical formula of praseodymium chloride, a number of oxygen atoms were detected in the XPS spectrum of praseodymium chloride (see Table II). The peak of O 1s was observed at 530.7 eV and the binding energy did not change after xenon ion-sputtering for 0.5 min while the peak intensity decreased. The peak of Pr 3d was present at 933.1 eV and a small shoulder was observed at 928 eV. After the sputtering, the main peak shifted to 932.5 eV and the shoulder at 928 eV appeared clearly. The peak of Cl 2p was observed at 198.2 eV while the peak shifted to 199.0 eV after the sputtering.

In the case of the praseodymium oxychloride produced from praseodymium oxide by the reaction in the presence of TCM (white-green portion), an O 1s peak at 529.3 eV with a shoulder at 531 eV was observed. The peak shifted to 528.9 eV after xenon ion-sputtering for 0.5 min. The peak of Pr 3d was similar to that for praseodymium chloride, that is, the main peak was observed at 932.9 eV with a shoulder at 928 eV which was more clearly defined than that in the spectrum for praseodymium chloride. The peak shifted to 932.5 eV after the sputtering and the shoulder at 928 eV intensified as seen in the spectra for praseodymium chloride. The peak of Cl 2p was present at 198.8 eV and the position of the peak did not change after the sputtering.

There were two peaks at 528.9 and 531.1 eV in the spectrum for O 1s of the praseodymium oxychloride taken from the inlet portion of the reactor after the reaction in the absence of TCM (the sample originated from praseodymium chloride heated at 750°C for 1 h). The peak at 531.1 eV disappeared after xenon-ion sputtering while the main peak was present at 529.1 eV. The spectra for Pr 3d before and after xenon-ion sputtering were similar to those for praseodymium oxychloride which originated from praseodymium oxide. Although the Cl 2p spectra for the

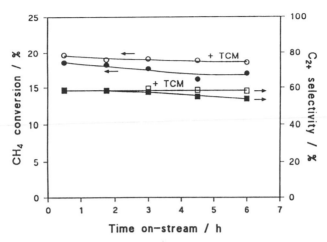

Figure 4   Conversion and C_{2+} selectivity for oxidative coupling of methane in the presence and absence of TCM over praseodymium chloride preheated in oxygen at 750°C for 15 h.

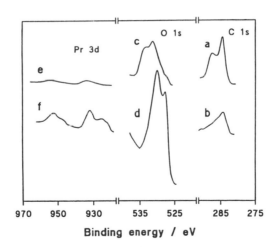

Figure 5   XPS bands of praseodymium oxide: (a) C 1s; (b) after sputtering; (c) O 1s; (d) after sputtering; (e) Pr 3d; (f) after sputtering.

Table II. Summary of XPS Analyses for Praseodymium Compounds

Sample	Molar ratio[a]		Binding energy[a,b]/eV		
	O/Pr	Cl/Pr	O 1s	Pr 3d	Cl 2p
Pr_6O_{11}	3.8 (1.7)	0.0 (0.0)	531.2 533.2 (530.0) (527.8)	933.8 (927.5) (932.0)	
$PrCl_3$	1.3 (0.6)	2.6 (1.4)	530.7 (530.7)	933.1 928s (932.5) (928s)	198.2 199.0
PrOCl produced from Pr_6O_{11} in the reaction with TCM	1.3 (0.9)	1.2 (0.7)	529.3 531s (528.9)	932.9 928s (932.5) (928s)	198.8 (198.8)
PrOCl produced from $PrCl_3$ in the reaction with no TCM	0.7 (0.6)	0.5 (0.4)	528.9 531.1 (529.1)	932.6 928s (932.5) (928s)	198.7 201s (198.9)
PrOCl produced from $PrCl_3$ in the reaction with TCM	0.8 (0.5)	0.8 (0.4)	529.0 (529.0)	932.9 928s (932.5) (928s)	198.9 200s (199.1)

[a] Values in parentheses refer to those obtained after xenon-ion sputtering for 0.5 min.
[b] Shoulder is symbolized as s.

former sample of praseodymium oxychloride had no shoulder, a shoulder at 201 eV was present on the main peak at 198.7 eV with this sample. The shoulder was not observed after the sputtering.

Praseodymium oxychloride produced from praseodymium chloride with a pretreatment in oxygen followed by methane conversion in the presence of feedstream TCM (as shown in Figure 3) displayed an O 1s peak at 529.0 eV but no shoulder. The peak position did not change after xenon-ion sputtering. The spectra for Pr 3d were similar to those for other praseodymium oxychloride samples. The binding energy of Cl 2p was 198.9 eV while the peak had a shoulder at 200 eV. The shoulder disappeared after the sputtering.

Discussion

The Role of TCM in the Reaction. Addition of TCM to the flow of methane conversion reactants produced a remarkable increase in the catalytic activity of praseodymium oxide. As evidenced by the XRD pattern for the white-green portion of the catalyst taken from the reactor after the reaction in the presence of TCM, praseodymium oxychloride is formed during the reaction. Although methyl chloride was detected in the methane reaction over praseodymium chloride preheated under either an oxygen or helium stream, no formation of methyl chloride was observed in the reaction over praseodymium oxide with TCM feed. The number of chlorine atoms trapped by praseodymium oxide can be estimated from the quantity of chlorine atoms trapped and is 1.1×10^{-3} mol after 6 h on-stream. Since the total number of chlorine atoms fed as TCM during the same period of time is 1.6×10^{-3} mol, the chlorine, in whatever form, in TCM was consumed mainly in the reaction between TCM and praseodymium oxide, and formation of methyl chloride is probably caused by the reaction between methane and chlorine atoms on the surface of the catalysts. At the initial stage of the reaction the catalytic activity of praseodymium chloride pretreated under a helium stream was low both in presence and in the absence of TCM, suggesting that praseodymium chloride is not active for methane coupling. Since the formation of praseodymium oxychloride was evident from the XRD patterns for the catalysts taken from the reactor after 6 h on-stream, the increase in methane conversion and C_{2+} selectivity after ca. 2 h on-stream (see Figure 2 and Table I) is evidently due to the formation of this compound in the catalyst. Since both the conversion of methane and the selectivity to C_{2+} hydrocarbons decreased with time-on-stream on praseodymium chloride pretreated under an oxygen stream at 750°C for 1 h, it can be concluded that the praseodymium oxychloride produced by this method was unstable without the addition of TCM to the feedstream (see Figure 3), although the catalyst still retained the crystalline structure of praseodymium oxychloride throughout the reaction. The observation of methyl chloride in the product stream suggests that the catalyst is dechlorinated. Although chlorine liberated from the catalyst surface and hence existing (in some form) in the gas phase may participate in the process, possibly in a manner similar to that of TCM, the number of chlorine atoms on the surface of the praseodymium oxychloride produced from the chloride by preheating in oxygen for 15 h is calculated as 2×10^{-5} mol g^{-1} from its density and the surface area, considerably smaller than that of chlorine fed as TCM (2×10^{-3} mol g^{-1} h^{-1} in Cl atoms). Moreover, the product distribution of the reaction catalyzed with the oxychloride was not affected by addition of TCM to the feedstream (see Figure 4 and Table I). Thus, the direct contribution of free chloride or TCM to the reaction is negligible and TCM mainly functions as a generator or stabilizer of praseodymium oxychloride in the reaction.

The Surface Properties of the Praseodymium Compound. Although the efficiency of catalysts in methane conversion has been ascribed to a variety of properties, a number of researchers have demonstrated the importance of basicity of the catalysts employed in this process (14-16). Thus estimates of basicity, such as may be obtained from the adsorption of carbon dioxide, are of some value in characterizing the catalysts. It is obvious that the surface state of a working catalyst at 750°C is different from that at room temperature. However, measurements of the adsorption of carbon dioxide at the latter temperature provide semiquantitative information on sites capable of donating electrons.

Irreversible adsorption of carbon dioxide was observed only on praseodymium oxide. In the XPS measurement for praseodymium oxide, there was a C 1s peak at 287.4 eV but no such peak was observed for other samples. Since adsorbed species are easily removed from the surface of the solid sample by ion sputtering, the peak can be attributed to carbon atoms in adsorbed species on the oxide. The binding energy of C 1s for carbon atoms bonded to oxygen is usually higher than 284.6 eV, hence, the peak at 287.4 eV can be attributed to carbon dioxide strongly adsorbed on the surface of praseodymium oxide. After etching the oxide surface the peak at 527.8 eV in the O 1s spectrum was observed. Since the binding energy is significantly lower than that of the oxygen atom normally observed (12), it appears that the electron density on the oxygen atom is relatively high. Thus, the oxygen atom responsible for the peak is probably a basic site, and it can adsorb carbon dioxide in air strongly. Although the main peak for O 1s at 531.2 eV is attributed to the lattice oxygen in praseodymium oxide, the O 1s peak for adsorbed carbon dioxide would overlap the peak because the 1s binding energy for oxygen in CO_3^{2-} is ca. 531 eV (9, 12).

Sites capable of adsorbing oxygen would presumably contribute to the activation of oxygen in the reaction process. Chemisorption of oxygen was also observed on the surface of praseodymium oxide. The number of the molecules adsorbed corresponds to 14% of the praseodymium atoms on the surface of the oxide estimated from the density of the oxide and the surface area of 7.9 m^2 g^{-1}. The O 1s peak at 533.2 eV was smaller than the main peak at 531.2 eV and it was removed by the xenon-ion sputtering. The binding energy is considerably higher than that for lattice oxygen in metal oxides, which suggests that the electron density of the oxygen responsible for the peak at 533.2 eV is low (12). Hence, the peak can be attributed to adsorbed species of oxygen on the oxide surface.

Since the binding energy of Pr 3d for PrO_2 is 935.2 eV and that for Pr_2O_3 is 933.2 eV, the main peak at 933.8 eV in the spectrum for praseodymium oxide (see Figure 5c) can be attributed to praseodymium ions with valence between 3+ and 4+ (17). After xenon-ion sputtering of the surface, the binding energy of Pr 3d was lower than that for Pr_2O_3, suggesting that the valence of the ion is smaller than 3+. This is consistent with the conclusion that a significant number of oxygen atoms was removed by the etching (see Table II). The shoulder at 927 eV in the spectrum

(Figure 5f) may be attributed to metallic praseodymium because, based on the binding energy, the electron density of the praseodymium atom is probably high.

No peak assignable to carbon dioxide was observed in the spectrum for C 1s for the samples containing chlorine. The peak for O 1s at 530.7 eV in the spectrum for praseodymium chloride can be attributed to oxygen atoms resembling the lattice oxygen in praseodymium oxide because the value falls between those observed for the lattice oxygen in the oxide and is significantly larger than that for the oxygen in praseodymium oxychloride. This suggests that the surface of the chloride sample resembles that of praseodymium oxide although a number of chlorine atoms are present near the surface (see Table II). This is supported by the observation that the initial activity of praseodymium chloride was inferior to that of the oxychloride (cf. Figures 2 and 3).

No adsorption of carbon dioxide or oxygen was observed on either praseodymium chloride or oxychloride. This finding is consistent with the XPS results. The main peaks at 529 eV in the spectra for praseodymium oxychloride samples are also attributed to the lattice oxygen of the oxychloride while the peaks at 531 eV are assignable to O 1s for praseodymium oxide, suggesting that the surfaces of the oxychloride samples are partially oxidized to praseodymium oxide. The 3d binding energy of 933 eV for praseodymium in the chloride and oxychloride implies that the valence of praseodymium is 3+, while the shoulder at 928 eV could be attributed to metallic praseodymium (*17*).

Since the Cl 2p binding energy of the main peak for the samples containing chlorine (198-199 eV) is similar to that for metal chlorides, most of the chlorine atoms in these samples are present as anions (*12*). The shoulder at 200 eV can be attributed to chlorine atoms with lower electron density than the anionic chloride (*12*), however, further identification is difficult.

Estrom and Lapszewicz reported that the lattice oxygen in praseodymium oxide can react with methane (*10*). The XPS results show that the praseodymium ion on the surface of the oxide is easily reduced to the lower valence ion. Hence, redox reaction can occur in the catalysis over praseodymium oxide; unfortunately, the reaction would mainly produce carbon oxides. The O 1s peak in the spectrum for the deactivated oxychloride catalyst has a shoulder attributed to the lattice oxygen in the oxide while no shoulder was observed in the spectrum for another oxychloride catalyst which originated from the chloride. Over the deactivated catalyst carbon dioxide and ethane were mainly produced, similarly to the result with praseodymium oxide (see Table I). Thus, the deactivation is believed to result from dechlorination of the oxychloride and partial formation of praseodymium oxide on the surface during the reaction. Addition of TCM to the feedstream would hinder the formation of the oxide on the surface. However, since there is a shoulder at 531 eV in the spectrum for O 1s, the praseodymium oxychloride sample originating from the oxide is believed to have lattice oxygens similar to those in the oxide. The surface concentrations of oxygen and chlorine are significantly larger than those for the oxychloride samples originating from the chloride while the molar ratio of O/Cl is about one, regardless of the samples (see Table II). This suggests that the samples originating from praseodymium oxide and chloride have different main crystalline faces on the surfaces and the praseodymium-oxide-like lattice oxygen would be rather stable on those crystalline faces which are dominant in the sample originating from praseodymium oxide.

The source of activity of praseodymium oxychloride is not clear. No strong basic sites or adsorption sites for oxygen were detected although these sites have been suggested as important in oxidative methane coupling (*14-16*). However, it can be speculated that chlorine ions on the oxychloride surface promote abstraction of hydrogen atom from methane and ethane because catalysts containing chlorine usually produce high C_2H_4/C_2H_6 ratio in methane coupling, as observed with praseodymium oxychloride (*2-7*).

From a simplistic viewpoint there are three possibilities for the interaction of methane with the surface of praseodymium oxychloride. The methane may interact with a surface oxygen species, as it may be expected to do in the case of praseodymium oxide, but with the oxychloride the presence of the chlorine modifies the electron density of the oxygen (Fig. 6a). The methyl radical produced in the scission process is released into the gas phase and an O-H bond is formed between the oxygen and the residual hydrogen. A second possibility involves the direct activation of methane by the surface chlorine atoms which are undoubtedly perturbed by the presence of the neighbouring oxygen atoms, but here the residual hydrogen forms a bond with the chlorine (Fig. 6b). Finally a two-site interaction between the methane and the oxychloride surface may occur with the formation and release to the gas phase of methyl chloride and the generation of a surface hydroxyl group (Fig. 6c).

Since the present work has shown that the chloride is less effective than the oxychloride it appears that the one-site process of Fig. 6a is probably the dominant scheme for methane activation and scission. However, since in the absence of feedstream TCM the catalyst deactivates, some participation of the schemes b and c seems likely. Further, the appearance, under some conditions, of methyl chloride provides additional evidence for scheme c.

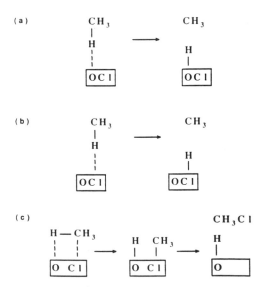

Figure 6   Schemes for the interaction of methane with praseodymium oxychloride.

Although generalization of the present results to other catalyst systems is tempting, the evidence currently available is undoubtedly insufficient to justify such an extrapolation. It is clear, however, from both the present and previous work in this laboratory, that the chlorine atoms interact with the surface of the catalyst. There is also strong evidence to support the contention that the effect produced by the introduction of TCM into the feedstream can be primarily attributed to a modification of the catalyst surface and not to a gas phase process. The present work demonstrates that, at least with praseodymium oxide, the oxychloride is produced on addition of TCM and further, the oxychloride is largely responsible for the beneficial effects. Since the enhancements observed with TCM have been shown (2, 4-7) to be related to the nature of the catalyst, it is conceivable that these effects, while dependent on the formation of the oxychloride, are also a function of the thermodynamic stability of the oxychloride. Further work is in progress.

Acknowledgement

The authors cordially thank Professor Satohiro Yoshida and Dr. Tunehiro Tanaka of Kyoto University for measurement of XPS. The financial support of the Natural Sciences and Engineering Research Council of Canada is gratefully acknowledged.

Literature Cited

1. Amenomiya, Y.; Birss, V. I.; Goledzinowski, M.; Galuszka, J.; Sanger, A. R. *Catal. Rev.-Sci. Eng.* **1990**, *32*, 163.
2. Ohno, T.; Moffat, J. B. *Catal. Lett.* **1991**, *9*, 23 and references therein.
3. Baldwin, T. R.; Burch, R.; Squire, G. D.; Tsang, S. C. *Appl. Catal.* **1991**, *75*, 153 and references therein.
4. Ahmed, S.; Moffat, J. B. *Catal. Lett.* **1988**, *1*, 141.
5. Ahmed, S.; Moffat, J. B. *J. Phys. Chem.* **1989**, *93*, 2542.
6. Ahmed, S.; Moffat, J. B. *J. Catal.* **1989**, *118*, 281.
7. Ahmed, S.; Moffat, J. B. *J. Catal.* **1990**, *125*, 54.
8. Gaffney, A. M.; Sofranko, J. A.; Leonard, J. J.; Jones, C. A. *J. Catal.* **1988**, *114*, 422.
9. Poirier, M. G.; Breault, R.; Kaliaguine, S.; Adnot, A. *Appl. Catal.* **1991**, *71*, 103.
10. Ekstrom, A.; Lapszewicz, J.A. *J. Phys. Chem.* **1989**, *93*, 5230.
11. Smith, J. V. *Index (Inorganic) to the Powder Diffraction File*; ASTM Publication No. PDIS-16i, Pa, 1966.
12. Wagner, C. D.; Riggs, W. M.; Davis, L. E.; Moulder, J. F.; Muilenberg, G. E. *Handbook of X-ray Photoelectron Spectroscopy*, Perkin-Elmer Corp., Eden Prairie, Minnesota, 1978.
13. Sensitivity Factors Determined by Perkin-Elmer-Phi.
14. Campbell, K.D.; Zhang, I.; Lunsford, J.H. *J. Phys. Chem.* **1988**, *92*, 750.
15. Lee. J.S.; Oyama, S.T. *Catal. Rev.-Sci. Eng.* **1988**, *30*, 249.
16. Choudhary, V. R.; Rane, V. H. *J. Catal.* **1991**, *130*, 411.
17. Sarma, D.D.; Rao, C.N.R. J. Electron Spectrosc. Relat. Phenom. **1980**, *20*, 25.

RECEIVED October 30, 1992

Chapter 26

Oxidative Coupling of Methane over Alkali-Promoted Simple Molybdate Catalysts

S. A. Driscoll, L. Zhang, and U. S. Ozkan

Department of Chemical Engineering, Ohio State University, Columbus, OH 43210

The effect of the addition of the alkali promoters Li, Na, and K to $MnMoO_4$ has been shown to increase the catalytic activity and selectivity for oxidative coupling of methane at 700 °C using a feed mixture of CH_4, O_2 and N_2 with a ratio of 2:1:2. The addition of K to $MnMoO_4$ provided the greatest increase in selectivity to C_2 hydrocarbons. The catalysts were characterized using laser Raman spectroscopy, X-ray diffraction, and TPR techniques. X-ray diffraction of reduced intermediates showed the reduction of pure $MnMoO_4$ and all three promoted catalysts by hydrogen to proceed through the formation of $Mn_2Mo_3O_8$ to Mo and MnO.

The study of various metal oxides and alkali promoted metal oxide catalysts has received much interest in recent years after the earlier reports of ethylene synthesis through oxidative coupling of methane (1), and of achieving high selectivities over a Li/MgO catalyst under methane and oxygen cofeed conditions (2). The addition of promoter ions to several oxide catalysts has been studied to determine the effect of the promoter ion on catalytic activity and selectivity (3 - 11). Alkali promoters were found to behave differently in different systems, causing an increase in selectivity while decreasing the catalytic activity for silica supported antimony oxides and nickel oxides (3, 11) while increasing both the activity and selectivity of samarium oxides (12). Alkali promoters have been found to affect the activity and selectivity in part due to a decrease in catalyst surface area, with additional selectivity changes due to the nature of the promoter ion (13-14). A decrease in product oxidation due to alkali

0097–6156/93/0523–0340$06.00/0

addition was given as an explanation for an observed selectivity increase during redox cycles by Jones et al. (*13*). The nature of the unpromoted metal oxide catalyst, the alkali metal ion used as a promoter, and the concentration of the promoter ion can all affect the activity of the catalyst. Burch et al. has suggested the function of alkali promoters may be to poison sites for complete oxidation on the surface for some metal oxide catalysts (*15*). Isotopic labelling of the feed gas has also been used to investigate the role of lattice oxygen and the residence time of carbon species through steady-state isotopic transient kinetic analysis using $^{18}O_2$ and $^{13}CH_4$ in the feed stream (*16-18*). Our work has focussed on the use of alkali promoters for a simple molybdate catalyst, $MnMoO_4$. Previous studies using molybdates dealt with unpromoted forms of these catalysts. A study of Na, Li, K, Mg, Ba, Mn, Co, Fe, Cu, Zn, and Ni molybdates by Kiwi et al. (*19*) showed that with the exception of $NiMoO_4$, the molybdates were stable for long periods of time under reaction conditions for oxidative coupling. At a conversion level of about 60 %, the molybdate selectivities ranged from 9.8 % to 16.6 %. The $MnMoO_4$ and K_2MoO_4 molybdates were the least selective catalysts. Another molybdate, $PbMoO_4$, was studied by Baerns et al., (*20*) with 19 % selectivity to C_2 hydrocarbons and an 11.4 % selectivity to formaldehyde at 1 % conversion. In this paper, we report the characterization and catalytic behavior of $MnMoO_4$ catalysts promoted with either Li, Na, or K in oxidative coupling of methane.

Experimental

Catalyst Preparation and Characterization. Simple molybdate catalysts have been prepared by a precipitation reaction as outlined previously (*21*). Alkali promoted catalysts containing either lithium, sodium, or potassium were prepared through wet impregnation of the molybdate with the alkali carbonate followed by drying in an oven overnight to drive off the water. The resulting catalysts were calcined in oxygen for 4 hours at 800 ºC. Two concentrations of each promoter were prepared such that the ratio of the moles of alkali metal to molybdenum was 1 % or 4 %. These catalysts were characterized through a number of techniques. BET surface area measurements were performed with a Micromeritics 2100E Accusorb instrument using krypton as the adsorbate. X-ray diffraction (XRD) patterns were obtained using a Scintag PAD V diffractometer with Cu K_α radiation as the incident X-ray source. X-ray photoelectron spectroscopy analysis was performed using a Physical Electronics/Perkin Elmer (Model 550)ESCA/Auger spectrometer, operated at 15kV, 20 mA. The X-ray source was MgK_α radiation (1253.6 eV). Laser Raman spectra were obtained using a Spex 1403 spectrometer equipped with a Datamate microprocessor for data collection and processing. The 514.5-nm line of a 5-W Ar ion laser (Spectra Physics) was used as the excitation Source. The laser power was 100 mW with a scanning rate of 1.0 cm^{-1}/sec.

Reaction Studies. A fixed-bed quartz reactor with 9 mm O.D. and 5 mm I.D. was used for the catalytic reaction experiments. The exit diameter was reduced to 2 mm at the exit of the catalyst bed to allow rapid exiting of the gas stream. The isothermal portion of the quartz tube was determined to be 20 mm, and the catalyst bed length was held constant at 15 mm by adding quartz chips to the catalyst. The total surface area of the catalyst was kept constant at 0.1 m^2. Blank studies using this reactor filled with quartz chips revealed negligible conversion under reaction conditions. The reaction feed gas consisted of methane, oxygen and nitrogen (ratio = 2:1:2). The gas composition was determined by mass flow controllers (Tylan), with a metering valve placed before the reactor to control the amount of gas to go through the reactor. The catalyst bed was heated from 650 °C to the reaction temperature of 700 °C while flowing the reaction mixture, and left overnight to reach steady-state conditions. The tubing and fittings following the reactor were heated to prevent condensation of products and the formation of paraformaldehyde. In one set of experiments, the flowrate was adjusted to attain a constant (10%) conversion of the entering methane. Additional flowrates were also examined for each catalyst system. A schematic of the analytical system used for steady-state, transient and steady-state isotopic labelling studies is shown in Figure 1. The 4-port Valco valve in the gas flow stream was used to switch the oxygen feed gas source from a $^{16}O_2$/He mixture to an $^{18}O_2$/He gas mixture. A second 4-port valve was used to isolate the reactor during feed gas analysis. Gas compositions were determined using an HP 5890 gas chromatograph. CH_4, CO_2, CO, N_2, O_2, formaldehyde, and coupling products that appeared in large concentrations were separated and analyzed by a thermal conductivity detector using an Haysep T column connected to a molecular sieve 5A column through a column isolation valve. Coupling products were also separated and analyzed through a second Haysep T column connected to a flame ionization detector. Transient pulse studies were analyzed by gas chromatography by placing the reactor in the carrier gas line. Isotopic labelling experiments were performed by leaking a small amount of the gas stream into the mass spectrometer detector (Hewlett Packard 5989A).

Temperature-Programmed Reduction Studies. The temperature programmed reduction (TPR) system, built in-house, is shown schematically in Figure 2. The same system was also used for temperature programmed desorption studies. The TPR profiles were determined by heating samples (50 mg) at a rate of 5 °C per minute from 50 °C to 950 °C. The temperature was then held at 950 °C until the reduction was completed. The reducing gas was 6 % H_2 in N_2, and the

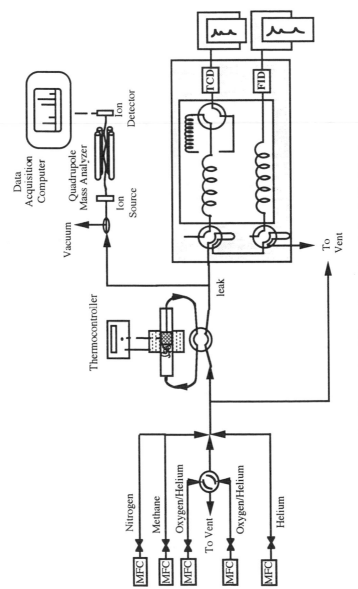

Figure 1. Schematic of Experimental and Analytical System.

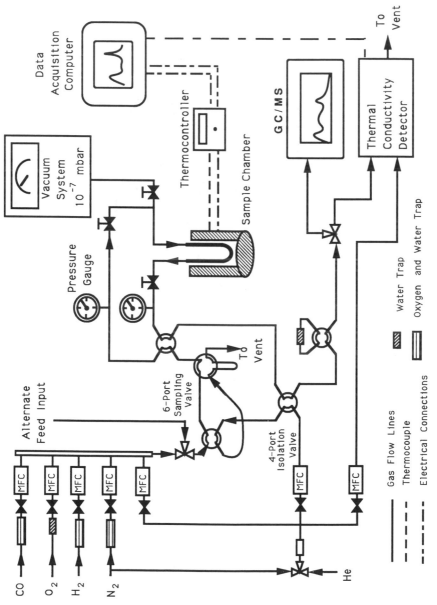

Figure 2. Schematic of Temperature Programmed Reduction System.

flow rate was 60 cm^3 (STP)/min.   X-ray diffraction and laser Raman spectroscopy were used to determine the nature of the reduction intermediates, and the final form.   To obtain the TPR intermediates for each of the samples, the TPR experiment was repeated, and terminated right after the first reduction peak. At the appropriate time, the hydrogen flow was stopped, and the sample temperature was quenched under nitrogen flow.   The TPR experiments were repeated   by replacing hydrogen with methane as the reducing agent. The reducing gas was 16.7 % methane in helium, and the total flowrate was maintained at 48 cm^3 (STP)/min.

Results and Discussion

The addition of the promoter ion to manganese molybdate (0.47 m^2/g) resulted in a decrease in BET surface area measurement   due to the addition of lithium (0.30 m^2/g), sodium (0.20 m^2/g), and potassium (0.27 m^2/g), with the largest decrease due to the addition of sodium.   The binding energies for fresh and spent catalysts obtained through X-ray photoelectron spectroscopy are reported in Table I. The metal binding energies were corrected for charging by the C 1s peak (BE=284.6 eV). The results indicated no major changes in the binding energies of Mn and Mo due to the addition of the promoter.   The signals from potassium and sodium were clearly visible, with binding energies for potassium at 292.6 eV for the 2p$_{1/2}$ electrons, and at 295.3 eV for the 2p$_{3/2}$ electrons, and for sodium at 1071.5 eV for 1s electrons in the promoted catalysts. Li was not detected through the XPS technique.   No major changes were observed in binding energies in post-reaction characterization of the catalysts.   Laser Raman spectra also revealed no major changes in the molybdate structure of the promoted catalysts compared to pure MnMoO$_4$, as shown in Figure 3. The effect of the promoter on the X-ray diffraction patterns compared to the pure MnMoO$_4$ revealed mainly changes in the relative intensities of the pattern, with no new phases observed.

The addition of a promoter ion to manganese molybdate affected the product distribution at equal conversion levels in favor of partial oxidation products as shown in Figure 4a. The selectivity was found to increase in the order of K > Na > Li. 1 % K/MnMoO$_4$ was found to be the most selective catalyst studied for C$_2$ production, while 1 % Na/MnMoO$_4$ was the most selective for HCHO production. Formaldehyde is generally thought to decompose under the reaction conditions contributing to the production of carbon oxides for oxidative coupling catalysts, and is often detected only in trace quantities. It is thought to occur through the formation of surface methoxide or peroxide species, a different route from the methyl radical formation leading to the gas phase coupling reaction (22). The formaldehyde observed here may be the result of the formation of some type of alkali molybdate structure. MacGiolla Coda and Hodnett

(23) studied formaldehyde production and have related it to the formation of crystalline Na_2MoO_4 through XRD examination of a sodium/ molybdenum impregnated silica support. The study examined the reaction at 500°C and 600°C using N_2O as an oxidant, where the observed increase in CO at the higher temperature was thought to be the result of HCHO decomposition. In the study of molybdate catalysts by Kiwi (20), higher temperatures and a more oxidizing atmosphere was used, with only trace amounts of formaldehyde reported. In our study, the catalytic activity of $MnMoO_4$ catalyst improved with the addition of promoter for equal residence times in each case, as shown in Figure 4 b. The addition of sodium increased the activity above both Li and K promoted catalysts, and the activity increased with the concentration of the sodium ion. The addition of Li was found to increase the catalytic activity more than the addition of K at equal promoter ion concentrations, however unlike sodium-promoted catalysts, the activity of catalysts promoted with lithium and potassium decreased as the promoter ion concentration increased from 1 % to 4 %.

TABLE I. X-Ray Photoelectron Spectroscopy Binding Energies (eV) for Fresh and Used Molybdate Catalysts

| Catalyst | Mn | | Mo | |
	$2p_{3/2}$	$2p_{1/2}$	$3d_{5/2}$	$3d_{3/2}$
$MnMoO_4$				
fresh	641.56	653.25	232.05	235.21
used	641.10	652.85	231.76	234.91
4% Li/$MnMoO_4$				
fresh	641.65	653.30	232.05	235.19
used	641.34	652.99	232.00	235.15
4% Na/MnMo				
fresh	641.40	653.08	231.86	235.01
used	641.17	652.86	231.88	235.04
4% K/$MnMoO_4$				
fresh	641.19	652.83	231.90	235.05
used	640.81	652.49	231.51	234.61

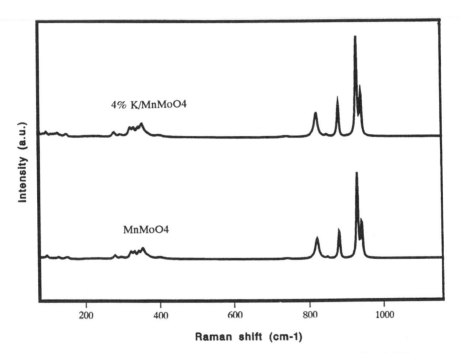

Figure 3. Laser Raman Spectra of MnMoO$_4$ and 4 % K/MnMoO$_4$ Catalysts.

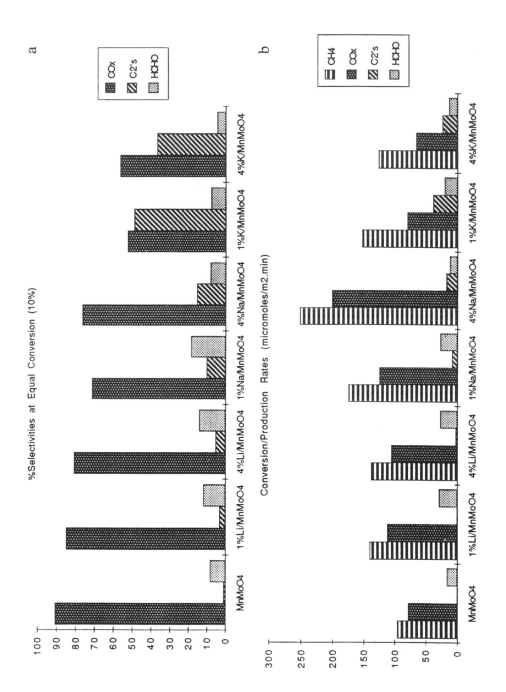

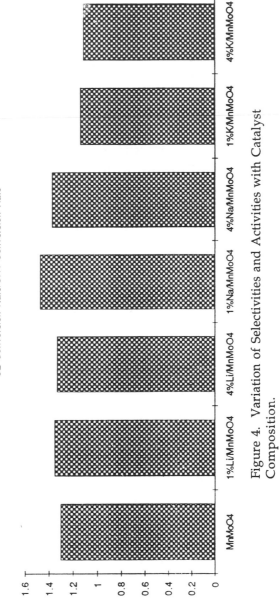

Figure 4. Variation of Selectivities and Activities with Catalyst Composition.

The highest formation rate for C_2 hydrocarbons was found over the K promoted catalysts. The ratio of oxygen conversion to methane conversion at equal residence times revealed an increased amount of oxygen utilized by both Li and Na catalysts over $MnMoO_4$, while the oxygen consumption for the potassium promoted catalyst decreased below that of $MnMoO_4$ catalyst (Fig. 4c).

The transient analysis using oxygen-free methane pulses over $MnMoO_4$ and 4 % $Na/MnMoO_4$ catalysts revealed no lattice oxygen reactivity at 700 °C. When the temperature was raised to 850 °C, both catalysts started to utilize lattice oxygen in reacting with methane, producing small amounts of CO_2, CO, and ethane. Under the conditions of the transient studies, the sodium promoted catalyst was found to be less active than pure $MnMoO_4$, but more selective for ethane formation. No reaction products were detected using a quartz filled reactor under the same reaction conditions. The initial results from steady-state isotopic labelling studies indicate significant differences in the way lattice oxygen is utilized over $MnMoO_4$ and promoted $MnMoO_4$ catalysts. Further studies using steady-state isotopic labelling technique are currently in progress.

The temperature programmed desorption (TPD) is a characterization technique used to examine the interaction of different reaction gases with the catalyst surface, and the adsorbed species remaining upon heating to reaction temperatures (23-26). Temperature programmed reduction and oxidation (TPR, TPO) are also characterization techniques that can provide complimentary information about the existence of different catalytic sites (27-29), and are often used as quality control indicators in catalyst production. The technique of TPD was used to examine $MnMoO4$ catalyst through room temperature adsorption of methane. No peaks were observed during the temperature ramp to 950 °C, indicating that any adsorption that occured was reversible at room temperature, or completely irreversible. TPR studies using methane as the reducing agent showed $MnMoO_4$ to be unreactive to methane in an oxygen free environment under the conditions employed below 900 °C. The results from the temperature programmed reduction (TPR) experiments using hydrogen as the reducing agent are shown in Figure 5 for pure $MnMoO_4$, (A) and 4 % K (B), 4% Na (C), and 4 % Li (D) promoted $MnMoO_4$ catalysts. The four TPR patterns are similar, with two peaks due to the reduction of $MnMoO_4$. X-ray diffraction patterns have identified the intermediate crystal structure after the first peak as $Mn_2Mo_3O_8$. The second peak represents the further reduction of $Mn_2Mo_3O_8$ to MnO and Mo. The peak maximum for the first reduction step has shifted to higher temperatures due to the addition of the promoter ion, with peak maximums for the first reduction occurring at 740 °C, 785 °C, 820 °C, and 840 °C for $MnMoO_4$, and the $MnMoO_4$ catalysts promoted with K, Na, and Li, respectively. The peak maximum for the second reduction occurred during the isothermal stage at 950 °C for all four catalysts. The 4 %

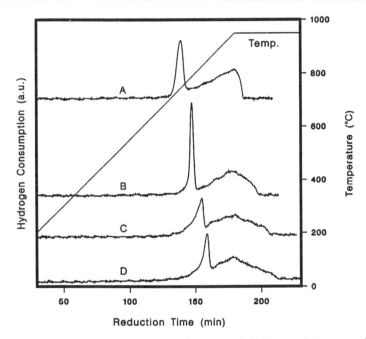

Figure 5. TPR Profiles of Manganese Molybdate and Promoted Manganese Molybdate Catalysts.

K/MnMoO$_4$ catalyst showed the sharpest peak for the first reduction step. The total hydrogen consumption for all samples was comparable, and was in agreement with the theoretical value of 13.96 mmol/g assuming the final species to be MnO and Mo.

The studies performed over promoted manganese molybdate catalysts have shown significant changes in catalytic behavior due to presence of the promoter. The preliminary results suggest that the pronounced differences observed in selectivity and activity may be related to the effect of the promoter cations on the reactivity of the lattice oxygen and the availability of adsorbed oxygen.

Acknowledgments

The financial support provided by the National Science Foundation through Grant CTS-8912247 is gratefully acknowledged.

Literature Cited

1. Keller, G.E., and Bhasin, M.M., *J. Catal.*, **1982**, *73*, 9.
2. Ito, T., and Lunsford, J.H., *Nature*, **1985**, *314*, 721.
3. Agarwal, S. K., Migone, R. A., and Marcelin, G., *Appl. Catal.*, **1989**, *53*, 71.
4. Bartsch, S., and Hofmann, H., In *New Developments in Selective Oxidation*; Centi, G. and Trifiro, F., Ed.; Elsevier: Amsterdam, **1990** , pp 353.
5. Gaffney, A.M., Jones, C.A., Leonard, J.J., and Sofranko, J.A., *J. Catal.*, **1988**, *114*, 422 .
6. Baronetti, G.T., Lazzari, E., Garcia, E.Y., Castro, A.A., Garcia Fierro, J.L., and Scelza, O.A., *React. Kinet. Catal. Lett.*, **1989**, *39*, 175.
7. Korf, S.J., Roos, J.A., Derksen, J.W.H.C., Vreeman, J.A., Van Ommen, J.G., and Ross, J.R.H., *Appl. Catal.*, **1990**, *59*, 291.
8. Lo, M.-Y., Agarwal, S.K., and Marcelin, G., *J. Catal.*, **1988**, *112*, 168 .
9. Matsuura, I., Utsumi, Y., Naki, M., and Doi, T., *Chem. Lett.*, **1986**, 1981.
10. Miro, E. E., Santamaria, J. M., and Wolf, E. E., *J. Catal.*, **1990**, *124*, 465 .
11. Otsuka, K., Liu, Q., and Morikawa, A., *Inorganica Chimica Acta*, **1986**, *118*, L23.
12. Otsuka, K., Liu, Q., Hatano, M., and Morikawa, A., *Chem. Lett.*, **1986**, 467.
13. Jones, C.A., Leonard, J.J., and Sofranco, J.A., *J. Catal.*, **1987**, *103*, 311.
14. Iwamatsu, E., Moriyama, N., Takasaki, T., and Aika, K., in *Methane Conversion*; Bibby, D.M., Chang, C.D., Howe, R.F., and Yurchak, S., Ed. Elsevier: Amsterdam, **1988**, pp 373.
15. Burch, R., Squire, G.D., and Tsang, S.C., *Appl. Catal.*, **1988**, *43*, 105.
16. Peil, K.P., Goodwin, J.G. Jr., and Marcelin, G.J., *J. Physical Chem.*, **1989**, *93*, 5977.
17. Peil, K.P., Goodwin, J.G. Jr., and Marcelin, G.J., *J. Catal.*, **1991**, *131*, 143.

18. Mirodatos, C., Holmen, A., Mariscal, R., and Martin, G.A., *Catal. Today*, **1990**, *6*, 473.
19. Carreiro, J.A.S.P., Follmer, G., Lehmann, L., and Baerns, M., in *Proceeding of the 9th International Congress on Catalysis*; Calgary, Phillips, M.J., and Ternan, M., Ed., Chemical Institute of Canada: Ottowa, **1988**, Vol. 2, pp 891.
20. Kiwi, J., Thampi, K.R., and Gratzel, M., *J. Chem. Soc., Chem. Commun.*, **1990**, 1690 .
21. Ozkan , U.S., Gill, R.C., and Smith, M.R., *J. Catal.*, **1989**, *116*, 171.
22. Peng, X.D., and Stair, P.C., *J. Catal.*, **1991**, *128*, 264 .
23. MacGiolla Coda, E., and Hodnett, B.K., in *New Developments in Selective Oxidation*; Centi, G., and Trifiro, F. Ed., Elsevier: Amsterdam, **1990**, pp 45.
24. Spinicci, R., and Tofanari, A., *Catal. Today*, **1990**, *6*, 473.
25. Ima, H., Tagawa, T., Kamide, N., and Wada, S., in *Proceedings of the 9th International Congress on Catalysis*; Calgary, Phillips, M.J., and Ternan, M., Ed., Chemical Institute of Canada: Ottowa, **1988**, Vol. 2; pp 952.
26. Korf, S.J., Roos, J.A., , Van Ommen, J.G., and Ross, J.R.H., *Appl. Catal.*, **1990**, *58*, 131.
27. Wendt, G., Meinecke, C.-D., and Schmitz, W.,*Appl. Catal.*, **1988**, *45*, 209.
28. Hicks, R.F., Qi, H., Young, M.L., and Lee, R.G., *J. Catal.*, **1990**, *122*, 295.
29. Larkins, F.P., and Nordin, M.R., *J. Catal.*, **1991**, *130*, 147.

RECEIVED October 30, 1992

Chapter 27

Methane-Selective Oxidation of Silica-Supported Molybdenum(VI) Catalysts

Structure and Catalytic Performance

Miguel A. Bañares and José Luis G. Fierro

Instituto de Catálisis y Petroleoquímica, CSIC, Campus UAM, Cantoblanco, E–28049 Madrid, Spain

Silica-supported molybdenum oxide catalysts have been prepared with variable surface concentrations and tested in the selective oxidation of methane. The nature of the supported molybdenum species has been studied by Laser Raman spectroscopy and their dispersion by FTIR of chemisorbed NO probe. It was found that polymolybdates are formed in the low Mo-content region (below 0.8 $Mo \cdot nm^{-2}$) while bulk MoO_3 dominates at higher Mo-contents; the former catalysts being more selective for the production of formaldehyde. The impregnation pH seems to play an important role in the genesis of catalysts as it controls both the nature and anchorage of molybdates species on the silica carrier. For instance, at pH values slightly below the zero point charge of the carrier (ca. 2.0) the anionic $Mo_7O_{24}^{6-}$ species interact more strongly than in neutral or basic media. Activity data also showed that formaldehyde and carbon dioxide are primary products while carbon monoxide is a secondary product. The deeper oxidation of formaldehyde to carbon monoxide can be reduced to a great extent by careful design of the reactor which minimizes the post reaction volume.

Considerable interest has been shown in the last decade in developing a relatively simple process for conversion of methane into C_1 oxygenates. A potential route that has received attention in the past, and in which there is now renewed interest, is the direct partial oxidation of methane (or natural gas). The work carried out in this area up to the eighties has been compiled in the review by Foster (1). A number of studies on the partial oxidation of methane to methanol and formaldehyde have been reported (2-6). The most interesting results for this reaction have been obtained by Liu et al. (7) and Iwamoto (8) while using nitrous oxide as oxidant. Also, steam has been added to the feed gases to inhibit the total oxidation of C_1 oxygenates produced. Liu et al. (7) also studied the catalyst MoO_3/SiO_2 and found that at 873 K and 1 bar, selectivities to CH_3OH and HCHO of 46,8 and 19,0 % at a CH_4 conversion of 16,4 % could be

0097–6156/93/0523–0354$06.00/0

obtained. More recently, these authors (9) have reported that they were unable to reproduce the high conversions obtained in their initial study.

Doubtless, Spencer´s work (10) on the partial oxidation of methane over MoO_3/SiO_2 catalyst is among the more succesful. For this system, the conversion-selectivity data could be plotted in a common curve, being independent of space velocity and only slightly temperature dependent. Moreover, for the Mo-containing catalysts attempts have been made to correlate catalyst structure with HCHO selectivity. For instance, Kasztelan and Moffat (11,12) and Barbaux et al. (13) have correlated increasing HCHO selectivity with formation of silicomolybdates on MoO_3/SiO_2 catalysts, and they attributed the increasing selectivity to stabilization on the silicomolybdate phase of the desirable O^{2-} species which acts selectively in the oxidation of methane. It is also recognized that the genesis of the anchored Mo-species markedly influences the catalytic behaviour (14). Many studies have been performed in order to elucidate the structure of silica supported molybdenum oxide (13, 20-22) looking for connections between molybdenum dispersion and catalyst preparation procedure. It is generally accepted that at low surface molybdenum loadings (<1.5 $Mo \cdot nm^{-2}$) the formation of molybdates is thought to be induced by the presence of Na or Ca impurities in the silica (23), which have a negative effect in the conversion of methane into formaldehyde (24). As the molybdenum loading increases, the polymeric species become the dominant species until the onset of the bulk molydenum oxide formation. The maximum of those amorphous species is reached right before the detection of XRD patterns (21). Accordingly, different catalytic behaviors could be expected from the different molybdenum species present on the surface of silica as has already been observed in propene metathesis (20, 21), propene oxidation (25), methanol oxidation (26), oxidative dehydrogenation of ethanol (21), ammonia oxidation to nitrogen (27) and methane oxidation (13, 28).

A tracer isotopic study of the reaction (Bañares, M. A.; Rodríguez-Ramos, I.; Guerrero-Ruíz, A. and Fierro, J. L. G., *10th International Congress on Catalysis*, to be published.) using oxygen-18 has shown that the oxygen is incorporated to the molecule of methane from the lattice molybdenum oxide and not from the gas phase molecular oxygen; the role of molecular oxygen (in general of the oxidant) is the reoxidation of the catalyst. Therefore, the different activities of the molybdenum species will be determined by the different ability that they have to release oxygen to the molecule of methane, which depends on the flexibility of the structures to accommodate the anion vacancy generated.

In addition, as the interaction of molybdenum with silica is through its silanol groups, such interaction of molybdenum with silica will be affected by the pH of the impregnating solution with respect to the isoelectric point of the support (29). In order to enlarge this view, the present study was undertaken to examine in greater detail the influence of Mo-concentrations and impregnation pH on the nature and exposure of surface molybdate species and their relevance for methane partial oxidation reaction. Particular emphasis was placed in characterizing these Mo-structures by Fourier Transform IR of chemisorbed NO probe, laser Raman spectroscopy and X-ray photoelectron spectroscopy. Moreover, the reactional pathways has been clarified by showing the importance of gas-phase reactions of formaldehyde and hence establishing optimal reactor geometry which maximizes formaldehyde yield.

Experimental

Catalyst Preparation. A silica, Aerosil 200, with a BET surface area of 180 m^2g^{-1}, particle size ca. 14 nm, and a composition SiO_2 >98,3 %, Al_2O_3 = 0,3-1,3 %, Fe < 0,01 % and Na_2O ≈ 0,05 % was used as carrier. The catalysts were prepared by impregnation with $(NH_4)_6Mo_7O_{24} \cdot 4H_2O$ (AHM) solutions at pH close to 4,5, whose concentrations were selected in order to achieve surface concentrations of 0.3, 0.8, 1.9 and 3.5 Mo atoms per nm^2 assuming complete dispersion. Two other preparations containing 0.8 Mo$\cdot nm^{-2}$ were prepared in a similar manner but starting with AHM solutions at pH values of 2.0 and 11.0, respectively. The impregnates were dried at 383 K for 16 h and then calcined in air in two steps: 623 K for 2 h and 923 K for 4 h.

Catalyst Characterization. The laser Raman Spectra (LRS) were recorded with a Jarrel Ash 25-300 spectrometer. The emission line at 514,5 nm from an Ar^+ laser (Spectra Physics, model 165) was used for excitation. The sensitivity for a given sample was adjusted according to the intensity of the Raman scattering. The wavenumbers obtained from the spectra are accurate to within about 2 cm^{-1}. The infrared spectra were recorded of very thin self-supporting wafers (8-10 mg$\cdot cm^{-2}$) of samples placed in a special infrared cell assembled with greaseless stopcocks and KBr windows, which allowed catalyst pretreatments either in vacuum or in controlled atmospheres. The samples were firstly purged in a He flow while heated up to 790 K, and subsequently reduced in a H_2 flow (60 cm^3min^{-1}) at this temperature for 1 h. After reduction the catalysts were outgassed under high vacuum for 0.5 h and then cooled down to room temperature. After this they were exposed to 4 kN$\cdot m^{-2}$ NO and the infrared spectra were recorded on a Nicolet ZDX Fourier Transform Spectrometer working with a resolution of 2 cm^{-1}.

The photoelectron spectra (XPS) were recorded with a Leybold LHS 10 spectrometer equipped with a hemispherical electron analyzer and a MgKα anode X-ray excitation source (hv = 1253.6 eV). The samples were turbopumped to ca. 10^{-3} N$\cdot m^{-2}$ before they were moved into the analysis chamber. The residual pressure in this ion pumped chamber was maintained below $5 \cdot 10^{-7}$ N$\cdot m^{-2}$ during data acquisition. Although surface charging was observed for all samples, accurate (±0.2 eV) binding energies (BE) could be determined by charge referencing with the Si2p peak for which a BE value of 103.4 eV was considered.

Activity Measurements. The activity meaurements were carried out at atmospheric pressure at temperatures 823-903 K in two quartz tube reactors consisting of 10 or 6 mm O. D., respectively, mounted vertically inside a tubular furnance having a 3 cm contant temperature zone. The reactant gases CH_4 (99.95% vol) and O_2 (99.98% vol) entered the reactor from the top, the flow being controlled by Brooks mass flow controllers. The catalysts were mounted in the constant temperature zone of the furnance and held in place by a quartz disc. The CH_4:O_2 molar ratio was adjusted to 11, and the methane residence time was selected to 1-6 seconds. The efluents of the reactor were analyzed by on-line GC using a Konik 3000 HR gas chromatograph fitted with a thermal conductivity detector. Chromosorb 107 and Molecular Sieve 5A packed columns using a column isolation analysis system were used in this study. The conversion of methane (X) is defined as the percentage converted to all products. The selectivity to product i (S_i) is defined as the amount of methane converted to product i divided by the amount of methane converted to all products and expressed as percentage. The yield to product i is defined as $X \cdot S_i/100$.

Results and Discussion

Infrared spectra of chemisorbed NO. Chemisorption of NO on partially reduced molybdena based catalysts has been shown to be a usefull technique in evaluating MoO_3 exposures on such a catalyst surface (see, e. g. (*16*)). However, the application of this technique to a series of catalysts which differ in MoO_3 loading, or even in preparation method, requires a carefull control of the degree of reduction. As NO appears to be chemisorbed mainly at Mo^{4+} sites, the catalysts must be quantitatively reduced according to the reaction:

$$MoO_3 + H_2 \rightarrow MoO_2 + H_2O \tag{1}$$

Previous microgravimetric experiments demonstrated that H_2-reduction proceeds rapidly at 790 K (*17*) almost reaching an equilibrium in reduction after one hour, which closely corresponds to the formation of stoichiometric MoO_2 phase. Following this procedure, the infrared spectra of NO chemisorbed on H_2-reduced catalysts were recorded. The infrared spectra in the region 1900-1650 cm^{-1} are given in Figure 1 B. It should be noted that the spectra were obtained by substracting the catalyst background from the overall spectra. This is an important requirement since the silica substrate shows intense overtones (Figure 1A) in the same spectral region where NO bands appear. All spectra show the characteristic doublet at ca. 1810 and 1710 cm^{-1} assigned to the symmetric and antisymmetric NO fundamental stretching vibrations, respectively of paired $(NO)_2$ molecules adsorbed as dinitrosyl or dimeric structures (*15,16*) on reduced Mo-sites, probably Mo^{4+}. The integrated intensities of the band at 1710 cm^{-1} as a function of Mo concentration and pH of impregnant AHM solutions are shown in Fig. 2. The intensity increases sharply with increasing Mo concentration up to 0.8 $Mo \cdot nm^{-2}$ and then levels off. As all catalyst samples have been reduced under the same experimental conditions, the observed intensities could be taken as a measure of molybdena exposure at the catalyst surface revealing the increasing importance of the clusters at molybdenum loadings higher than 0.8 $Mo \cdot nm^{-2}$. The opposite trend is observed however for the dependence of impregnant pH on the intensity of NO bands. The higher intensity of the NO band for the lowest pH and its progressive decay at higher pH values indicates, in agreement with the above reasoning, a parallel decrease in Mo dispersion. This can be interpreted in terms of the equilibrium:

$$Mo_7O_{24}^{6-} + 4 H_2O \Leftrightarrow 7 MoO_4^{2-} + 8 H^+ \tag{2}$$

which is shifted to the left hand side in acid media and to the right hand side in basic media. As the isoelectric point of silica is close to 2, one would expect that positively charged silica surface at pH values about 2 should be able to adsorb $Mo_7O_{24}^{6-}$ ions by electrostatic forces in a well dispersed manner as the silica becomes uniformly

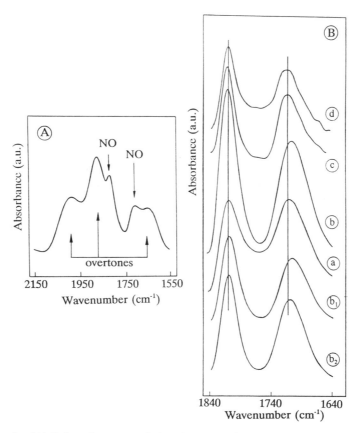

Figure 1. (A) Infrared spectra of chemisorbed NO on hydrogen prereduced 0.3 Mo•nm^{-2} catalyst without background substraction; (B), Net infrared spectra of adsorbed NO on hydrogen prereduced catalysts: (a)0.3 Mo•nm^{-2}; (b) 0.8 Mo•nm^{-2}; (c) 1.9 Mo•nm^{-2}; (d) 3.5 Mo•nm^{-2}; (b$_1$)0.8 Mo•nm^{-2}; impregnated at pH close to 2.0; (b$_2$) 0.8 Mo•nm^{-2} impregnated in basic medium (pH = 11.0)

charged. Conversely, in basic media the surface would appear negatively charged, therefore, no chance exists to adsorb $Mo_7O_{24}^{6-}$ or MoO_4^{2-} ions, only molybdena being deposited at the end of impregnation by molybdate precipitation over the silica surface. An intermediate situation to these extreme cases is expected to occur for medium pH values, thus explaining the observed tendency of IR intensities.

Laser Raman Spectra. Laser Raman Spectroscopy provides detailed information about the structure of supported molybdenum oxide phases. The Raman bands of calcined MoO_3/SiO_2 catalysts are summarized in Table I. For comparative purposes, those of the reference MoO_3 sample are also included in the same table. The peaks due to crystalline MoO_3 were present on supported catalysts for Mo contents of 0.8 Mo•nm^{-2} and above. For the 0.8 Mo•nm^{-2} catalyst additional weak bands at 943 and 677 cm^{-1} were observed, the former being the only one detected for the 0.3 Mo•nm^{-2} catalyst, where no bands due to MoO_3 were detected. As the 943 cm^{-1} band is in the range of vibrational frequencies corresponding to Mo=O double bond stretching, but it is on the lower frequencies side, it can reasonably be attributed to a dioxo structure but not a monoxo species (*18*). Williams et al. (*23*) observed Raman bands of hydrated silica-supported molybdenum oxide at 965-943, 880, 390-380 and 240-210 cm^{-1} and attributed these to octahedrally coordinated surface molybdate species, interacting weakly with the silica surface, similar in structure to the aqueous $Mo_7O_{24}^{6-}$ anion (*34*). Several aqueous polymolybdate anions composed of edge sharing MoO_6 octahedra display an intense Mo-O symmetric stretching vibration between 943 and 965 cm^{-1}, concordant with a Raman peak in 0.3 and 0.8 Mo•nm^{-2} samples. For these two samples the absence of other Raman bands in the lower energy region can result not only of their lower intensity but also to the silica background. The formation of polymolybdates on the surface of silica is also expected from the equation 2, since at the pH of impregnation, somewhat above zero point charge (ZPC) of the silica (ca. 2.0 pH), octahedral Mo(VI) clusters containing a discrete number of Mo cations, usually 7 as observed for slightly acid solutions of molybdate anions, are formed. This interpretation is consistent with literature findings. For instance, Jeziorowski et al. (*30*) assigned Raman bands at 956-970 and 884 cm^{-1} to an analogue of Mo-polyanion

Table I. Vibrational Bands in Silica-Supported Molybdate Catalysts

Catalyst	Mo=O str.	Mo-O-Mo as, str	Mo-O-Mo sy, str	Mo=O bend	Mo-O-Mo defor[c]
MoO$_3$[a]	996	821	473	338 367	218 246
			667	397	284 290
3.5Mo•nm^{-2}	995	818	473	337 363	216 245
			665	378	282 290
1.9Mo•nm^{-2}	995	818	473[b]	336 363[b]	216 244
			665	378	283 289
0.8Mo•nm^{-2}	995	818	665	336 363[b]	216 244
	943[b]		677	378	283 289
0.3Mo•nm^{-2}	943[b]				

[a] Commercial Sample; [b] Weak band; [c] Only the bands above 200 cm^{-1} were considered

chemically bonded to silica surface. Less structural detail was observed, however, by Rodrigo et al. (*31*) who observed a single broad band at ca. 950 cm^{-1}, as in our case, and suggested that it was indicative of a highly dispersed molybdate. Similar Raman bands for MoO_3/SiO_2 catalysts have already been reported by Cheng and Schrader (*32*) at 950-980 and 855-870 cm^{-1} and by Kakuta et al. (*33*) at 950 and 880 cm^{-1}, and which were attributed to aggregated molybdenum species and to several types of octahedrally coordinated molybdates.

From these results and the large body of research existing on MoO_3/SiO_2 catalysts it appears that ZPC of the silica carrier determines the nature of adsorbed molybdate species during aqueous impregnation with AHM. Because of the low ZPC of silica, molybdena assumes stable structures in an acid aqueous environment, those being the $Mo_7O_{24}^{6-}$ (or even its parent $Mo_8O_{26}^{4-}$) polyanion.

X-ray Photoelectron Spectroscopy. XP spectra of the catalysts showed the characteristic Mo3d doublet whose resolution was found to increase with increasing MoO_3 content, and the binding energy of the Mo3d peaks were slightly shifted towards lower values. These findings suggest that for the catalysts below 1.9 Mo•nm^{-2} the surface molybdena species become partly photoreduced under X-ray irradiation, which is also consistent with the intense blue color of samples after analysis. In order to obtain an estimate of molybdena dispersion, the intensity of the molybdenum signal relative to that of Si, corrected for the changes in surface area, have been calculated. The results are summarized in Table II and ploted in Figure 3 as a function of Mo content. The increase of this ratio up to 0.8 Mo•nm^{-2} indicates, in agreement with the infrared data, development of the disperse molybdate phase on the silica surface. However, above 0.8 Mo•nm^{-2} the intensity ratio starts to decrease, which means that molybdate becomes aggregated as MoO_3 crystallites, as Raman bands have also confirmed. It can also be noted in Figure 3 the almost constant values of the intensity ratios for the other two catalysts impregnated in acid (pH = 2.0) and basic (pH = 11.0) media. This latter tendency seems, in principle, anomalous since, as stated above, poorer MoO_3 dispersion in basic media should be expected. The appearance of broad features in the XRD pattern, characteristic of MoO_3 crystallites (d≈ 5 nm), and the absence of polymolybdates, as expected from equation 2, in the catalyst prepared at pH = 11.0 indicate clearly a change in the anchorage mechanism of molybdenum oxide. It appears that in the basic media MoO_3 precipitates at the end of impregnation step on the outer silica surface while maintaining essentially uncovered the inner surface.

Catalytic Behavior.

Influence of Mo content. The reaction of CH_4 with O_2 on MoO_3/SiO_2 catalysts under specific reaction conditions leads mainly to the formation of formaldehyde and carbon oxides, with minor amounts of dimerization products and methanol. Activity and selectivity data obtained for a given set of experimental parameters were practically the same when the experiment was duplicated. For all catalysts the increase in reaction temperature resulted, as expected, in an increase in the CH_4 and O_2 conversions but in a decrease of the selectivity towards selective oxidation products in favor of deeper oxidation (CO_x) and dimerization (C_2) products. A similar

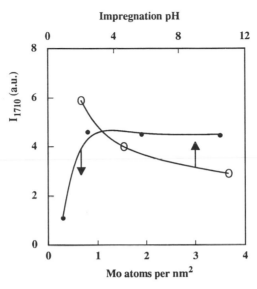

Figure 2. Dependence of molybdenum content (•) and impregnation pH of AHM for the 0.8 Mo•nm^{-2} catalyst (O) on the integrated intensity of the band at ca. 1710 cm^{-1} of NO chemisorbed on prereduced catalyts

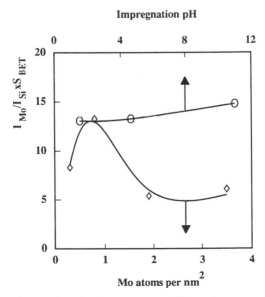

Figure 3. Dependence of molybdenum content (◊) and impregnation pH of AHM solutions for 0.8 Mo•nm^{-2} catalysts (O) on the XPS Mo/Si intensity ratios of fresh catalysts.

tendency was also observed for catalysts when increasing the CH_4 residence time while maintaining the reaction temperature. In agreement with literature findings (10, 17), HCHO decreased monotonically with methane conversion indicating that HCHO is a primary reaction product. Selectivity to carbon monoxide was zero at the limit of zero methane conversion and increased with methane conversion implying that carbon monoxide is a secondary product. Activity and selectivity data for the selective oxidation of methane at fixed experimental conditions: reaction temperature = 863 K, contact time = 3 s and $CH_4:O_2 = 11$ (molar), are summarized in Table II as a function of Mo-content and impregnation pH. To better illustrate the changes in catalytic behavior, Figures 4 and 5 show the dependence of both Mo-loading and impregnation pH, respectively, on CH_4 conversion and HCHO selectivity. Apart from these general tendencies, HCHO selectivity has been obtained at constant methane conversion (1%). On comparing the data in Table II and Figure 4 with those obtained using IR (Figure 2), a close correlation between molybdate dispersion and catalyst activity for Mo-loadings below 0.8 Mo•nm^{-2} is revealed. The reaction mechanism for the homogenous gas-phase reaction at moderate pressures leading to C_1-oxygenates is well understood (36) although kinetic data for a few steps require further refinement. The initiation of the reaction implies breaking down of the C-H bond of CH_4 by molecular oxygen:

$$CH_4 + O_2 \rightarrow CH_3^\bullet + HO_2^\bullet \tag{3}$$

Thereafter, the propagation reactions proceed controlling the overall chemistry. Recent models include more than 160 radical reactions (37), of which those accounting for formaldehyde formation are:

$$HO_2^\bullet + CH_4 \rightarrow H_2O_2 + CH_3^\bullet \tag{4}$$
$$CH_3^\bullet + O_2 \rightarrow CH_3O_2^\bullet \tag{5}$$
$$CH_3O_2^\bullet \rightarrow HCHO + HO^\bullet \tag{6a}$$
$$CH_3O^\bullet \rightarrow HCHO + H^\bullet \tag{6b}$$
$$CH_3O^\bullet + O_2 \rightarrow HCHO + HO_2^\bullet \tag{6c}$$

However, formaldehyde can be rapidly destroyed through radical reactions or even decomposed by oxygen at very short residence times. This is one of the main reasons why formaldehyde selectivity is low at high conversions. Apart from this, it is difficult to compare the gas-phase and heterogenous reactions because there is a pressure gap and temperature gap, i. e., moderately high pressure and low temperatures in the purely gas-phase oxidation and high temperature and atmospheric pressure in the heterogenous oxidation.

For the heterogenous process, it is generally believed that the methyl radicals, equation 3, may be produced relatively easily on a catalyst surface and that they may desorb to react further in the gas phase. However, if we try to take into account reactions at the catalyst surface, involving longer lived species, a precise understanding of the processes confined to the catalyst surface is required. Without doubt, the amount of mechanistic work on the selective oxidation of methane is more abundant for Mo-containing catalysts. Work by Yang and Lunsford (35) demonstrated that MoO_3/SiO_2 catalysts are very active for methanol oxidation with moderate temperatures, i. e.

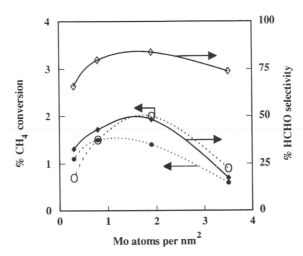

Figure 4. Influence of molybdenum content on methane conversion (circles) and formaldehyde selectivity (rhombus) at 873 K, a contact time of 3 sec and $CH_4:O_2$ molar ratio of 11. Filled symbols for reactor 1 and open symbols for reactor 2.

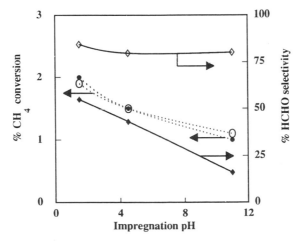

Figure 5. Influence of pH impregnation of AHM solutions for 0.8 $Mo \cdot nm^{-2}$ catalysts on methane conversion (circles) and formaldehyde selectivity (rhombus). Reaction conditions as in Figure 4.

TABLE II. Activity and Product Distribution of Catalysts

Mo/nm^2	Conversion (%)		Selectivity (% molar)					%Yield
	CH_4	O_2	CO	CO_2	C_2	HCHO	CH_3OH	HCHO
0.8A	1.9	15.2	23.0	5.0	-	71.9	0.1	1.4
0.8I	1.5	17.4	23.1	5.4	-	71.5	0.1	1.1
0.8B	1.1	9.0	19.0	5.0	-	75.9	0.1	0.9
0.3I	0.7	9.3	22.7	2.4	2.7	71.7	0.5	0.5
0.8I	1.5	17.4	23.1	5.4	-	71.5	0.1	1.1
1.9I	2.0	26.1	39.1	6.9	-	54.0	-	1.1
3.5I	0.9	9.5	18.3	4.3	-	77.3	-	0.7

Reaction conditions were as follows: reaction temperature = 863 K; contact time 3 sec; CH_4:O_2 molar ratio = 11. A, B and I refer to acid, basic and intermediate AHM impregation pH media, respectively. Notice that 0.8 Mo•nm^{-2} is common for the two catalysts series. Reactor 2.

350° C, so it is not surprising that we did not observe methanol as an intermediate at temperatures close to 600°C in the oxidation of CH_4 to HCHO. These authors also suggested a mechanism whereby a switch in valence between Mo^{IV}/Mo^{VI} was responsible for the oxidation of methanol to formaldehyde, the active species being a bridging-oxygen, formally corresponding to O^{2-}:

The methoxy species would then decompose to formaldehyde. Using oxygen-18 tracer we have recently demonstrated (Bañares, M. A.; Rodríguez-Ramos, I.; Guerrero-Ruíz, A. and Fierro, J. L. G., *10th International Congress on Catalysis*, to be published.) that the O-atom of formaldehyde arises from the lattice oxygen of molybdenum oxide, hence confirming the model proposed by Yang and Lunsford. Consistent with this model, the increase in selectivity is observed in those catalysts which develop a larger proportion of polymolybdates, i. e., molybdenum concentrations below 0.8 Mo•nm^{-2}. Hence the selectivity decrease observed at higher Mo-contents can be associated with the decrease in the active O^{2-} sites of the polymolybdates which occurs when the crystalline MoO_3 phase is developed as shown in the Raman spectra.

On the other hand, the data in Figure 5 show that for virtually the same Mo compositions, both activity and HCHO selectivity decrease with impregnation pH. This tendency can be explained in terms of equilibrium 2. At pH values close to isoelectric point of silica (ca. 2), molybdate appears to be uniformly adsorbed by electrostatic forces, however at medium and, especially, at higher pH, crystalline MoO_3 is deposited during the impregnation step.

The influence of the reactor. The product distribution is strongly affected by the characteristics of the reactor, since the smaller the section the reactor has, the better the selectivity to HCHO obtained. This effect is illustrated in Table III, indicating clearly that the time the reaction products stay in the hot zone must be minimized, since they may undergo successive reactions to give CO_X. Additionally, the data presented in Table III show that the methane conversion is almost unaltered by using the same reaction conditions in both reactors, but the product distribution is affected in the sense that partial oxidation product, HCHO, becomes most important when the volume to surface ratio of the reactor is smaller, while the selectivity to CO_2 is only slightly affected. This indicates, again, that in the oxidation of methane, CO_2 and HCHO are primary products. Carbon monoxide should be produced from the successive decomposition of formaldehyde in homogenous phase.

TABLE III. Effect of the characteristics of the reactor

Reactor	% Conversion		% Selectivity				% Yield
	CH4	O2	C O	CO2	HCHO	CH3OH	HCHO
1	2.0	30.6	59.6	7.1	33.0	0.2	0.66
2	1.9	15.2	23.0	5.0	71.9	0.1	1.36

Reaction conditions: catalyst 0.8A; reaction temperature = 863 K; $CH_4:O_2 = 11$ molar.

The sets of product distributions, obtained either with different reactors and the same catalyst (Table III) or with the same reactor and various catalysts (Table II), indicate that HCHO is the primary product of the methane and oxygen reaction, CO being a secondary product along with traces of methanol. The origin of CO_2 is much more complex. It may be directly formed from CH_4 since it is observed at very low methane conversions, in agreement with Spencer´s findings for the selective oxidation of methane on catalysts with comparable Mo-loadings (*28*). Using an isotopic labeling technique we have previously shown that CO_2 comes directly from the interaction of methane with molybdena (Bañares, M. A.; Rodríguez-Ramos, I.; Guerrero-Ruíz, A. and Fierro, J. L. G., *10th International Congress on Catalysis*, to be published.), which suggests that more than one oxidation site is formed on the molybdenum oxide phase. Moreover, carbon oxide can be produced from the oxidation of CO with the unreacted oxygen and also from the Water Gas Shift (WGS) reaction of CO. In fact, according to the amounts of H_2 present, ca. 20% of the CO_2 detected probably originates via WGS reaction. Apart from this, the extent

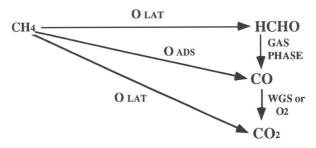

of carbon dioxide formation does not appear to be very important, as the close link between CO and HCHO indicates the latter as the main source of CO. As a whole, the reactional pathway, as depicted in the simplified scheme, is that lattice oxygen will yield selective and complete oxidation products while gas phase molecular oxygen will be mainly responsible for the oxidation of formaldehyde into CO (cf. Table III). Recent dynamic isotopic tracing studies (38) reveal that the residence times of HCHO and CO_2 are virtually identical, suggesting that they arise from a common pool of intermediates, while CO is obtained after extended residence times.

Conclusions

Supported molybdate catalysts are active and selective for methane partial oxidation. However, the surface Mo-loading and impregnation pH are important parameters in the genesis of catalysts as these influence to a great extent the nature and the relative abundance of the particular molybdenum species, whose intrinsic activities differ. The appearance of Raman band at 943 cm^{-1} in the catalysts with low Mo-loading (below 0.8 Mo•nm^{-2}), is indicative of the formation of polymolybdates with structures similar to those in aqueous media. These structures appear to be the most selective Mo-species for the partial oxidation of methane. At higher Mo concentrations the MoO_3 phase is the major phase at catalyst surfaces and acts less selectively. Finally, minimizing the postreaction volume by proper design of the reactor can be a method of increasing the HCHO yield, by reducing the extent of gas-phase reactions leading to combustion products.

Acknowlegments

This research was supported by the Comisión Interministerial de Ciencia y Tecnología, Spain (Grants MAT88-039 and MAT91-0494). Thanks are due to Dr. J. A. Anderson for critical reading of the manuscript.

Literature Cited

(1)   Foster, N. R., *Appl. Catal.*, **1985**, *19*, 1
(2)   Leonov,V.K. ; Ryzhak, I.A.; Kalinichenko, L.M.; Vysotskii, Yu L.; Semikova, L.E. and Shuster, Yu A.,*Kinet. Katal.*, **1977**, *15*, 6
(3)   Otsuka, K.and Hatano, M., *J.Catal.* **1987**, *108*, 252
(4)   Kastanas, G.; Tsigdinos, G. and Schwank, J., *Spring National AIChE Meeting*, Houston, Texas, **1989**, *April* ,Paper 52nd.
(5)   Sinev, M. Yu ; Korshak, V. N. and Krylov, O.V., *Russ. Chem. Rev.*, **1989**, *58*, 22
(6)   Brown, M.J. and Parkins, N.D., *Catal. Today*, **1991**, *8*, 305
(7)   Liu, R.S., Iwamoto, M. and Lunsford J.H., *J. Chem. Soc., Chem. Commun.*, **1982**, 78
(8)   Iwamoto, M., *Jpn. Pats. 58 92.629* (**1983**); *58 92.630* (**1983**)
(9)   Liu, H. F.; Liu, R. S.; Liew, K.Y.; Johnson, R. E. and Lunsford, J. H., *J. Am. Chem. Soc.*, **1984**, *106*, 4117
(10)   Spencer, N.D., *J. Catal*, **1988**, *109*, 187
(11)   Kasztelan, S. and Moffat, J. B., *J. Catal.*, **1987**, *106*, 512
(12)   Moffat, J. B. and Kasztelan, S., *J. Catal.*, **1988**, *109*, 206
(13)   Barbaux, Y.; Elmrani, A. R.; Payen; E. Gengembre; L. Bonnelle, J. B. and Gryzbowska, B., *J. Catal.*, **1988**, *44*, 117

(14) Rocchiccioli-Deltcheff, C.; Amirouche, M.; Che, M.; Tatibouet, J. and Fournier, M.,*J. Catal.*, **1990**, *125*, 292

(15) Yao, H. C. and Rothschild, G. W. in Proceedings of the *Climax Fourth International Congress on the Chemistry and Uses of Molybdenum*, Barry., H. F. and Mitchell, P. C. H.,Eds.), The Climax Molybdenum Co., Ann. Arbor, Michigan, 1982, p. 31.

16)   Fierro, J. L. G. in Spectroscopic Characterization of Heterogeneous Catalysts. Part B: Chemisorption of Probe Molecules, Fierro, J. L. G., Ed., Elsevier, Amsterdam, 1990, Vol. 57B, p. 367

(17) Bañares, M. A. and Fierro, J. L. G., *An. Quím.*, **1991**, *87*, 223

(18) Desikan, A. N., Huang, L. and Oyama, S. T., *J. Phys. Chem.*, **1991**, *95*, 10050

(19) Nakamoto, K., *Infrared and Raman Spectra of Inorganic and Coordination Compounds*, 3rd ed., Wiley, New York , 1978, p. 145

(20) Liu, T.-Ch.; Forissier, M.; Coudurier, M. and Vèdrine, J. C., *J. Chem. Soc., Faraday Trans. I*, **1989**, *85*, 1607

(21) Ono, T.; Anpo, M. and Kubokawa, Y., *J. Phys. Chem.*, **1986**, *90*, 4780

(22) Che, M.; Louis, C. and Tatibouët, J. M., *Polyhedron*, **1986**, *5*, 123 (1986)

(23) Williams, C. C.; Ekerdt, J. G.; Jehng, J. M.; Hardcastle, F. D.; Turek, A. M. and Wachs, I. E., *J. Phys. Chem.*, **1991**, *95*, 8781

(24) Spencer, N. D.; Pereira, C. J. and Grasselli, R. K., *J. Catal.*, **1990**, *126*, 546

(25) Giordano, N.; Meazzo, M.; Castellan, A.; Bart, J. C. and Ragaini, V., *J. Catal.*, **1977**, *50*, 342

(26) Louis, C.; Tatibouët, J. M. and Che, M., *Polyhedron*, **1986**, *5*, 123

(27) Boer, M. de; van Dillen, A. J.; Konigsberger, D. C.; Janssen, F. J. J. G.; Koerts, T. and Geus, J. W. In Procs. *"New Developments in Selective Oxidation"*, Lovain-la-Neuve (Belgium), April, 8-10, 1991

(28) MacGiolla Coda, E.; Mulhall, E.; van Hoeck, R. and Hodnett, B. K., *Catal. Today*, **1989**, *4*, 383

(29) Ismail, H. M.; Zaki, M. I.; Bond, G. C. and Shukri, R., *Appl. Catal.*, **1991**, *72*, L1

(30) Jeziorowski, H.; Knözinger, H.; Grange, P. and Gajardo, P., *J. Phys. Chem.*, **1980**, *82*, 2002

(31) Rodrigo, L.; Marcinkowska, K.; Adnot, A.; Roberge, P. C.; Kaliaguine, S.; Stencel, J. M.; Makovsky, L. E. and Diehl, J. R., *J. Phys. Chem.*, **1986**, *90*, 2690

(32) Cheng, C. P. and Schrader, G. L., *J. Catal.*, **1979**, *60*, 276

(33) Kakuta, N.; Tohji, K. and Udagawa, Y., *J. Phys. Chem.*, **1988**, *92*, 2583

(34) Greenwood, N. N. and Earnshaw, A. *"Chemistry of the Elements"*, Pergamon Press, 1984, p. 1177

(35) Yang, T. J. and Lunsford, J. H., *J. Catal.*, **1987**, *103*, 55

(36) Baldwin, T. R.; Burch, R.; Squire, G. D. and Tsang, S. C., *Appl. Catal.*, **1991**, *74*, 137

(37) Bedeneev, V. I.; Gol'denberg, M. Ya.; Gorban', N. I. and Teiltel'boim, *Kinet. Katal.*, **1988**, *29*, 7

(38) Mauti, R. and Mims, C. A., *ACS Symp. Ser., Div. Petr. Chem.*, **1992**, *37(1)*, 65

RECEIVED October 30, 1992

Chapter 28

Partial Oxidation of Ethane over Silica-Supported Molybdate Catalysts

A. Erdőhelyi, J. Cserényi, and F. Solymosi

Institute of Solid State and Radiochemistry, University of Szeged,
P.O. Box 168, H−6701 Szeged, Hungary

The partial oxidation of ethane was investigated on Rb_2MoO_4, $ZnMoO_4$, $MgMoO_4$ and $MnMoO_4$ catalysts at 773-823 K using N_2O and O_2 as oxidants. Kinetic measurements were carried out in a fixed-bed continuous-flow reactor. Additional measurements included temperature programmed reduction, pulse experiments and the study of the catalytic decomposition of N_2O and C_2H_5OH. For comparison, data for MoO_3/SiO_2 obtained under the same conditions are also determined. The main product of the selective oxidation of ethane were ethylene and acetaldehyde. The best catalytic performance as regards conversion and selectivity was exhibited by Rb_2MoO_4. Mo^{6+}-O^- surface species are considered to be the active centers in the partial oxidation and the formation of $C_2H_5\cdot$, and C_2H_5-O^- surface complexes are assumed to play an important role. The high activity and selectivity of Rb_2MoO_4 as compared with other catalysts are correlated with its ready reduction and with its high number of basic sites

The selective oxidation of alkanes is currently one of the most widely studied classes of catalytic reactions. This work mainly concentrates on the oxidative dehydrogenation of methane, with some attention paid to the partial oxidation of the product of this reaction, ethane. As regards the latter reaction, higher yields of partial oxidation products (acetaldehyde and ethylene) were achieved when N_2O was used instead of O_2 (1-6).

In a program to develop an efficient and selective catalyst in the partial oxidation of ethane, we found that the addition of various compounds of potassium to V_2O_5/SiO_2 *(5,6)* and to MoO_3/SiO_3 *(7)* greatly enhanced the conversion of ethane and advantageously influenced the product distribution. However, a better catalytic performance was obtained in both cases when KVO_3 or K_2MoO_4 was used as catalyst *(6-9)*. In the present work, the studies have been extended to silica-supported Rb_2MoO_4 and some divalent metal molybdates, $MgMoO_4$, $ZnMoO_4$ and $MnMoO_4$, several properties of which differ from those of the alkali metal molybdates. For comparison, we also present data for Mo_3/SiO_2. As oxidants, we use O_2 and N_2O.

Experimental

The ethane reaction was studied in a fixed bed continuous-flow reactor made of quartz (100 mm x 27 mm o.d.).The reacting gas mixture consisted of 20% C_2H_6, 40% N_2O (or O_2) and He as diluent. The flow rate of the reactants was usually 50 ml/min, and the space velocity was 6000 h^{-1} .

Analysis of the reaction products was carried out with a Hewlett-Packard 5780 gas chromatograph. Details on the experimental set up and on the analysis of the products are given in our previous papers *(5,8)*. The number of basic sites was determined by adsorption of CO_2 by dosing at 300 Torr and room temperature *(10,11)*, while the number of acidic sites was measured by n-butylamine titration *(12)*. Conversions are defined as moles of ethane reacted per min/mol ethane fed per min. Selectivity S is defined as mol of products formed per min/mol ethane reacted per min. Yield is given as mol products formed per min/ mol ethane fed per min.

Molybdates were prepared from $ZnCl_2$, $MgCl_2$ and $MnCl_2$ and Na_2MoO_4 following the method described by Sleight and Chamberland *(13)*. Rb molybdate was prepared from Rb_2CO_3 and MoO_3 *(14)*. The MoO_3 was stirred in hot alkali carbonate solution containing the calculated amount of carbonate until it dissolved. The solution was filtered, and the Rb molybdate was crystallized. The supported molybdates were prepared by impregnating the SiO_2 support (Cab-O-Sil) with a solution of these molybdate salts to yield a nominal 2% loading of MoO_3. The suspension was dried at 373 K and calcined at 873 K for 5 h. Before the catalytic measurements, each catalyst sample was oxidized in a O_2 stream at 773 K in the reactor; the catalysts were then flushed with He and heated to the reaction temperature in flowing He.

Results.

Characterization of Catalysts. The BET surface areas of the divalent metal molybdate catalysts were practically the same as that of the MoO_3/SiO_2, and they remained almost unchanged after the high-temperature reaction. A much lower value was measured for Rb_2MoO_4/SiO_2. The number of acidic sites was about 10% higher for the divalent metal molybdates than that for the Rb salt, and significantly lower than that for silica-supported molybdenum oxide. As expected, the number of basic sites was highest for the Rb compound, followed by molybdenum oxide and then the divalent metal molybdates. Data are given in Table I.

The reducibility of the catalysts was investigated by temperature-programmed reduction at 300-1100 K. Characteristic TPR spectra of the metal molybdates are shown in Figure 1, with that of MoO_3/SiO_2 for comparison. The reduction of MoO_3 starts above 800 K, and occurs in a rather broad temperature range peaking at 946 K. Reduction of the metal molybdates proceeds above 650-700 K. The onset temperature of the main reduction stage and the temperature of the TPR maximum, increase in the sequence Rb, Zn, Mn and Mg. The difference between the lowest and highest peak temperatures is about 140 K. In the reduction of the divalent molybdates one peak was obtained, while the reduction of the Rb salt occurred in two steps. It appears that most of these molybdates are reduced at lower temperatures than MoO_3/SiO_2. The consumption of hydrogen indicated that the average valency of molybdenum after reduction varied between 1.37 and 1.97. Results are listed in Table I.

Oxidation with N_2O.

Divalent Metal Molybdates, $MMoO_4$. The oxidation of ethane was first followed at 823 K2. The initial conversion of ethane was about 3-5%, which decreased slightly during the conditioning period (Figure 2). A steady-state activity was attained after a reaction time of 40-50 min. The main product of the partial oxidation was ethylene (7-20%), but small amounts of methane, acetaldehyde and ethanol (1-5%) were also formed. The selectivity and the yield for ethylene production decreased in time, except for $MnMoO_4/SiO_2$, where the 20% level was maintained for several hours. For comparison, data for MoO_3/SiO_2 obtained under the same conditions are likewise plotted in Figure 2. Some characteristic data for the catalytic behavior of metal molybdates are given in Table II.

With lowering of the temperature, the selectivities for ethylene, acetaldehyde and ethanol were significantly enhanced. The highest selectivity for ethylene in the case of $MnMoO_4$ was 28%, while in the other two cases they were 22-25%. The selectivity for acetaldehyde was the highest (8%) on $MnMO_4$ (Figure 3).

The effects of the reactant concentrations were determined on $MnMoO_4$ at 823 K after the steady state had been attained. With increase of the C_2H_6 concentration from 5% to 20%, the acetaldehyde production increased, but no change occurred in the production of ethylene. When the N_2O concentration was increased from 12% to 75%, the conversion of ethane increased and the selectivities of ethylene and acetaldehyde formation decreased.

Table I

Some characteristic data of silica supported MoO_3 and metal molybdates

Catalysts	BET surface m²/g	Acidic sites umol/g	Basic sites umol/g	T_i K	T_p K	H_2 consumption[a] umol/g	$MO \cdot MoO_x$[a]
MoO_3/SiO_2	166	360	16.3	842	946	202	1.54
Rb_2MoO_4/SiO_2	73	239	34.0	695	833, 944	172	1.70
$ZnMoO_4/SiO_2$	144.9	263	18.7	762	868	143	1.97
$MnMoO_4/SiO_2$	163.1	293	9.6	800	900	248	1.37
$MgMoO_4/SiO_2$	145.3	286	13.4	861	975	164	1.81

T_i: the onset temperature of the reduction. T_p: the temperature of TPR peak maxima (from fig. 1)

a the composition of molybdates (M metal ion) after the TPR, calculated from H_2 consumption

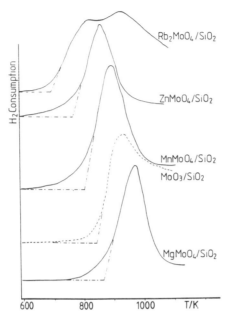

Figure 1. Temperature programmed reduction of silica supported MoO_3 and metal molybdates.

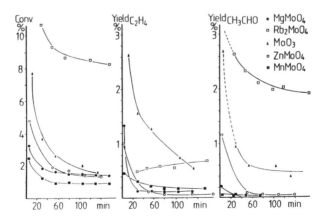

Figure 2. The conversion of ethane and the yields of acetaldehyde and ethylene in the $C_2H_6 + N_2O$ reaction on silica supported MoO_3 and metal molybdates at 823 K.

Table II.

Some characteristic data for the oxidation of ethane on different metal molybdates at 823 K^x

Catalyst	N_2O					O_2				
	Conv. %	Selectivity %				Conv. %	Selectivity %			
		CH_4	C_2H_4	CH_3CHO	C_2H_5OH		CH_4	C_2H_4	CH_3CHO	C_2H_5OH
MoO_3/SiO_2	1.1	0.6	34.1	16.9	–	31.6^{xxx}	0.74	0.06	0.01	–
Rb_2MoO_4/SiO_2	7.2	–	13.2	15	0.4	8.9	0.49	45.7	7.13	0.26
$ZnMoO_4/SiO_2$	1.25	4.2	7.3	1.43	2.16	2.85	0.8	51.4	6.9	0.18
$MnMoO_4/SiO_2$	0.86	3.2	20.1	2.7	3.6	1.9	0.33	34.7	0.22	0.24
$MgMoO_4/SiO_2$	1.35	1.32	7.9	0.07	1.85	1.42	0.75	50.2	4.8	0.34

^{x}Data refer to the steady state.

^{xx}For comparison data obtained for MoO_3/SiO_2 are also shown.

^{xxx}At 783 K, the conversion was 20 %, the selectivity for ethylene was 0.7%.

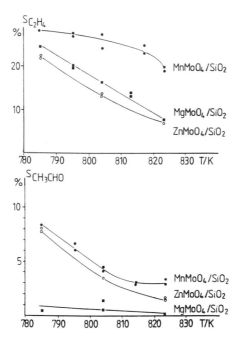

Figure 3. Effects of temperature on the selectivities of ethylene and acetaldehyde formation in the $C_2H_6 + N_2O$ reaction on silica supported metal molybdates.

Rb$_2$MoO$_4$ This catalyst was found to be much more active and selective than the divalent metal molybdates. As shown in Figure 2, very little decay was observed in the conversion of ethane. In contrast with the previous catalysts, acetaldehyde was the main product of partial oxidation: at 823 K it was formed with a selectivity of 23-24%. The selectivity for ethylene was 10-13%. As it appears from Figure 2, the yield of acetaldehyde formation was about 5 times higher than on MoO$_3$/SiO$_2$ catalyst.

Variation of the contact time showed that the conversion of ethane and the rate of product formation linearly increased with increasing contact time. A higher selectivity for acetaldehyde production could be attained at higher space velocities. This result clearly indicates the occurrence of secondary reactions of acetaldehyde, particularly at high contact times. The space velocity change did not influence the selectivity of ethylene formation. Some experiments with the pulse method were performed with this catalyst. The purpose of these measurements was to evaluate the interaction of ethane with Rb molybdate and to compare it with that observed for MoO$_3$. Results are plotted in Figure 4. It appears that Rb$_2$MoO$_4$ is much more reactive towards ethane than MoO$_3$. In the latter case only the nonoxidative dehydrogenation of ethane to ethylene occurred.

Oxidation with O$_2$.

Divalent Metal Molybdates. Except for the first point, the conversion of ethane did not change in the conditioning period at 823 K and it lay in the same range as for the N$_2$O as oxidant. The main product was ethylene; the selectivity of its formation markedly exceeded that obtained with N$_2$O as oxidant. Acetaldehyde was formed with 4.8% and 6.8% selectivity on the Mg and Zn salts. As a result, the yields for ethylene and acetaldehyde were much higher than in the case of N$_2$O oxidation (Figure 5). Other hydrocarbons and alcohols were also detected in very small concentrations, with less than 1% selectivity.

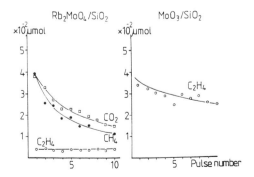

Figure 4. The amounts of products formed in the interaction of C$_2$H$_6$ with Rb$_2$MoO$_4$/SiO$_2$ and MoO$_3$/SiO$_2$ catalysts at 823 K as a function of pulse number. One pulse contains 32.6 μmol ethane. The amount of the catalyst was 0.3 g.

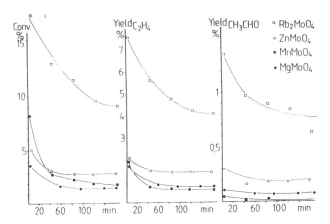

Figure 5. The conversion of ethane and yield of ethylene and acetaldehyde in the $C_2H_6 + O_2$ reaction on silica supported metal molybdates at 823 K.

With decrease of the reaction temperature, no change was experienced in the selectivity for ethylene, but there was an increase in the selectivity of acetaldehyde formation. The oxidation of ethane has been also investigated on MoO_3/SiO_2. Under these conditions, this catalyst was found to be very active for the total oxidation of ethane. At 510 K, the conversion of ethane was 21%, the products of partial oxidation were formed only in trace amount.

Rb_2MoO_4 This catalyst was found to be much more active than the divalent metal molybdates. Although the catalyst underwent a significant activity loss, even in the steady state the conversion was about 9.0%. The selectivities for ethylene and acetaldehyde formation in this state was 46% and 7.1%. Some data for the oxidation of ethane on molybdates with O_2 are also listed in Table II.

Decomposition of Ethanol.

As previous studies had suggested that the selective oxidation of ethane might occur through the formation and further reaction of ethoxide, it seemed useful to investigate the effects of these molybdate catalysts in the decomposition of ethanol. The decomposition of ethanol at 603 K yielded acetaldehyde (64-69%), ethane (25-26%), ethylene (3-5%) and small amounts of methane and CO. A decay in catalytic activity was observed for all catalysts. At the steady state, neither the activity nor the selectivity differed significantly for these molybdates.

A somewhat higher conversion was measured for Rb_2MoO_4 and the selectivity of acetaldehyde formation was also higher (76.8%). Interestingly, no ethylene or methane formation was detected.

Decomposition of N_2O.

In a closed circulation system the decomposition of N_2O on the divalent metal molybdates (activated in vacuum at 700 K) was measurable above 600 K. The initial decomposition was very slow on oxidized samples, but occurred more rapidly if the metal molybdates were reduced before the reaction at the same temperature. The rate of the decomposition declined, however, and reached the value measured for the sample activated in vacuum.

The activity sequence for molybdate catalysts was Mg - Mn - Zn. The Rb salt behaved similarly, but it was less active than the above metal molybdates.

Discussion

Comparison of the Catalytic Performances of Supported MoO_3 and Metal Molybdates

Oxidation with N_2O. In the evaluation of the catalytic data, we first compare the catalytic behavior of the metal molybdates with that of MoO_3/SiO_2. Data obtained for this catalyst earlier *(7)* under the present conditions are also included in Figure 2 and the appropriate tables. The conversion of ethane on MoO_3/SiO_2 at 823 K was initially 8 %, which decreased to about 1% at the steady state. The selectivity of acetaldehyde formation was about 21% and that of ethylene production was 31%. Depending on the preparation of the catalysts, Mendelovici and Lunsford *(3)* measured corresponding selectivities of 7-36% and 35-51%, but they studied the reaction in the presence of water and used a larger amount of catalyst and a lower space velocity.

Different features were observed for Rb_2MoO_4/SiO_2. In this case we attained a significantly higher rate for ethane oxidation (especially if we took into account the BET areas), and a high rate and selectivity of acetaldehyde formation (Figure 2). This was particularly pronounced at the steady state, when the yield of acetaldehyde production was five times larger than that measured for MoO_3/SiO_2.

Such advantageous catalytic properties were not exhibited by the divalent metal molybdates: the ethane conversion was low and was not connected with higher selectivity. Ethylene was the main product of oxidative dehydrogenation, but its selectivity and yield were much less than on the above catalysts. Acetaldehyde was produced in only small amounts, with low selectivity. A decrease was observed in the selectivities for C_2H_4 and CH_3CHO with increasing conversion, which is a common trend in the oxidative dehydrogenation of alkanes *(15)*. We point out here that the nonoxidative dehydrogenation of ethane on Rb and on divalent metal molybdates was extremely slow in the temperature range used, and its contribution to the ethylene formation is negligible.

Oxidation with O_2. In this case there was a dramatic difference in the catalytic behavior of supported MoO_3 and metal molybdate catalysts. Whereas in harmony with previous studies *(16)*, only the complete oxidation of ethane occurred on MoO_3/SiO_2 in the temperature range 500-550 K, on metal molybdates, ethylene was formed with selectivities (40-50%) which were considerably in excess of the values obtained even with N_2O. The selectivity of acetaldehyde formation was also higher. These improvements were achieved at somewhat higher ethane conversion (with the exception of $MnMoO_4$).

The best catalytic performance as regards conversion and selectivity was exhibited by Rb_2MoO_4 in this case also.

Mechanism of the Oxidation of Ethane

Previous studies indicate that O^- ion is the oxidizing agent in the partial oxidation of ethane on supported MoO_3 and V_2O_5 catalysts, and this is the surface species which can activate the ethane molecule (3). O^- species can readily be formed in the decomposition of N_2O on oxide surfaces (17), their formation being facilitated by the presence of Mo^{5+} surface centers. These Mo^{5+} centers can be generated by the surface decomposition of MoO_3 and molybdates at high reaction temperatures and/or by the reduction of Mo^{6+} ions by ethane, very probably in a non-selective step. From previous studies it was concluded that this active form of Mo in MoO_3/SiO_2 comprises only a very small fraction of the total molybdenum present (3). It appears that when O_2 is used as oxidant on this catalyst the surface concentration of other oxygen species active in complete oxidation is so high that the total oxidation of ethane predominates.

Accordingly, the key step in the selective oxidation is the removal of a hydrogen atom by O^- to give the ethyl radical:

$$C_2H_6 + O^- = C_2H_5{}^{\bullet} + OH^- \qquad (1)$$

The subsequent step is the formation of ethoxide:

$$C_2H_5{}^{\bullet} + O^- = C_2H_5\text{-}O^- \qquad (2)$$

which reacts with OH^- groups to give acetaldehyde:

$$2\ Mo^{6+} + C_2H_5\text{-}O^- + OH^- = CH_3CHO + H_2O + 2\ Mo^{5+} \quad (3)$$

or decomposes to ethylene:

$$C_2H_5\text{-}O^- = C_2H_4 + OH^- \qquad (4)$$

The formation of ethyl radical in the reaction $C_2H_6 + N_2O$ has been confirmed by ESR using the matrix isolation method (3,18). The kinetic results and the product distribution suggest that this reaction route dominates in the oxidation of ethane on V_2O_5 (5) and alkali metal vanadates (6,8), and very probably in the present case too. Attempts to identify $C_2H_5\text{-}O^-$ during the reaction were hampered by the low transmittance of the silica supported catalysts in the region 1000-1200 cm^{-1}. However, the formation of small amounts of ethanol can be regarded as evidence of the above reaction sequence. As mentioned before, the decomposition of ethanol on these molybdates is quite fast and the main product is acetaldehyde. As concerns the formation of CO_2, the product of full oxidation, there are several routes: direct oxidation of ethane, decomposition and oxidation of acetaldehyde, etc..

The high activity in ethane oxidation of Rb_2MoO_4 as compared with the other catalysts is very probably connected with its ready reduction (Figure 1). Under the given reaction conditions, this process results only in a higher surface concentration of Mo^{5+} (or Mo^{6+}-0$^-$) species, and not in the further reduction of molybdenum, which would involve the occurrence of a two-electron transfer.

The selectivity of the catalyst may be correlated with the numbers of acidic and basic sites. There are no significant differences in the acidities of the catalysts, but a marked variation may be observed in the numbers of basic sites (Table I). It may be assumed that the high number of basic sites on Rb_2MoO_4 as compared with MoO_3, and particularly divalent metal molybdates, may be responsible for the formation of acetaldehyde from surface ethoxide species (step 3).

One of the most interesting features of the metal molybdates is that the oxidative dehydrogenation of ethane (on divalent metal molybdates) and the formation of an oxygenated compound (on Rb_2MoO_4) occurred even with O_2. This indicates that the nature of the oxygen species and/or the state of the Mo during the catalytic reactions on these metal molybdates are markedly different from those on MoO_3/SiO_2. Experiments are in progress to exploit these features.

Literature Cited

1. Thorsteinson, E.M.; Wilson, T.P.; Young, F.G.; Kasai, P.H.; *J. Catal.*, **1978**, *52*, 116.
2. Iwamoto, M.; Taga, T.; Kagawa, S.; *Chem. Lett.,* **1982**, 1496.
3. Mendelovici, L.; Lunsford, J.H.; *J. Catal.* **1985**, *94*, 37.
4. Iwamatsu, E.; Aika, K.; Onishi, T.; *Bull. Chem. Soc. Japan,* **1986**, *59*, 1665.
5. Erdőhelyi, A.; Solymosi, F.; *J. Catal.,* **1990**, *123*, 31.
6. Erdőhelyi, A.; Solymosi, F.; *Appl. Catal.* **1988**, *39*, L11.
7. Erdőhelyi, A.; Máté, F.; Solymosi, F.; *Catal. Lett.,* **1991**, *8*, 229.
8. Erdőhelyi, A.; Solymosi F.; *J. Catal.* **1991**, *129*, 497.
9. Erdőhelyi, A.; Máté, F.; Solymosi, F.; *J. Catal.* **1992**, *135*, 563.
10. Borodin, V.N.; *Zhur. Fiz. Khim.* **1977**, *51*, 928.
11. Ai, M.; *J. Catal.* **1978**, *54*, 223.
12. Tanabe, K.; *Solid Acid and Bases;* Tokyo Kodansha and New York Academic Press 1970.
13. Sleight, A.W.; Chamberland, B.L.; *Inorg. Chem.,* **1968**, *7*, 1672.
14. Retgers, J.M.; *Z. Phys. Chem.* **1891**, *8*, 6.
15. Kung, M.C.; Kung, H.H.; *J. Catal.,* **1991**, *128*, 287.
16. Murakami, Y.; Otsuka, K.; Wada, Y.; Morikawa, A.; *Bull. Chem. Soc. Japan,* **1990**, *63*, 340.
17. Iwamatsu, E.; Aika, K.; Onishi, T.; *Bull. Chem. Soc. Japan,* **1986**, *59*, 1665.
18. Martir, N.; Lunsford, J.H.; *J. Amer. Chem. Soc.,* **1981**, *103*, 3728.

RECEIVED October 30, 1992

Chapter 29

Selective Oxidative Dehydrogenation of Propane on Promoted Niobium Pentoxide

R. H. H. Smits, K. Seshan, and J. R. H. Ross[1]

Faculty of Chemical Technology, University of Twente, P.O. Box 217, 7500 AE Enschede, Netherlands

It has been shown that the activity of niobia for the oxidative dehydrogenation of propane can be increased by adding vanadium or chromium, while maintaining a high selectivity towards propylene. The vanadium-containing materials have been characterised further by XRD, TPR, laser Raman spectroscopy and XPS in combination with Ar^+-sputtering. The influence of various catalyst pretreatments on the catalytic performance has also been investigated.

There is currently an increasing interest in the catalytic activation of lower alkanes. Following the initial work of Chaar et al. (1) on n-butane, a number of papers have appeared on the oxidative dehydrogenation of propane (2-10) over V-Mg-O catalysts. Propane is usually investigated because oxidative dehydrogenation may be an attractive alternative route for the production of propylene, compared to conventional dehydrogenation and cracking. Oxidative dehydrogenation of isobutane is another interesting application as the isobutylene produced could be used in the synthesis of MTBE. Research on this subject has focused on discussions of the nature of the active phase: magnesium orthovanadate or pyrovanadate (1, 5, 6).

We attempted to improve the V-Mg-O catalyst system by doping it with various elements (8, 11), but these attempts were not successful. We have therefore focused our attention on niobium pentoxide. In previous papers (9,10) we have shown that high selectivities towards propylene are possible using pure niobia, but that the conversions were low.

In this paper we present the results of experiments which show that the activity of niobia can be improved by adding other suitable elements, while maintaining high selectivity. These results bring the proces of oxidative dehydrogenation of propane closer to practical application.

[1]Current address: University of Limerick, Plassey Technological Park, Limerick, Ireland

0097–6156/93/0523–0380$06.00/0

Experimental

Niobia-based catalysts containing approximately 5 mol% of a high-valency element oxide were made by three different methods. A number of elements of which the oxides are soluble in both oxalic acid and ammonia (P, V, Cr, Ge and Mo) were incorporated by evaporation of the water from an oxalic acid solution of niobium and the oxide using a rotary evaporator. Elements of which the oxides are soluble in oxalic acid but not in ammonia (Ti, Zr and Sn), were incorporated using coprecipitation: a mixed oxalic acid solution of the two elements was slowly added to an ammonium oxalate/ammonia buffer (pH 10) at 60 °C, the pH being kept around 10 by addition of ammonia. Elements of which the oxides are not soluble in oxalic acid or ammonia (Sb, Pb and Bi) were incorporated by simultaneous slow addition of a nitric acid solution of the element and an oxalic acid solution of niobium to an ammonium oxalate/ammonia buffer (pH 10) at 60 °C. The resulting precipitates were filtered and washed on a glass filter. In all cases, the powders were then dried overnight at 80 °C, followed by calcination for 5h at 630 °C. This temperature was chosen because it appeared to be optimal for unpromoted niobia (*9, 10*). The chemicals used in the different methods were $(NH_4)_2HPO_4$, $TiO(AcAc)_2$, NH_4VO_3, $Cr(NO_3)_3.9H_2O$, GeO_2, $ZrOCl_2.8H_2O$, $(NH_4)_6Mo_7O_{24}.4H_2O$, $SnCl_2.2H_2O$, $Sb(OAc)_3$, $Pb(NO_3)_2$ and $BiO(NO_3)$. The niobium oxalate was provided by Niobium Products Company, Inc., USA (batch AD/651).

Because of the interesting results obtained with the addition of vanadium, catalysts containing 0.25, 1 and 10 wt% V_2O_5 were prepared by evaporation as described above. In addition, two other preparation methods were also used. The first of these involved dry-mixing of the hydrated niobia (Niobium Products Company batch AD/628, 10g), NH_4VO_3 and oxalic acid dihydrate (40g), and heating the resultant mixture at 120 °C overnight. This causes the oxalic acid to dissolve in its water of crystalisation, and then to react with the hydrated niobia and the vanadium to form niobium oxalate and vanadium(IV) oxalate. A large part of the oxalates formed dissolves; the oxalic acid evaporates slowly and the oxalates then decompose, the result being a very fine powder (V1Nb melt).

The other method involved wet impregnation of the vanadium by evaporation of an aqueous NH_4VO_3 solution in the presence of a niobia support (14.3 m^2/g) prepared by calcination of hydrated niobia for 5h at 600 °C (V1Nb imp.).

The catalysts prepared in these ways were characterised by X-ray powder diffraction (for phase composition), Temperature Programmed Reduction (TPR, for reducibility), Ar or N_2 adsorption (for total BET area), and X-ray fluorescence (for chemical composition). TPR experiments were performed using a TCD detector, a gas flow of 20 ml/min of 5% H_2 in argon, a heating rate of 5 °C/min, and 100 mg of sample (0.3-0.6 mm grains). The materials were tested for catalytic activity using a flow reactor system as described previously (*10*), using a feed consisting of 10 vol% O_2, 30 vol% C_3H_8 and 60 vol% He, a gas flow of 140 ml/min, and 600 mg of catalyst (0.3-0.6 mm grains). The tests were carried out by heating the sample in a series of seven steps of 25 °C in the required temperature range, and carrying out a sequence of measurements during each temperature step; each temperature was maintained for two hours. Laser Raman spectra were taken using a Specs Triplemate spectrometer equipped with an argon laser (wavelength 514 nm), and XPS in combination with Ar^+-sputtering was done on a Kratos XSAM 800 (sputtering rate 1-2 Å/min, vanadium concentrations were calculated from the corrected areas of the peaks of V2p and Nb3d).

Results

The results of XRF, BET surface area measurements and catalytic testing are summarized in Table I, while the results of catalytic testing are also shown in Figure 1. The rates displayed in Table I are the rates of reaction of propane per unit surface area as calculated from the interpolated propane conversion. Propane conversion was 3% or less, while oxygen conversion was less than 30%.

Table I. Results of XRF, BET Surface Area Measurements and Catalytic Testing for Niobia Catalysts Containing Various Additives

Sample	Amount (mol%[a])	Surface area (m^2/g)	Rate at 550 °C ($\mu mol/m^2$.s)	Temperature (at 20% O_2 conversion)	Selectivity
Nb_2O_5[b]	0	9.9	2.7	581	72
P5Nb	nd	35.7	2.8	551	71
Ti5Nb	4.4	28	3.0	544	66
V5Nb	4.6	11.3	8.6[c]	353	75
Cr5Nb	4.3	43.2	1.4[c]	372	78
Ge5Nb	4.7	41	2.2	542	48
Zr5Nb	5.9	21	5.8	542	70
Mo5Nb	nd	15.7	2.0[d]	498	88
Sn5Nb	nd	16	2.9	539	18
Sb5Nb	nd	20	0.8	-	-
Pb5Nb	4.9	17	3.2	-	-
Bi5Nb	8.7	18	2.2	551	14

[a] Element relative to Nb; [b] Prepared by calcination of hydrated niobia for 5h at 650 °C; [c] At 350 °C; [d] At 450 °C.

The BET surface areas fall in the range 10-45 m^2/g, with Cr5Nb and Ge5Nb having the highest areas, and V5Nb having the lowest surface area. Most surface areas are higher than that of pure niobia also when calcined at 600 °C (14.3 m^2/g). The XRD results showed that mixed oxides were formed with V5Nb ($V_2Nb_{23}O_{62}$, (12)), Sb5Nb ($SbNbO_4$) and Pb5Nb ($PbNb_2O_6$); only in the case of Ge5Nb (and possibly Pb5Nb) the pure dopant oxide was found. TT-Nb_2O_5 was found to be the major phase in all cases. Any other phase present in the catalysts was either amorphous or present in a quantity too small to be detected.

In order to compare the activities of the catalysts, the temperature at which 20% oxygen conversion is reached is displayed in Table I, while the oxygen conversion as a function of temperature is shown in Figure 1a. The increase in activity caused by the addition of vanadium is remarkable; good results were also obtained for the chromium-containing material. In both cases, the temperature at which reaction started was 200 °C below that required for the unpromoted Nb_2O_5. The addition of molybdenum also increased the activity, but molybdenum was lost

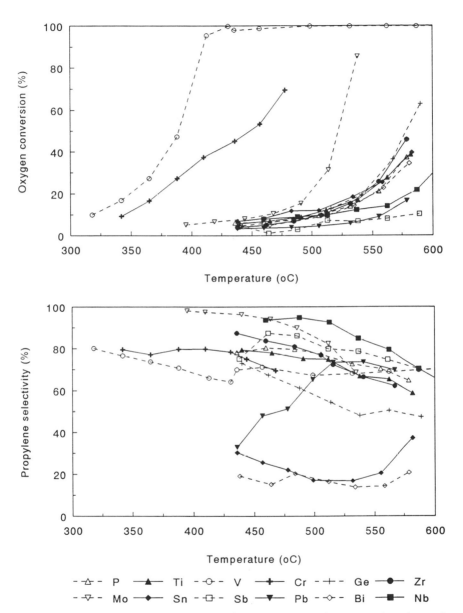

Figure 1 Oxygen conversion and propylene selectivity as a function of temperature for niobia with various additives.

from the catalyst, and deposited at the end of the reactor (thus resulting in an unstable catalyst). Most of the other additives had a negligible effect on the activity compared with pure niobia, while Sb and Pb caused a decrease in oxygen conversion (they did not reach 20% oxygen conversion in the temperature range investigated). The curves for the conversion of propane have the same shape as the oxygen conversion curves, while the amount of propane converted depends on the efficiency of the use of oxygen, i.e. on the selectivity towards total oxidation. The propane conversion may reach 16% at 100% oxygen conversion. The same conclusions can be drawn from the rates of reaction of propane per unit surface area shown in Table I, with the possible exception of Zr5Nb, which is slightly more active than the others, and Pb5Nb, which shows a normal reactivity towards propane.

The effect of the additives on the selectivity towards propylene was more pronounced. Figure 1b shows that both V5Nb and Cr5Nb, in addition to having a high activity, have high selectivities towards propylene. A V-Mg-O catalyst prepared according to the method described by Chaar et al. (1) and tested under our circumstances reached a selectivity towards propylene of less than 60% at a temperature which was 50 °C higher. Other products produced by these catalysts were CO (5-20% sel), CO_2 (10-15% sel) and small amounts (less than 1%) of C_2H_4 and C_2H_2. In contrast, the addition of Bi or Sn was detrimental to the selectivity towards propylene, while addition of Pb or Ge also caused a decrease in selectivity over part of the temperature range. The other additives did not affect the selectivity towards propylene, although they tended to produce more total oxidation products (CO and CO_2) than cracking products (C_2H_4 and CH_4).

Table II. Results of XRF, BET Surface Area Measurements and Catalytic Testing for Vanadia-Niobia Catalysts

Sample code	Amount (mol%)	Surface area (m²/g)	Rate at 350 °C (μmol/m².s)	Temperature (at 50% O₂ conversion)	Selectivity
V0.25Nb	0.26	11.0	8.5[a]	565	77
V1Nb	0.93	33.2	2.0	423	77
V5Nb	4.58	11.3	8.6	390	70
V10Nb	8.64	9.4	10.3	408	75
V1Nb 1. fresh	0.99	16.1	4.6	423	76
melt 2. used			1.5	448	79
3. oxidized			4.0	420	76
4. reduced			1.3	452	80
5. no further treatment			1.4	448	79
V1Nb imp.	1.47	8.5	5.9	455	75

[a] At 500 °C.

The results of the characterisation of the series of catalysts with different vanadium loadings are presented in Table II. The XRD results showed that at higher loading of vanadium, at least one new phase is formed. In V10Nb, the major phase was ß$(V, Nb)_2O_5$ (13), while TT-Nb_2O_5 was only a minor phase. The same conclusion can be drawn from the laser Raman spectrum of V10Nb shown in Figure 2. The broad band at 600 cm^{-1}, which is typical for Nb_2O_5 (10), has become much smaller, and a new band at 950 cm^{-1} appears; this may be attributed to V=O bonds. In laser Raman spectra of samples with a lower vanadium concentration, only the broad band at 600 cm^{-1} was visible.

It appears that the catalyst containing 0.26 mol% V_2O_5 is much less active than the other samples. The other catalysts are more or less equally active when oxygen conversion is considered. On a per unit surface area basis, V1Nb is less active than V5Nb and V10Nb, but this sample also has a much higher surface area. Results not given here showed that the minimum value of selectivity, which usually occurs when the oxygen conversion reaches 100%, became lower for the more active catalysts.

Table II also gives results for the effect of various pretreatments on a V1Nb melt catalyst. After a catalyst of this type had gone through the usual temperature cycle used to test the catalyst (100% oxygen conversion is reached during the last part of this temperature cycle), its activity during a second cycle was lower than that of the fresh catalyst. However, when the catalyst which had been used once was given an oxygen treatment (1h in flowing oxygen at 600 °C), the original behaviour of the fresh catalyst was restored. When the catalyst was then reduced by hydrogen (1h in 5% H_2/Ar at 600 °C), the activity decreased. When this catalyst was then tested again without any further treatment, the behaviour of the used catalyst was restored. From the TPR results for this catalyst shown in Figure 3 it follows that the reduction treatment mentioned above is sufficient to reduce vanadium to V^{3+}. The consumption of hydrogen during the TPR was equivalent to the reduction of all the vanadium in two equal steps from V^{5+} to V^{3+}. A TPR of the same catalyst after use showed the same pattern, except that the first reduction peak at 506 °C was smaller. The niobium reduction peak at 890 °C corresponds to 8% of the Nb being reduced to Nb^{4+}. This reduction peak also occured during TPR of undoped TT- and T-Nb_2O_5, but not in the case of H-Nb_2O_5 (14).

The results of XPS combined with Ar^+-sputtering on a V1Nb imp.-sample are shown in Figure 4. In this figure, sputtering profiles are shown for a fresh catalyst and for one which had been used for 300h at 510 °C (at this temperature 100% oxygen conversion was reached, no deactivation was observed). Since XRF showed that no vanadium was lost from the catalyst, it can be concluded that in the fresh catalyst, vanadium is present in a homogeneous layer which is at least 40-80 Å thick. Upon use, sub-surface vanadium has diffused into the bulk of the catalyst, while the surface concentration of vanadium remains the same. The vanadium at the surface was found to be mostly V^{5+} in both the fresh and the used catalyst. Because of the low concentrations of vanadium at the surface and interference from the O1s peak, the presence of small amounts of reduced vanadium cannot be excluded.

Discussion

Since little is known about crystal structures of the low-temperature mixed oxides of niobia, it is difficult to explain from a theoretical point of view the good results obtained for doping with vanadium or chromium. It appears that the vanadium and chromium are present in the catalyst in a unique environment, which makes

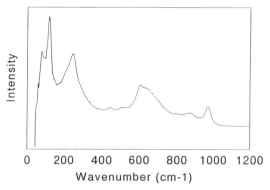

Figure 2 Laser Raman spectrum of V10Nb.

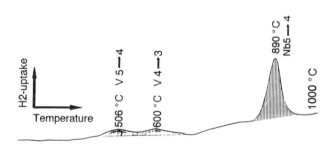

Figure 3 TPR of V1Nb melt.

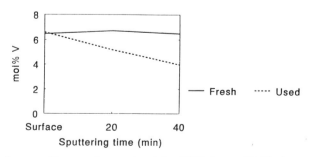

Figure 4 Ar$^+$-sputtering profiles as determined by XPS for a V1Nb imp.-catalyst. Sputtering rate 1-2 Å/min.

them selective for the oxidative dehydrogenation of propane. It is known that vanadium is a selective catalyst for this reaction only when it is present in a certain environment; in other environments (e.g. in the bulk oxide or when present on a support either as a monolayer or in multiple layers) it causes total oxidation (5). The use of chromium in selective oxidation catalysts usually results in total oxidation, although examples of selective chromium-containing oxidation catalysts exist (15).

The results obtained with the variation of the vanadium content suggest that when the amount of vanadium is increased until 1 mol%, a compound is initially formed (possibly a solid solution of V in niobia) in which the vanadium is present in sites where it is not particularly active. Upon addition of more vanadium (1 mol% or more), another compound is formed which contains vanadium in an environment in which it is active and selective for the oxidative dehydrogenation of propane. This compound probably has a fixed composition, and is preferably segregated at the surface of the catalyst. Addition of more vanadium causes this layer to grow thicker, without a change in surface structure or in surface vanadium concentration and thus in activity (Figure 5). It is possible that this compound is one of the phases found in XRD of V10Nb, i.e. $V_2Nb_{23}O_{62}$ (8 mol% V, half of which is V^{4+}) or ß(V, Nb)$_2O_5$ (10 mol% V) (12, 13). These phases have structures similar to those of VNb_9O_{25} and H-Nb_2O_5, with vanadium replacing niobium in the tetrahedral sites at the junction of the octahedral blocks (16). The results for the V1Nb-catalyst suggest that this catalyst contains the same amount of active phase exposed at the surface per gram (T for 50% oxygen conversion for 600 mg catalyst is similar, see Table II), but that because the surface area is larger, more inactive "solid solution" is exposed at the surface, resulting in a lower activity per unit surface area.

The effects of the various pretreatments to a V1Nb-catalyst, in combination with TPR results, give evidence that V becomes reduced to V^{4+} upon use, causing a lower activity and a higher selectivity towards propylene. The presence of V=O bonds in the V10Nb catalyst (Figure 2) does not harm the selectivity towards propylene. This is in contrast to the suggestion made by Chaar et al. (1, 2), who explain the selectivity of magnesium orthovanadate for the oxidative dehydrogenation of propane and butane from the fact that in this compound, V=O bonds are absent. The XPS results of the fresh catalyst show that vanadium is not present as a monolayer, since a decrease in vanadium concentration just below the surface would be expected in this case. The fact that in the used catalyst, vanadium at the surface is present in the same concentration and oxidation state as in the fresh material suggest that the compound mentioned

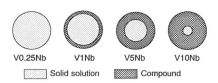

Figure 5 Model for the formation of a layer of the active compound on a vanadia-niobia catalyst.

above has formed in a thin layer at the surface of the catalyst during the cooling down in the reaction mixture after the temperature cycle used to test it. The XPS results also show that upon use, a redistribution of the vanadium occurs, possibly because of the destruction of the compound due to reduction.

Acknowledgements

We wish to thank E.M. Kool, P.H.H. in 't Panhuis and A. Toebes for carrying out a substantial part of the experimental work. The hydrated niobia and the niobium oxalate used in the preparation of the catalysts were kindly provided by the Niobium Products Company, Inc., USA. One of the authors (R.H.H. Smits) thanks the Dutch Foundation for Chemical Research (SON) for financial assistance.

References

1 Chaar, M.A.; Patel, D.; Kung, M.C.; Kung, H.H. *J. Catal.* **1987,** *105,* 483.
2 Chaar, M.A.; Patel, D.; Kung, H.H. J. Catal. **1988,** *109,* 463.
3 Nguyen, K.T.; Kung, H.H. J. Catal. **1990,** *122,* 415.
4 Patel, D.; Andersen, P.J.; Kung, H.H. *J. Catal.* **1990,** *125,* 132.
5 Siew Hew Sam, D.; Soenen, V.; Volta, J.C. *J. Catal.* **1990,** *123,* 417.
6 Guerrero-Ruiz, A.; Rodriguez-Ramos, I.; Fierro, J.G.L.; Soenen, V.; Herrmann, J.M.; Volta, J.C. *Stud. Surf. Sc. Catal.* **1992,** *72,* 203.
7 Corma, A.; López-Nieto, J.M.; Paredes, N.; Pérez, M.; Shen, Y.; Cao H.; Suib, S.L. *Stud. Surf. Sci. Catal.* **1992,** *72,* 213.
8 Seshan, K.; Swaan, H.M.; Smits, R.H.H.; Van Ommen, J.G.; Ross, J.R.H. *Stud. Surf. Sc. Catal.* **1990,** *55,* 505.
9 Smits, R.H.H.; Seshan, K.; Ross, J.R.H. *J. Chem. Soc., Chem. Comm.* **1991,** *8,* 558.
10 Smits, R.H.H.; Seshan, K.; Ross, J.R.H. *Stud. Surf. Sci. Catal.* **1992,** *72,* 221.
11 Smits, R.H.H.; Seshan, K.; Van Ommen, J.G.; Ross, J.R.H. unpublished results.
12 Waring, J.L.; Roth, R.S. *J. Res. Nat. Bur. Stand.* **1965,** *69A (2),* 119.
13 Goldschmidt, H.J. *Metallurgia* **1960,** *62,* 211.
14 Smits, R.H.H.; Seshan, K.; Ross, J.R.H. to be published.
15 Loukah, M.; Coudurier G.; Vedrine, J.C. *Stud. Surf. Sci. Catal.* **1992,** *72,* 191.
16 Wadsley, A.D.; Andersson, S. In *Perspectives in Structural Chemistry;* Dunitz, J.D.; Ibers, J.A.; John Wiley & Sons, 1970, Vol. 3; 19-26.

RECEIVED November 6, 1992

Chapter 30

Factors That Determine Selectivity for Dehydrogenation in Oxidation of Light Alkanes

H. H. Kung, P. Michalakos, L. Owens, M. Kung, P. Andersen, O. Owen, and I. Jahan

Ipatieff Laboratory and Department of Chemical Engineering, Northwestern University, Evanston, IL 60208–3120

The product distributions in the oxidation of butane over supported vanadium oxide catalysts and orthovanadates of cations of different reducibilities, and C_2 to C_6 alkanes over Mg orthovanadate, Mg pyrovanadate, and vanadyl pyrophosphate were compared. From these product distribution patterns, the factors that determine selectivity for dehydrogenation versus formation of oxygen-containing products (which include oxygenates and carbon oxides) were identified. It was proposed that in the formation of the primary products, the selectivity for dehydrogenation decreased with increasing probability of a surface alkyl species (or adsorbed alkene formed from the alkyl) to react with a surface lattice oxygen. This probability was higher for vanadates that had low differential heats of reoxidation or that contained cations of high reduction potential, and for vanadates whose active sites could bind the surface species at least two different carbon atoms so as to bring the hydrocarbon species close to the surface reactive lattice oxygen. These factors could be used to interpret the different selectivities for dehyrogenation in the oxidation of butane over SiO_2-supported vanadia catalysts of different vanadia loadings.

It has always been desirable to obtain high yields of the desired products in chemical production processes. High yields reduce raw material cost, capital cost, and operating costs which include those for separation, recycling, and removal of environmentally unacceptable byproducts. In hydrocarbon oxidation reactions, the byproducts are often carbon oxides and oxygen-containing organic molecules such as acids, ketones, aldehydes. In terms of ease of operation, since the formation of carbon oxides is a highly exothermic reaction, nonselective reactions leading to their production could result in hot spots in the reactor that could be very dangerous. In terms of environmental acceptability, many of the organic wastes are pollutants that are unacceptable even when they are present in trace quantities. Thus there is a strong desire to have highly selective oxidation reactions.

Over the years, there have been many proposals regarding factors that determine selectivity in hydrocarbon oxidation [1-6]. Basically, the formation a certain product can be categorized by two factors: how many oxygen atoms are incorporated into the reactant and where in the molecule are they incorporated. The ability to control these two factors determines whether one can control the observed selectivity.

Various investigators have suggested factors that control the number of oxygen atoms that are incorporated into the molecule [6-8]. It is reasonable to expect that this is determined by: (i) the residence time of the molecule on the surface, (ii) the number of oxygen atoms available at the active site during the residence of the molecule on the surface, and (iii) the reactivity of the oxygen at the active site. These factors may in turn be affected by other properties of the catalyst and the surface species. For example, the number of oxygen atoms available at the active site could be affected by the diffusivity of the lattice oxygen as well as the atomic arrangement of the active site and the size of the surface species with respect to the size of the site. The reactivity of the oxygen may depend on the reducibility of the cation, or more broadly speaking, the metal–oxygen bond strength.

The idea that the reactivity of the oxygen at the active site is important has received substantial attention. It has long been proposed that the reactivity of the lattice oxygen can be measured with either the heat of evaporation of lattice oxygen [7] or the rate of change of this heat with the extent of reduction of the oxide [8]. One can extend these concepts to illustrate the relationship among the number of oxygen atoms available at the active site, the size of the site and the surface intermediate, and the reactivity of the oxygen using the following hypothetical example.

Consider two oxides each containing one of the following two type of active sites. One site consists of a MO_4 tetrahedron (Fig. 1a) and another consists of a M_2O_7 unit which is two corner-sharing MO_4 tetrahedra (Fig. 1b). The reactivity of the lattice oxygen in these sites can be represented by the heat of removal of the lattice oxygen. It is reasonable to expect that this heat increases (that is, the removal of oxygen becomes increasingly difficult) with increasing number of oxygen atoms removed from the site, that is, the degree of reduction of the oxide. This is illustrated in Fig. 2a. If the selectivity is determined by the oxide property shown in this figure, i.e. the number of oxygen removed per cation, which is the common definition of degree of reduction, the two catalysts containing these two sites would be expected to show similar selectivities since they show similar dependence of heat of reduction versus the degree of reduction of the oxide.

On the other hand, consider another case that each of the two sites adsorbs one surface intermediate species. The reaction of this intermediate is such that

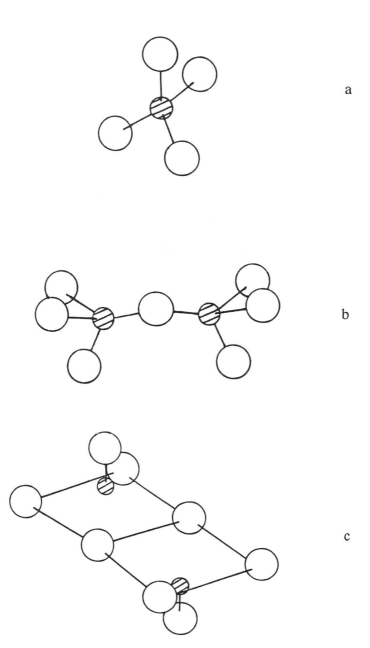

a

b

c

Figure 1. Schematic drawing of the active sites: a: VO_4 unit in $Mg_3(VO_4)_2$, b: V_2O_7 unit in $Mg_2V_2O_7$, and c: V_2O_8 unit in $(VO)_2P_2O_7$. Open circles are O ions and filled circles are V ions.

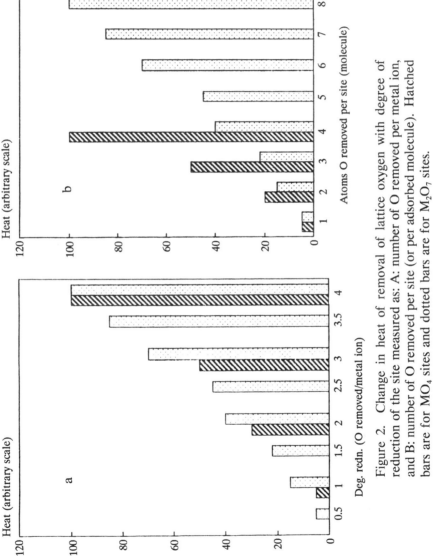

Figure 2. Change in heat of removal of lattice oxygen with degree of reduction of the site measured as: A: number of O removed per metal ion, and B: number of O removed per site (or per adsorbed molecule). Hatched bars are for MO_4 sites and dotted bars are for M_2O_7 sites.

it can acquire oxygen atoms up to a certain heat of reduction. In this case, as is shown in Fig. 2b, this intermediate would be able to take up a larger number of oxygen atoms from the M_2O_7 site than from the MO_4 site. Thus the oxide with the M_2O_7 sites would be less selective than the one with the MO_4 sites. This example demonstrates that detailed information about the interaction of the surface intermediate with the active site is very helpful in understanding changes in selectivity patterns in oxidation catalysis.

As to the position at which the reactive oxygen reacts with the surface intermediate, it has been suggested that this is determined by the electron density of the oxygen [5,9]. A nucleophilic (high in electron density) oxygen species would tend to attack the carbon atom to form C–O bonds, whereas an electrophilic (low in electron density) oxygen species would tend to attack the region of high electron density of the molecule (such as C=C bonds) leading to breaking of the carbon skeleton and eventually to degradation products. In addition to the nucleophilicity and electrophilicity, whether the adsorbed hydrocarbon species is situated favorably to react with the surface lattice oxygen should also be a factor [10].

We have been investigating the selective oxidation reaction of light alkanes on a number of vanadium oxide-based catalysts, which include orthovanadates of various cations, Mg pyrovanadate ($Mg_2V_2O_7$), vanadyl pyrophosphate (($VO)_2P_2O_7$), and vanadium oxide supported on silica and on alumina. The alkanes include ethane, propane, 2-methylpropane, butane, pentane, and cyclohexane. The different catalysts were found to produce a wide variety of products, ranging from oxidative dehydrogenation products, to various oxygen-containing organic products, to carbons oxides. It has been of great interest to attempt to understand the observed selectivity pattern with respect to the properties of the catalysts.

This paper is a summary of our current understanding of this system. In particular, we will be discussing the observations in terms of selectivity with respect to the availability of reactive lattice oxygen. The organization of the paper is as follows. First, the general features of the reaction scheme for alkane oxidation on vanadate catalysts will be presented. This is followed by a discussion of results on the effect of ease of removal of oxygen from the lattice on the selectivity, and then a discussion on the importance of the atomic arrangement of the active sites.

General Features of C_2 to C_6 Alkane Oxidation on Vanadates

In general, the products in reactions of alkanes with oxygen on vanadates fall into three categories as shown in Scheme I: oxidative dehydrogenation products (whence referred to simply as dehydrogenation products), oxygen-containing organic products (oxygenates), and combustion products (CO_x).

In the formation of any of these products, the first step of the reaction involves activation of the alkane molecule. Experimental results point to the fact that in general, breaking of the first C–H bond is a rate-limiting step. For example, on a V-P-O catalyst, a sizable deuterium kinetic isotope effect was observed in the oxidation of butane to maleic anhydride for deuterium substitution at the secondary carbon of the alkane molecule [11]. Thus breaking of the secondary C–H bond in butane is involved in the rate-limiting step in this system.

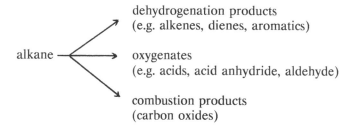

dehydrogenation products
(e.g. alkenes, dienes, aromatics)

alkane

oxygenates
(e.g. acids, acid anhydride, aldehyde)

combustion products
(carbon oxides)

C–H bond breaking initial
~ r.d.s.

Scheme I

A similar conclusion applies to a Mg-V-O catalyst in which Mg₃(VO₄)₂ is the active component. The relative rates of reaction for different alkanes on this catalyst follow the order: ethane < propane < butane ~ 2-methylpropane < cyclohexane (Table I) [12-14]. This order parallels the order of the strength of C–H bonds present in the molecule, which is primary C–H > secondary C–H > tertiary C–H. Ethane, which contains only primary C–H bonds, reacts the slowest, whereas propane, butane, and cyclohexane react faster with rates related to the number of secondary carbon atoms in the molecule, and 2-methylpropane, with only one tertiary carbon and the rest primary carbons, reacts faster than propane which contains only one secondary carbon. Similar to a Mg-V-O catalyst, the relative rates of oxidation of light alkanes on a Mg₂V₂O₇ catalyst follow the same order (Table I).

Table I. Relative Rates of Reaction of Alkane with Oxygen on V-Mg-O and Mg₂V₂O₇ Catalysts

Alkane	V-Mg-O[a]	Mg₂V₂O₇[b]
Ethane	0.19	0.16
Propane	0.56	0.70
2-Methylpropane	1.29	0.9
Butane	1	1
Cyclohexane	~ 6	

a: Reaction at 540 C.
b: Reaction at 500 C.

After the first C–H bond is broken, a surface alkyl species is formed. There are at least three possible reactions for this alkyl species that lead to different products. Dehydrogenation products would be formed if the alkyl species reacts by breaking another C–H bond at the ß-position:

possible mechanism

$$\text{M–O–M} + *\text{–CH}_2\text{–CH}_2\text{–R} \rightarrow \overset{\overset{\textstyle H}{\textstyle |}}{\text{M–O–M}} + \text{CH}_2\text{=CHR} + * \qquad (1)$$

In this equation, $*$ represents a surface ion in the active site, and is assumed to be a surface V ion. Alternatively, if the alkyl species reacts not only by breaking a C–H bond, but also by forming a C–O bond, then an oxygen-containing organic product could be produced, such as the one shown in equation (2).

$$\text{M–O–M} + *\text{–CH}_2\text{CH}_2\text{R} + * \rightarrow \text{M } \square \text{ M} + * + *\text{–H} \qquad (2)$$

$$\overset{\overset{\textstyle HC\text{–CH}_2R}{\textstyle |}}{\underset{\overset{\textstyle |}{\textstyle O}}{}}$$

An oxygen-containing product can also be formed by insertion of a lattice oxygen into a C–C bond to form C–O bonds. This reaction would most likely lead to combustion products:

$$2* + \text{M–O–M} \rightarrow *\text{–CH}_2\text{CH}_2\text{R} \rightarrow \text{M } \square \text{ M} + 2*\text{–H} + \overset{\overset{\textstyle H}{\textstyle |}}{*\text{–C–O–CHR}} \qquad (3)$$

Since it is likely that the formation of a C–O bond is irreversible, depending on whether the surface alkyl species reacts by forming a C–O bond or not, oxygen containing (including products of both equations 2 and 3) or dehydrogenation products would be formed. Thus one can view the reaction of the alkyl species as a selectivity-determining step, and the ease of removal of an oxygen atom from the lattice to form a C–O bond with the surface intermediate should be an important factor that determines the selectivity of the reaction. By grouping all oxygen-containing products together, Scheme I is simplified to Scheme II:

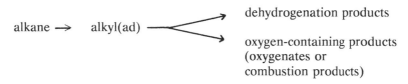

Scheme II

It should be stated here that the discussion to be followed would remain the same if the selectivity-determining step is the reaction of an adsorbed alkene formed from a surface alkyl. In that case, dehydrogenation products are observed if the alkene desorbs, and oxygen-containing products are observed if it further reacts with a lattice oxygen. (*See* Scheme III.)

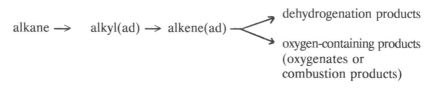

Scheme III

The $*-CH_2CH_2R$ species in eqns (1) to (3) will be replaced by an adsorbed alkene, but the selectivity remains a function of the ease of removal of a lattice oxygen to form a C–O bond.

In assigning the reaction of the surface intermediate derived from alkane to be the selectivity-determining step, it is implicitly assumed that the discussion is confined to the regime of low conversions of alkanes. Thus the selectivity of the products are not affected by secondary reactions, such as readsorption and further reaction of desorbed primary products.

Effect of the Reactivity of Lattice Oxygen

As mentioned earlier, we postulated that the probability for the surface intermediate in the selectivity-determining step to react with an oxygen species to form a C–O bond is an important factor. For the catalysts under consideration, we will assume that the oxygen involved in the selectivity-determining step is a lattice oxygen. In the case of $(VO)_2P_2O_7$, oxygen isotopic labeling experiments have confirmed the participation of lattice oxygen in the oxidation of butane [11]. In the case of $Mg_3(VO_4)_2$, it has been found that the selectivity in butane oxidation was nearly the same whether the reaction was conducted in a flow reactor with oxygen in the gas phase or in a pulse reactor using butane pulses without oxygen [15]. These data are consistent with the assumption that lattice oxygen is involved in the selectivity-determining step.

It follows from this assumption that the easier it is to remove a lattice oxygen from the active site, the more likely it is to form oxygen-containing products, assuming that the lattice oxygen would otherwise react in a similar manner with the surface intermediate. The ease of removal of lattice oxygen can be measured in at least two different ways: one as the metal-oxygen bond strength, and another as the ease of reduction of the cation in the solid.

Selectivity and Metal-Oxygen Bond Strength. The metal-oxygen bond strength could be measured as the differential heats of reoxidation of the reduced catalysts. Conceptually, the heat of reoxidation measures the energy released when two metal–oxygen bonds are formed in the solid together with all associated processes, the latter include structural rearrangement and changes in oxide-support interaction. Although a quantitative measure of the heats of the associated processes is not available, it is likely that their magnitudes are smaller than the metal–oxygen bond energy, and a meaningful qualitative trend can still be derived directly from the reoxidation heat.

We have measured the differential heats of reoxidation and the reaction characteristics of a number of V_2O_5/γ-Al_2O_3 catalysts. The chemical transformation corresponding to the heat measurement in this system is:

$$x/2 O_2 + V_2O_{5-x}/\gamma\text{-}Al_2O_3 \xrightarrow{\Delta H_{ox}} V_2O_5/\gamma\text{-}Al_2O_3 \qquad (4)$$

For these measurements, the samples were first reduced with H_2 to an extent equivalent to roughly the conversion of all V^{5+} to V^{4+}. Then they were incrementally reoxidized with doses of O_2. The heat released during reoxidation was measured with a microcalorimeter and the amount of O_2 consumed was measured volumetrically [16]. From such measurements, the heats of reoxidation were determined for two samples of 8.2 and 23.4 wt% V_2O_5 loadings, which were equivalent to surface coverages of 2.9 and 8.2 V/nm^2, respectively, as a function of the degree of reduction, ϕ, which was defined as the number of oxygen atoms removed per vanadium ion. It was found that the differential heats of reoxidation were functions of ϕ, vanadium loading, and temperature. Fig. 3a shows the data at 400°C for the 8.2 V/nm^2 sample, and Fig. 3b shows those at 500°C for the 2.9 V/nm^2 sample. For both samples, the heats were found to be high at large ϕ and decreased quite rapidly as the sample became close to being fully reoxidized. In addition, the heat at large ϕ's was somewhat higher for the sample of lower loading.

These two samples were then tested for the oxidation of butane in a pulse reactor. In these studies, pulses of butane (0.05 ml at 1 atm) were passed in a He carrier over 0.05 g of 8.2 V/nm^2 catalyst or 0.75 g of 2.9 V/nm^2 catalyst [17]. The products were collected in a trap at 77 K and later flashed into a gas chromatograph for analysis. Since no gaseous oxygen was present, the oxygen consumed by the reaction had to be originated from the lattice. Therefore, from the product selectivities and conversions of the butane pulses, the degrees of reduction of the catalysts could be calculated.

For the 8.2 V/nm^2 sample, the products observed for the pulse reaction at 400°C consisted of only dehydrogenation products (butenes and butadiene) and carbon oxides. No oxygenates were observed, and the carbon balance for each pulse was satisfied within experimental error. The selectivity for dehydrogenation is shown in Fig. 3a as a function of ϕ. It shows that the selectivity was very low when the catalyst was in a nearly fully oxidized state, but increased rapidly when the catalyst was reduced beyond $\phi = 0.15$. It should be noted that the dependence of selectivity for dehydrogenation on ϕ shown in the figure was not

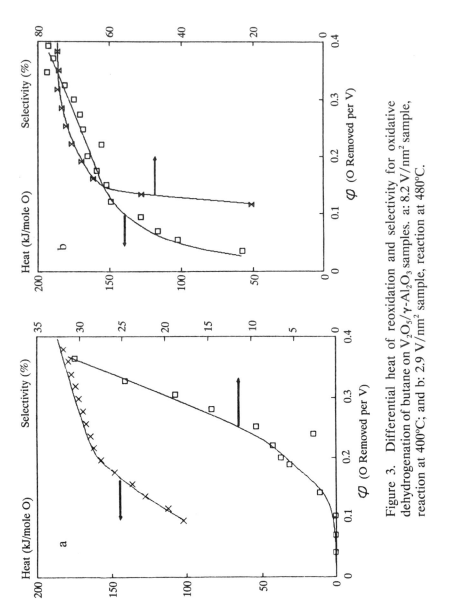

Figure 3. Differential heat of reoxidation and selectivity for oxidative dehydrogenation of butane on $V_2O_5/\gamma\text{-}Al_2O_3$ samples. a: 8.2 V/nm^2 sample, reaction at 400°C; and b: 2.9 V/nm^2 sample, reaction at 480°C.

a result of changes in conversion of butane in the pulse since these data were for experiments of about the same conversion.

The data for the 2.9 V/nm^2 catalyst is shown in Fig. 3b. The activity of this catalyst was lower and the experiments had to be performed at 480°C. Over this catalyst, a fraction of the butane which reacted in the pulse remained adsorbed. The fraction decreased from 60% in the first pulse to 30% to 40% after a few pulses. To calculate the degree of reduction of the catalyst, the adsorbed material was assumed to be elemental carbon, and the hydrogen that was not accounted for in the detected products was assumed to have reacted with oxygen of the vanadia phase to form water which subsequently desorbed. Upon reoxidation of the catalyst, the adsorbed carbon was recovered as carbon oxides.

Fig. 3b shows that the selectivity for dehydrogenation (based on detected products) was very low at low values of ϕ, but increased rapidly as the catalyst was reduced. On this catalyst, small amounts of crotonaldehyde and maleic anhydride were also detected. These amounts decreased slowly with increasing ϕ.

From the data in Figures 3a and 3b, one could conclude that there is a strong correlation between the selectivity for dehydrogenation and the heat of reoxidation of the catalyst. The selectivity is low when the heat is low and increases rapidly when the heat increases rapidly. Since the heat of reoxidation is a measure of the metal–oxygen bond strength, this observation is consistent with the model that the ease of removal of lattice oxygen is an important factor that determines selectivity for dehydrogenation versus formation of oxygen-containing products.

Reducibility of the Cations. The removal of lattice oxygen from the active site is accompanied by the reduction of the neighboring cation. Thus another way to view the ease of removal of lattice oxygen is to look at the reduction potential of the cations at the active site. This concept was tested using a series of orthovanadates of the formula $M_3(VO_4)_2$, where M = Mg, Zn, Ni, and Cu, and MVO_4, where M = Fe, Sm, Nd, and Eu. The cations in these two series were chosen to span a range of (aqueous) reduction potentials from -2.40 V to +0.77 V. Orthovanadates are made up of isolated VO_4 units that are separated from each other by MO_x units. Thus there are only M–O–V bonds in these structures and no V–O–V bonds, and the difference in the ease of removal of a lattice oxygen should depend on the difference in the reduction potential of the M ion.

The oxidation of butane on these orthovanadates were tested at 500°C in a flow reactor using a butane:oxygen:helium ratio of 4:8:88. The observed products were isomers of butene, butadiene, CO, and CO_2. The carbon balance in these experiments were within experimental errors, thus the amount of any undetected product if present should be small. The selectivity for dehydrogenation (butenes and butadiene) was found to depend on the butane conversion and be quite different for different orthovanadates. Fig. 4 shows the selectivity for dehydrogenation at 12.5% conversion of butane [15,18,19]. Its value ranged from a high of over 60% for $Mg_3(VO_4)_2$ to a low of less than 5% for

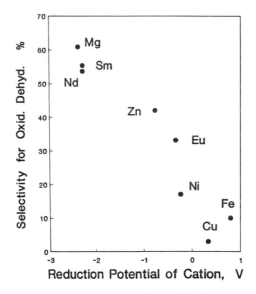

Figure 4. Dependence of selectivity for oxidative dehydrogenation of butane over orthovanadates on the reduction potential of the cations. Reaction conditions: 500°C, butane/O_2/He = 4/8/88, butane conversion = 12.5%.

$Cu_3(VO_4)_2$. The figure shows that there exists a correlation between the reduction potential of the cation and the selectivity: the higher is the reduction potential, that is, the more easily reduced is the cation, the lower is the selectivity.

This correlation can be explained by the model presented thus far that there exists a selectivity-determining step. Depending on whether or not the surface intermediate in this step reacts with a lattice oxygen to form a C–O bond, oxygen-containing or dehydrogenation products were formed.

This interpretation of the correlation assumes that lattice oxygen is the species that reacts with the surface intermediate. This assumption is supported by the observation that a similar correlation between selectivity for dehydrogenation and cation reducibility was observed in pulse reactions of butane without oxygen on Mg, Zn, Ni, Fe, and Cu orthovanadates [15]. Thus gaseous oxygen is not involved in the selectivity-determining step in these systems, although it might be involved in subsequent steps, particularly in the formation of carbon oxides.

Effect of Atomic Arrangement of an Active Site

The data in Figs. 3 and 4 show that the ease of removal of a lattice oxygen, which can also be expressed in terms of the reducibility of the neighboring cations, has a strong effect on the selectivity for oxidative dehydrogenation of butane. If this is the only factor that determines selectivity, then a catalyst that is selective for dehydrogenation of butane, such as $Mg_3(VO_4)_2$, will be selective for other alkanes as well. Likewise, any catalyst that contains V^{5+}–O–V^{5+} bonds will not be selective for any alkanes and produce large amounts of oxygen-containing products.

When the data for the oxidation of C_2 to C_6 alkanes on three vanadate catalysts were examined, it became apparent that although ease of removal of lattice oxygen (or reducibility of the cations) explains most of the data , it is insufficient to explain all of them. This is illustrated by the data in Table II which shows some representative data obtained over these catalysts: $Mg_3(VO_4)_2$, $Mg_2V_2O_7$, and $(VO)_2P_2O_7$ [10]. $Mg_2V_2O_7$ is made up of V_2O_7 units connected by MgO_6 units, and $(VO)_2P_2O_7$ is made up of V_2O_8 units of edge-sharing VO_5 units (see Fig. 1). Assuming that these VO_x units are the active sites, then the active sites on both $Mg_2V_2O_7$ and $(VO)_2P_2O_7$ contain V^{5+}–O–V^{5+} bonds, which is absent in $Mg_3(VO_4)_2$.

The data in the table show that $Mg_3(VO_4)_2$ is indeed a selective dehydrogenation catalyst for the alkanes studied except ethane, whereas $Mg_2V_2O_7$ and $(VO)_2P_2O_7$ produce mostly oxygen-containing products. However, contrary to expectation, propane reacted on $Mg_2V_2O_7$ with a high dehydrogenation selectivity, whereas ethane reacted with high dehydrogenation selectivity on $(VO)_2P_2O_7$.

To explain the data for propane, we propose that in addition to the necessary condition of possessing readily removable lattice oxygen in the active site, the formation of oxygen-containing products is enhanced if the hydrocarbon

Table II. Typical Product Distributions for Alkane Oxidation

Reactant	Rxn T (°C)	Alkane conv.%	% Selectivity[a]						

Over VPO catalyst

			CO	CO_2	C_2H_4				
C_2H_6	305	3.8	12	7	81				

			CO	CO_2	C_2H_4	C_3H_6	Ac	Ar	
C_3H_8	300	8	62	26	4	0	6	2	

			CO	CO_2	C_2H_4	C_3H_6	Ac	Ar	MA
C_4H_{10}	300	7	16	10	2	1	7	2	64

			CO	CO_2	MA	PA			
C_5H_{12}	325	7	22	15	19	44			

Over VMgO catalyst

			CO	CO_2	C_2H_4				
C_2H_6	540	5.2	28	49	24				

			CO	CO_2	C_3H_6				
C_3H_8	500	8.4	14	24	62				

			CO	CO_2	C_4H_8				
iC_4H_{10}	475	4.0	7	22	71				

			CO	CO_2	C_4H_8	C_4H_6			
C_4H_{10}	475	4.1	9	22	55	14			

			CO	CO_2	C_6H_{10}	C_6H_6			
cC_6H_{12}	484	8.4	3	14	47	36			

Over $Mg_2V_2O_7$ catalyst

			CO	CO_2	C_2H_4				
C_2H_6	540	3.2	49	21	30				

			CO	CO_2	C_3H_6				
C_3H_8	475	10	27	18	56				

			CO	CO_2	C_4H_8				
iC_4H_{10}	502	6.8	39	36	25				

			CO	CO_2	C_4H_8	C_4H_6			
C_4H_{10}	500	6.8	33	33	31	2			

a) Ac = acetic acid, Ar = acrylic acid, MA = maleic anhydride,
   PA = phthalic anhydride.

intermediate in the selectivity-determining step can be bonded to the two vanadium ions of the linked VO_x units such that the hydrocarbon species is being held close to the reactive surface lattice oxygen. This latter requirement could be met only if the molecule is sufficient large to do so. Whether this is satisfied or not can be examined by comparing the molecular size with the separation of the vanadium ions in the solid derived from crystallographic data. For the V_2O_7 unit in $Mg_2V_2O_7$, the separation is 0.339 nm [20]; for the V_2O_8 unit in $(VO)_2P_2O_7$, it is 0.319 nm [21]. The separation of V ions in adjacent VO_4 units in $Mg_3(VO_4)_2$ on a low index plane is 0.37 nm [22]. Using a value of 0.072 nm for the ionic radius of V^{5+} [23] and a covalent radius of 0.07 nm for C [24], the V—C bond length is estimated to be 0.14 nm. Using this V—C bond distance, the V—V separation in an active site needed to bond with a surface 1,2-diadsorbed C_2 and a 1,3-diadsorbed C_3 species can be estimated to be 0.244 nm and about 0.327 nm, respectively. From these values, it can be seen that a C_3 species (such as a 1,3-diadsorbed allylic species) could readily bond with the two V ions in the active site of $(VO)_2P_2O_7$, but could do so only with difficulty with those in $Mg_2V_2O_7$. Thus one would argue that propane would react on $Mg_2V_2O_7$ as if the active sites are isolated VO_4 units, like in $Mg_3(VO_4)_2$, whereas it would react like the larger hydrocarbons on $(VO)_2P_2O_7$. This is indeed observed.

The manner in which a 1,3-diadsorbed C_3 carbon chain is expected to bond to the two vanadium ions in the active site could explain the much larger production of acetic acid from 2-methylpropane than from propane on $(VO)_2P_2O_7$ (see Table II) [10]. As shown in Scheme IV, when a 2-methylallyl species bridges the adjacent VO_5 units in $(VO)_2P_2O_7$, the methyl group in the middle carbon is not interacting with the surface. This configuration is conducive to the formation of acetate group when the two C—C bonds of the C_3 unit are cleaved by reaction with the lattice oxygen. This may explain the substantial production of acetic acid observed on this catalyst. (See Scheme IV.)

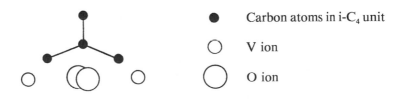

●	Carbon atoms in i-C_4 unit
○	V ion
○	O ion

Scheme IV

The difference between propane and 2-methylpropane on $Mg_2V_2O_7$ cannot be explained by the size of the C_3 chain. We propose that although with difficulties, a C_3 species still has a finite probability to interact with both vanadium ions in a V_2O_7 unit in the catalyst to lead to the formation of combustion products. This probability is twice as large for 2-methylpropene (or

2-methylpropyl) than propene (or propyl), which may account for the lower selectivity for dehydrogenation for 2-methylpropane on this catalyst.

The behavior of ethane is different from the other alkanes. It is the only alkane that undergoes significant dehydrogenation on the VPO catalyst, as well as the only one for which combustion is the predominant reaction on VMgO. However, an ethyl species is too small to interact with two V ions simultaneously on any of the three catalysts. A phenomenological explanation of this behavior of ethane has been suggested [10]. In this explanation, the possible reactions of ethyl, propyl, and 2-methylpropyl species were compared by statistically counting the number of various types of bonds in each species:

$$
\begin{array}{ccc}
C^{\beta}H_3 & H & C^{\beta}H_3 \\
| & | & | \\
C^{\alpha}H_2 & C^{\beta}H_3-C^{\alpha}-C^{\beta}H_3 & C^{\beta}H_3-C^{\alpha}-C^{\beta}H_3 \\
| & | & | \\
* & * & *
\end{array}
$$

In these species, a reaction in which a $C^{\beta}-H$ bond is broken would lead to dehydrogenation. However, breaking a $C^{\alpha}-H$ bond or cleaving a $C^{\alpha}-C^{\beta}$ bond would lead to degradation products. The statistical probabilities of these three processes are proportional to the number of these bonds in the species, which are shown in Table III. They show that these species only differ in the relative number of $C^{\beta}-H$ bonds. If the $C^{\alpha}-H$ and the C–C bond react with equal probability, whereas the $C^{\beta}-H$ bond reacts somewhat faster, combustion would be more likely for ethane than for the other alkanes. Since the reaction conditions, especially temperatures, for the data on $Mg_2V_2O_7$ and $Mg_3(VO_4)_2$ were similar, this argument would account for the low dehydrogenation selectivity observed on these two oxides.

Table III. Number of Various Types of Bonds in an Alkyl Species

Species	Number			Normalized value		
	C^{β}-H	C^{α}-H	C-C	C^{β}-H	C^{α}-H	C-C
ethyl	3	2	1	6	4	2
propyl	6	1	2	6	1	2
2-methylpropyl	9	0	3	6	0	2

For the VPO catalyst, for some unknown reasons, which may be related to the much lower reaction temperature, the $C^{\beta}-H$ bond breaks much more readily than the C^α–H bond, and dehydrogenation becomes the dominant reaction. For the higher alkanes, the formation of 1,3-diadsorbed species is very favorable that dehydrogenation is not an important reaction any more.

Oxidation of Butane on SiO₂-Supported Vanadium Oxide

It is encouraging that the selectivity for dehydrogenation versus the formation of oxygen-containing products could be explained by the simple idea of availability of reactive oxygen, which depends on the atomic arrangement of the active sites and the metal–oxygen bond strength. There are different ways to change the availability of reactive oxygen. One is by using oxides of different structures and therefore different active sites as shown above, another is to use highly dispersed supported oxides.

It is well established that vanadium oxide when present in very low loadings on silica does not form crystalline V_2O_5. Instead, a highly dispersed phase is formed [25-28]. This highly dispersed phase is structurally different from crystalline V_2O_5. For example, crystalline V_2O_5 shows peaks in a Raman spectrum at approximately 997, 703, 526, 480, 404, 304, and 284 cm⁻¹. However, the spectrum of a 1 wt% V_2O_5 on SiO_2 only shows a peak at 1040 cm⁻¹ and none of these other peaks. Peaks of crystalline V_2O_5 begin to appear when the vanadia loading is increased, and are clearly discernable in the spectrum of a sample of 10 wt% V_2O_5 loading (Fig. 5). It has been proposed that the 1040 cm⁻¹ peak is due to the stretching vibration of the V=O bond in a VO_4 species on silica.

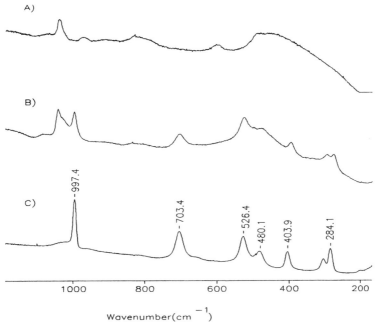

Figure 5. Laser Raman spectrum of a A: 1 wt% V_2O_5/SiO_2, B: 10 wt% V_2O_5/SiO_2 and C: V_2O_5.

If the structural interpretation of the species on silica is correct, low loading SiO_2-supported vanadia samples should be a more selective catalysts for oxidative dehydrogenation of alkanes than high loading ones, because the former catalysts would contain only active sites that have small numbers (or none at all) reactive lattice oxygen atoms (those bonded between two easily reducible cations). Table IV shows the selectivity for dehydrogenation in the oxidation of butane for a 1 wt.% and a 10 wt.% V_2O_5/SiO_2 catalyst. The data show clearly that the lower loading sample was much more selective than the higher loading sample. This difference was not due to the effect of impurities in the support because the data were obtained on an acid-washed Davison 62 silica which contained less than 10 ppm of Na and 200 ppm of Ca. The same effect was observed on catalysts prepared with Cabosil silica which contained no detectable impurities. We have found that, however, if the Davison 62 silica was used without acid-wash such that it contained 600 ppm Na impurity, a 10 wt% V_2O_5 sample became about as selective as a 1 wt% sample. Whether the improvement for the 10 wt% sample was a result of Na interacting with the vanadia or changes in the silica due to acid-washing needed to be investigated.

Table IV. Oxidation of Butane over V_2O_5/SiO_2 Catalysts

SiO_2[a]	V_2O_5 wt%	Catalyst Wt.(g)	Conv.[b] %C_4H_{10}	CO	CO_2	1-B	t-2-B	c-2-B	BD	TD
1	1.0	0.5	15	19	10	27	14	13	12	66
1	10.0	0.1	11	55	30	4	2	2	0	9
2	1.0	0.5	16	15	10	30	12	11	8	60
2	10.0	0.1	18	61	29	5	3	2	0	10

Selectivity (%)[c]

a: 1: acid-washed Davison 62; 2: Cabosil silica.
b: Feed: $C_4H_{10}/O_2/He$ = 4/8/88, total flow rate = 100 ml/min, temperature = 520°C.
c: 1-B = 1-butene, t-2-B = t-2-butene, c-2-B = c-2-butene, BD = butadiene, TD = total dehydrogenation; the balance was C_2 and C_3 hydrocarbons.

Conclusion

This paper summarized our current understanding of the factors that determine selectivity for dehydrogenation versus formation of oxygen-containing products in the oxidation of light alkanes. From the patterns of product distribution in the oxidation of C_2 to C_6 alkanes obtained with supported vanadium oxide, orthovanadates of cations of different reduction potentials, and vanadates of different bonding units of VO_x in the active sites, it was shown that the selectivities can be explained by the probability of the surface alkyl species (or the surface alkene formed from the alkyl) to react with a reactive surface lattice oxygen. Catalysts for which this occurs with a high probability would show low selectivities. This probability increases for vanadates that have low heats of removal of lattice oxygen, which are those that contain easily reducible cations in the active sites, and for vanadates whose active sites can bind the surface alkyl species (or alkene) in a way that bring the surface intermediate close to the reactive lattice oxygen.

Acknowledgment

The portion of this work on the vanadates has been supported by the Department of Energy, Basic Energy Sciences, Division of Chemical Sciences, and that on the heats of reoxidation by the National Science Foundation. PM acknowledges support by the Battelle's NASA Advanced Materials Center for the Commercial Development of Space, and LO acknowledges fellowship support from the 3M Company and a Faculty Minority Internship from the Monsanto Company.

References

1. Dadyburjor, D.B., Jewur, S.S., Ruckenstein, E., Catal. Rev.-Sci. Eng., **1979**, 19, 293.
2. Kung, H.H., Ind. Eng. Chem. Proc. Res. Dev. **1986**, 25, 171.
3. Kung, H.H., "Transition Metal Oxides: Surface Chemistry and Catalysis," Elsevier Science Publ., Amsterdam, **1989**.
4. Sokolovskii, V.D., Catal. Rev.-Sci. Eng., **1990**, 32, 1.
5. Bielanski, A., and Haber, J., "Oxygen in Catalysis," Marcel Dekker, Inc., NY, **1991**.
6. Grasselli, R., and Burrington, J., Adv. Catal., **1980**, 30, 133.
7. Morooka, Y., Morikawa, Y., and Ozaki, A., J. Catal., **1967**, 7, 23; **1966**, 5, 116.
8. Sachtler, W.M.H., Dorgelo, G.J.H., Fahrenfort, J., and Voorhoeve, R.J.H., Rec. Trav. Chim. Pays-Bas, **1970**, 89, 460.
9. Haber, J., in "Solid State Chemistry in Catalysis," Grasselli, R.K., and Brazdil, J.F. ed., Amer. Chem. Soc. Symp. Series no. 279, **1985**, p.1.
10. Michalakos, P., Kung, M.C., Jahan, I., and Kung, H.H., submitted.
11. Pepera, M.A., Callahan, J.L., Desmond, M.J., Milberger, E.C., Blum, P.R., and Bremer, N.J., J. Amer. Chem. Soc., **1985**, 107, 4883.
12. Patel, D., Kung, M., and Kung, H., Proc. 9th Intern. Congr. Catal., Calgary, 1988, M.J. Phillips and M. Ternan, Eds., Chem. Institute of Canada, Ottawa, **1988**, 4, 1553.
13. Kung, M., and Kung, H., J. Catal. **1991**, 128, 287.
14. Kung, M.C., and Kung, H.H., J. Catal., **1992**, 134, 688.
15. Owen, O.S., PhD Thesis, Northwestern University, **1991**.
16. Andersen, P.J., and Kung, H.H., J. Phys. Chem., **1992**, 96, 3114.
17. Andersen, P.J., and Kung, H.H., Proc. 10th Intern. Congr. Catal., Budapest, **1992**.

18.   Owen, O.S., Kung, M.C., and Kung, H.H., Catal. Lett., **1992**, 12, 45.
19.   Owen, O.S., and Kung, H.H., submitted.
20.   Gopal, R., and Calvo, C., Acta. Cryst., **1974**, B30, 2491.
21.   Gorbunova, Yu. E., and Linde, S.A., Sov. Phys. Dokl., **1979**, 24, 138.
22.   Krishnamachari, N., and Calvo, C., Canad. J. Chem., **1970**, 49, 1629.
23.   Shannon, R., Acta. Cryst., **1976**, A32, 751.
24.   Atkins, P.W., Physical Chemistry, W.H. Freeman and Company, 3rd ed.,
       New York, N.Y., **1986**.
25.   Oyama, S.T., Went, G.T., Lewis, K.B., Bell, A.T., and Somorjai,
       G.A., J. Phys. Chem., **1989**, 93, 6786.
26.   Eckert, H., and Wachs, I.E., J. Phys. Chem., **1989**, 93, 6796.
27.   Roozeboom, F., Mittelmeijer-Hazeleger, M.C., Moulijn, J.A.,
       Medema, J., de Beer, V.H.J., and Gellings, P.J., J. Phys. Chem.,
       1980, 84, 2783.
28.   Deo, G., and Wachs, I.E., J. Catal., **1991**, 129, 307.

RECEIVED October 30, 1992

STATE-OF-THE-ART ENGINEERING CONCEPTS IN SELECTIVE OXIDATION

Chapter 31

Optimal Distribution of Silver Catalyst for Epoxidation of Ethylene

Asterios Gavriilidis and Arvind Varma

Department of Chemical Engineering, University of Notre Dame, Notre Dame, IN 46556

The optimal distribution of silver catalyst in α-Al_2O_3 pellets is investigated experimentally for the ethylene epoxidation reaction network, using a novel single-pellet reactor. Previous theoretical work suggests that a Dirac-delta type distribution of the catalyst is optimal. This distribution is approximated in practice by a step-distribution of narrow width. The effect of the location and width of the active layer on the conversion of ethylene and the selectivity to ethylene oxide, for various ethylene feed concentrations and reaction temperatures, is discussed. The results clearly demonstrate that for optimum selectivity, the silver catalyst should be placed in a thin layer at the external surface of the pellet.

The effects of non-uniform distribution of the catalytic material within the support in the performance of catalyst pellets started receiving attention in the late 60's (cf. *1-4*). These, as well as later studies, both theoretical and experimental, demonstrated that non-uniformly distributed catalysts can offer superior conversion, selectivity, durability, and thermal sensitivity characteristics over those wherein the activity is uniform. Work in this area has been reviewed by Gavriilidis et al. (*5*). Recently, Wu et al. (*6*) showed that for any catalyst performance index (i.e. conversion, selectivity or yield) and for the most general case of an arbitrary number of reactions, following arbitrary kinetics, occurring in a non-isothermal pellet, with finite external mass and heat transfer resistances, the optimal catalyst distribution remains a Dirac-delta function.

The ethylene epoxidation reaction network, occurring in a Dirac-type catalyst, has previously been studied theoretically (*7-8*). Both studies showed that the selectivity to ethylene oxide is maximized when the catalyst is located at the external surface of the pellet, i.e. for an egg-shell type catalyst. A systematic experimental investigation of the performance of such catalysts for this industrially important reaction network has recently been reported (*9*). A summary of this work, as well as some new results, are presented in this paper.

Experimental

Ethylene reacts over supported silver catalyst with oxygen to give ethylene oxide. Important side reaction is the complete combustion of ethylene, and to a lesser extent that of ethylene oxide, according to the scheme:

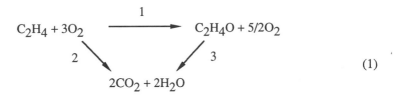

$$C_2H_4 + 3O_2 \xrightarrow{1} C_2H_4O + 5/2O_2$$

(1)

$$2CO_2 + 2H_2O$$

Although this reaction network has been studied extensively, its mechanism is still under debate (*10*). In this study, a single-pellet reactor was used, and the pellet was prepared mechanically by pressing the active catalyst layer between two alumina layers. In this way a step-type catalyst pellet was produced, which approximated a Dirac-type catalyst distribution.

Catalyst Pellet. The silver catalyst was prepared by impregnation of -325 mesh α-alumina powder (Alcoa A-17) with silver lactate, using a procedure similar to that of Klugherz and Harriott (*11*). In order to decompose the silver lactate, the catalyst was heated at 500 °C, under nitrogen, for 5 hours. After cooling, it was subjected to two oxidation-reduction cycles, with 20% O_2 in Ar and 20 % H_2 in Ar respectively, at 350°C. This treatment was necessary in order to get a catalyst with stable activity. In this way the *active powder* was obtained, and its loading was determined by atomic absorption spectroscopy. Blank runs conducted with pellets made from α-alumina alone, showed that it was somewhat reactive towards ethylene oxide combustion and isomerization to acetaldehyde. In order to minimize this activity, the α-alumina was impregnated with sodium hydroxide (0.3g NaOH per 100g α-alumina) following the procedure of Lee et al. (*12*). This sodium impregnated α-alumina support is the *inert powder*.

The catalyst pellet with step-distribution of the active layer was prepared by pressing together two inert powder layers and one active layer in a two-piece die. The three layers were first pressed at 4000 psi, and then the pellet was pressed at 7000 psi. The pellet was cylindrical, with 20.4 mm diameter, and contained 152 mg of silver in *all* cases. Three types of pellets were prepared as follows. *Type 1*, in which 0.45 g of active powder containing 33.8 wt.% silver and 4.55 g of inert powder were used as such; the thickness of the active layer was 0.45 mm while the thickness of the pellet was 7.1 mm. *Type 2*, in which 0.45 g of the same active powder as above was mixed with a portion of inert powder to form the active layer. These pellets had variable width of the active layer, ranging from 0.45 mm to 3.6 mm, while the total thickness was 3.6 mm. *Type 3*, in which active powders with loadings 8.9, 33.8 and 52.6 wt. % silver were used as such. The amounts of active powders used were such that the pellets always contained 152 mg of silver. These pellets had a total thickness of 7.1 mm. (For ease of notation, pellets with the active layer at the top surface and below the top surface are referred to later as *surface* and *subsurface* catalyst pellets, respectively). The curved surface of the pellet was wrapped with flexible Teflon tape and pressed fit into a sample holder made of Teflon. Similarly, the bottom surfaces of type 2 pellets were also covered with flexible Teflon tape. The pellet was thus thermally insulated from the surroundings, and the reaction gases would diffuse in and out of the pellet *only* through the top surface.

Apparatus. The gases, oxygen (Linde, Zero grade) and ethylene (Linde, CP grade), were passed through gas purifiers for removing any trace of water. The gas flowrates were controlled using mass flow controllers. The gases were mixed and delivered to the reaction section. The reaction unit included a preheater, which was used not only for preheating the reaction gases, but also to promote mixing of the recycle stream with the fresh feed. The reactor was heated by two semi-cylindrical heating elements. A small portion of the gas exiting the reactor was delivered to the analysis section, while the majority was recycled using a diaphragm pump. The recycle ratio was ca. 170, so CSTR conditions (i.e. a well-mixed fluid phase) were ensured.

The reactor was a specially designed single-pellet reactor, shown in Figure 1. It consisted of two stainless steel tubes of rectangular cross-section. The tubes were constructed such that one fit tightly into the other, leaving a channel of rectangular cross-section, through which the reaction gases passed. The sample holder contained the catalyst pellet in such a way that its top surface was flush with the reaction gases that flowed parallel to this surface. In this arrangement external transfer resistances were present, and could become important under certain conditions (9). The length of the tube upstream of the pellet was sufficient for the flow to develop fully. The flange at one end of the reactor had a feedthrough, which was used to insert a thermocouple that recorded the temperature of the reaction gas just above the pellet.

The gases exiting the reactor passed through a Beckman 565 infrared CO_2 analyzer, which continuously monitored the production of carbon dioxide. Gas composition analysis was performed on-line using a Hewlett Packard 5890 II gas chromatograph, equipped with both a thermal conductivity and a flame ionization detector and a Porapak-Q column. Additional experimental details are given elsewhere (9).

Results and Discussion

The catalyst pellets showed stable activity with time. Conversion and selectivity remained essentially constant over the duration of the experiment, which for each pellet lasted several days. Furthermore, by scanning the temperature range several times, in both increasing and decreasing order, it was determined that the reproducibility was very good and that multiple steady states were not present.

As noted earlier, previous theoretical studies (7-8) have shown that the selectivity to ethylene oxide is maximized, when the active material is located at the external surface of the pellet. This behavior results primarily from the fact that the main undesired reaction 2 has a higher activation energy than the desired one. Therefore, intraphase temperature gradients are detrimental to the selectivity. Indeed, in Figure 2, where results for Type 1 pellets are presented, it is shown that for all the temperatures studied, selectivity decreases when the active layer is located deeper inside the pellet. This behavior was observed for all the inlet ethylene concentrations investigated.

A convenient way to illustrate the influence of active layer location on catalyst performance is to examine the selectivity vs. conversion behavior. For five Type 1 catalyst pellets with different active layer locations and for the case where the inlet ethylene concentration is 7.2%, this behavior is shown in Figure 3. For each active layer location, the conversion increase was attained by increasing the reaction temperature (from 210°C to 270°C). This causes the selectivity to decrease, which is typical for the ethylene epoxidation reaction network. This figure shows clearly that conversion and selectivity are always higher for the surface catalyst pellet. If, for example, an ethylene conversion of 20% is desired, then the surface pellet operated at about 215°C will give ethylene oxide selectivity of about 72%, while the first subsurface pellet (s = 0.64 mm) must be operated at 225°C and will give a selectivity of only about 60%. When the active layer is placed deeper within the pellet, the reaction has to be carried out at progressively higher temperatures and gives progressively lower selectivities.

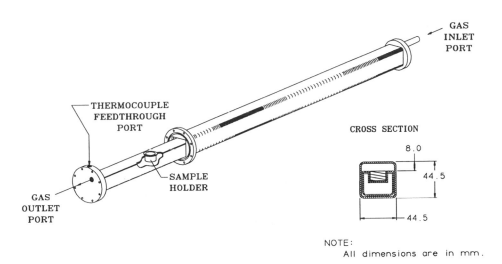

NOTE:
All dimensions are in mm.

Figure 1. Single-pellet reactor. (Reproduced with permission from ref. 9. Copyright 1992 AIChE.)

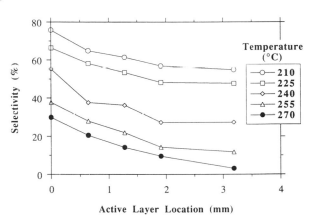

Figure 2. Selectivity as a function of location of the active layer, for various bulk fluid temperatures: inlet ethylene concentration = 7.2%.
(Reproduced with permission from ref. 9. Copyright 1992 AIChE.)

A set of experiments was performed to investigate the effect of *width* of the active layer, which was changed by mechanically mixing the active powder with the appropriate amount of inert powder before pelletizing (i.e., Type 2 pellets). In Figure 4, the selectivity vs. conversion behavior for three surface pellets with different active layer widths is shown. The pellet with a 3.6-mm active layer thickness corresponds to a *uniform* pellet with 6.1 wt.% overall silver loading (for this pellet *all* the inert powder was mixed with the active powder). From Figure 4, it is evident that when the silver is concentrated in thinner regions closer to the top surface (in other words, going from uniform to thin egg-shell pellets), pellet performance improves; for the same conversion, selectivity increases. These results verify that the appropriate Dirac distribution gives better performance than more uniform distributions.

Another method to control silver content in the active layer is to use active powders with different silver loadings. Even though this method introduces additional complexities since the intrinsic reactivities of these active powders may be different, a

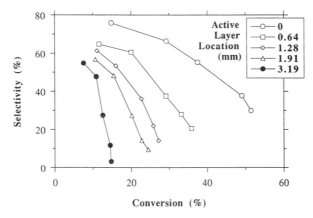

Figure 3. Selectivity as a function of conversion for various active layer locations: inlet ethylene concentration = 7.2%.
(Reproduced with permission from ref. 9. Copyright 1992 AIChE.)

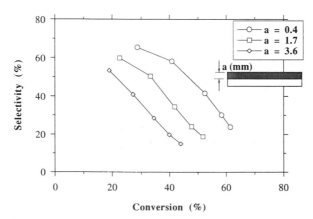

Figure 4. Selectivity as a function of conversion for various active layer widths: inlet ethylene concentration = 4.6%.

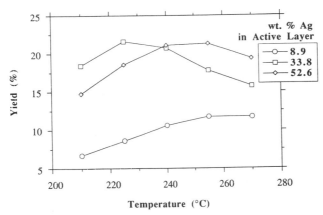

Figure 5. Yield as a function of reaction temperature for various active layer loadings: inlet ethylene concentration = 2.7%.

set of experiments with Type 3 pellets was performed. In Figure 5, the dependence of yield on reaction temperature is shown for three surface pellets with 8.9, 33.8 and 52.6 wt. % silver, for the case where the inlet ethylene concentration is 2.7 vol.%. It is interesting that maxima of yield with temperature are observed for the catalyst pellets with 33.8 and 52.6 wt.% silver in the active layer, and that these maxima have similar values. For the lower loading, the maximum is encountered at a lower reaction temperature. The above results show that when the silver is concentrated towards the external surface by using more heavily loaded active powder, an *optimum loading* of the active powder exists. In this context, it should be emphasized that the pellet with 8.9 wt.% silver in the active layer, which is closer to industrial catalysts, performed poorly for all ethylene feed concentrations considered. Additional results about this behavior are described elsewhere (*13*).

Literature Cited

1. Michalko, E. *U.S. Patent 3,259,589*, July 5, **1966**.
2. Kasaoka, S.; Sakata, Y. *J. Chem. Eng. Japan*, **1968**, *1*, 138.
3. Minhas, S.; Carberry, J. J. *J. Catal.*, **1969**, *14*, 270.
4. Roth, J. F.; Reichard, T. E. *J. Res. Inst. Catalysis, Hokkaido Univ.*, **1972**, *20*, 85.
5. Gavriilidis, A.; Varma, A.; Morbidelli, M. In *Computer Aided Design of Catalysts and Reactors;* Pereira, C. J.; Becker, E. R., Eds.; Chemical Industries Series; Marcel-Dekker, in press.
6. Wu, H.; Brunovska, A.; Morbidelli, M.; Varma, A. *Chem. Eng. Sci.*, **1990**, *45*, 1855.
7. Morbidelli, M.; Servida, A.; Paludetto, R.; Carra, S. *J. Catal.*, **1984**, *87*, 116.
8. Pavlou, S.; Vayenas, C. G. *J. Catal.*, **1990**, *122*, 389.
9. Gavriilidis, A.; Varma, A. *A.I.Ch.E.J.*, **1992**, *38*, 291.
10. Van Santen, R. A.; Kuipers, H. P. C. E. *Adv. Catal.*, **1987**, *35*, 265.
11. Klugherz, P. D.; Harriott, P. *A.I.Ch.E.J.*, **1971**, *17*, 856.
12. Lee, J. K.; Verykios, X. E.; Pitchai, R. *Appl. Catal.*, **1988**, *44*, 223.
13. Gavriilidis, A.; Varma, A., manuscript in preparation.

RECEIVED October 30, 1992

Chapter 32

Synthesis Gas Formation by Direct Oxidation of Methane over Monoliths

D. A. Hickman[1] and L. D. Schmidt

Department of Chemical Engineering and Materials Science, University of Minnesota, Minneapolis, MN 55455

The production of H_2 and CO by catalytic partial oxidation of CH_4 in air and O_2 at atmospheric pressure has been examined over Pt- and Rh-coated monoliths at residence times of less than 10^{-2} sec, with a Rh-coated foam monolith giving H_2 and CO selectivities as high as 0.90 and 0.96 respectively. Studies of several catalyst configurations including Pt-10% Rh woven gauzes and Pt- and Rh-coated ceramic foam and extruded monoliths show that better selectivities are obtained by operating at higher temperatures, maintaining high rates of mass transfer through the boundary layer at the catalyst surface, and using high metal loadings. For both metals, the high selectivities to H_2 and CO strongly suggest that the primary surface reaction is methane pyrolysis, $CH_4 \rightarrow C + 4H$, and Rh catalysts are superior to Pt because the formation of H_2O via OH is energetically less favorable on Rh .

Much recent research (*1-5*) has been devoted to converting methane to products that are more easily transported and more valuable. Such more valuable products include higher hydrocarbons and the partial oxidation products of methane which are formed by either direct routes such as oxidative coupling reactions or indirect methods via synthesis gas as an intermediate. The topic of syngas formation by oxidation of CH_4 has been considered primarily from an engineering perspective (*1-5*). Most fundamental studies of the direct oxidation of CH_4 have dealt with the $CH_4 + O_2$ reaction system in excess O_2 and at lower temperatures (*6-10*).

Synthesis gas is usually produced from methane by steam reforming:

$$CH_4 + H_2O \rightarrow CO + 3 H_2, \qquad \Delta H = +49.2 \text{ kcal/mole} \qquad (1)$$

This endothermic reaction is driven by heating the reactor externally or by adding oxygen to the feed to provide the necessary energy by highly exothermic combustion reactions. A typical steam reformer operates at 15 to 30 atm and 850 to 900°C with a Ni/Al$_2$O$_3$ catalyst and a superficial contact time (based on the feed gases at STP,

---

[1]Current address: The Dow Chemical Company, Midland, MI 48674

0097–6156/93/0523–0416$06.00/0

standard temperature and pressure) of 0.5 to 1.5 seconds (*11*), which corresponds to a residence time of several seconds. The CO/H_2 ratio of the reformer product gases is often modified by the water-gas shift reaction:

$$CO + H_2O \leftrightarrow CO_2 + H_2, \qquad \Delta H = -9.8 \text{ kcal/mole} \qquad (2)$$

A high temperature water-gas shift reactor ($\sim 400°C$) typically uses an iron oxide/chromia catalyst, while a low temperature shift reactor ($\sim 200°C$) uses a copper-based catalyst. Both low and high temperature shift reactors have superficial contact times (based on the feed gases at STP) greater than 1 second (*12*).

The direct oxidation of CH_4 is an alternate route for synthesis gas production:

$$CH_4 + \frac{1}{2} O_2 \rightarrow CO + 2 H_2, \qquad \Delta H = -8.5 \text{ kcal/mole.} \qquad (3)$$

While this reaction is exothermic, the oxidation reactions which produce H_2O from methane are much more exothermic,

$$CH_4 + \frac{3}{2} O_2 \rightarrow CO + 2 H_2O, \qquad \Delta H = -124.1 \text{ kcal/mole,} \qquad (4)$$

$$CH_4 + 2 O_2 \rightarrow CO_2 + 2 H_2O, \qquad \Delta H = -191.8 \text{ kcal/mole.} \qquad (5)$$

Oxidation reactions are much faster than reforming reactions, suggesting that a single stage process for syngas generation would be a viable alternative to steam reforming.

The objective of the research described here is to explore synthesis gas generation by direct oxidation of CH_4 (reaction 3). A reactor giving complete conversion to a 2/1 mixture of H_2 and CO would be the ideal upstream process for the production of CH_3OH or for the Fischer-Tropsch process. As discussed above, currently implemented or proposed processes utilize a combination of oxidation and reforming reactions to generate synthesis gas from CH_4 and O_2. In this work, we seek a faster, more efficient route of syngas generation in which H_2 and CO are the *primary products* of CH_4 oxidation. It is expected that this may be difficult because the reactions $H_2 + O_2 \rightarrow H_2O$ and $CO + \frac{1}{2} O_2 \rightarrow CO_2$ are extremely fast (*13-15*), either heterogeneously or homogeneously, while CH_4 activation is quite slow except at high temperatures.

In this paper, we summarize results from a small scale methane direct oxidation reactor for residence times between 10^{-4} and 10^{-2} seconds. For this work, methane oxidation (using air or oxygen) was studied over Pt-10% Rh gauze catalysts and Pt- and Rh-coated foam and extruded monoliths at atmospheric pressure, and the reactor was operated autothermally rather than at thermostatically controlled catalyst temperatures. By comparing the steady-state performance of these different catalysts at such short contact times, the direct oxidation of methane to synthesis gas can be examined independent of the slower reforming reactions.

Apparatus

The details of the experimental apparatus and procedures are outlined in another paper (*16*). The reactor consisted of a quartz tube with an inside diameter of 18 mm which held the monolith or gauze pack. The reactor was operated at a steady state temperature which is a function of the heat generated by the exothermic reactions and the heat losses from the reactor. Thus, the temperatures reported in these experiments are autothermal steady-state temperatures, which are a function of the heat generated by the exothermic oxidation reactions. To better approximate adiabatic operation, the catalyst was immediately preceded and followed by inert alumina monoliths which acted as radiation shields, and the outside of the tube at the reaction zone was usually insulated. When higher autothermal temperatures were desired, the feed gases were preheated by heating the upstream section of the reaction tube externally. Bypass of the reactant gases around the annular space between the catalyst sample and the quartz tube was minimized by sealing the catalyst sample with a high temperature alumina-silica cloth.

The feed gas flow rate was monitored and controlled by mass flow controllers. Product gases were fed through heated stainless steel lines to a sample loop in an automated gas chromatograph. The GC analysis was performed using two isothermal columns (80°C) in series, a Porapak T and a Molecular Sieve 5A column. When necessary, a second GC analysis using a temperature programmed Hayesep R column was used to separate and detect small hydrocarbons (such as ethylene and ethane) and H_2O.

Although direct oxidation of methane to synthesis gas occurs at feed compositions outside the flammability limits, all of these experiments were conducted in a reactor inside a fume hood. The Tygon feed gas lines were not clamped too tightly so that a sudden pressure rise of a few psig would disconnect the tubing and stop the flow of gases to the reactor. Furthermore, the mixing point of the feed gases was inside the hood to prevent propagation of a flame through tubing outside the fume hood. In addition, all of the product gases were burned in an incinerator so that any toxic and otherwise dangerous products would be converted to CO_2 and H_2O before venting to the atmosphere.

Monoliths

Three basic types of catalysts were studied in these experiments: Pt-10% Rh gauzes, foam monoliths, and extruded monoliths. The gauze catalysts were 40 mesh (40 wires per inch) or 80 mesh Pt-10% Rh woven wire samples which were cut into 18 mm diameter circles and stacked together to form a single gauze pack 1 to 10 layers thick. Gauze catalysts are used industrially in the oxidation of NH_3 to NO for HNO_3 production and in the synthesis of HCN from NH_3, CH_4, and air.

The foam monoliths were α-Al_2O_3 samples with an open cellular, sponge-like structure. We used samples with nominally 30 to 80 pores per inch (ppi) which were cut into 17 mm diameter cylinders 2 to 20 mm long. A 12 to 20 wt.% coating of Pt or Rh was then applied directly to the alumina by an organometallic deposition.

The cordierite extruded monoliths, having 400 square cells/in^2, were similar to those used in automobile catalytic converters. However, instead of using an alumina washcoat as in the catalytic converter, these catalyst supports were loaded directly with 12 to 14 wt.% Pt in the same manner as the foam monoliths. Because these extruded monoliths consist of several straight, parallel channels, the flow in these monoliths is laminar (with entrance effects) at the flow rates studied.

To a good approximation, we believe that all of the supported catalysts behave as Pt or Rh metal surfaces. The loadings are sufficient to deposit uniform films ~1 μm thick on the Al_2O_3 or cordierite ceramics. Scanning electron microscopy (SEM) micrographs of these monoliths before use revealed that the catalyst formed large crystallites on the support with the metal completely covering the support surface. The absence of significant changes in reaction selectivity over several hours of use and the similar performance of all the monoliths confirms that the performance of these catalysts is not strongly influenced by the support, even though SEM showed that the metal formed large crystalline grains during reaction, thus exposing ceramic surfaces.

For all types of monoliths, experiments were run on many samples. We shall show results from only a few of these, but the results were nominally reproducible for all samples.

Reaction Stoichiometry and Equilibrium

Before presenting the results of our experiments, we briefly describe the essential roles of stoichiometry and thermodynamics in this system.

The performance is governed by the conversions of CH_4 and O_2 and the selectivities in producing H_2 and CO:

$$S_{H_2} = \frac{0.5\, F_{H_2}}{F_{CH_4 in} - F_{CH_4 out}} = \frac{F_{H_2}}{F_{H_2} + F_{H_2O}} \tag{6}$$

$$\text{and} \quad S_{CO} = \frac{F_{CO}}{F_{CH_4 in} - F_{CH_4 out}} = \frac{F_{CO}}{F_{CO} + F_{CO_2}} \tag{7}$$

where F_i is the molar flow rate of species i. Because there are only 3 atomic species in this reaction system (C, H, and O), these selectivities are related by the equation

$$S_{H_2} + \frac{1}{2} S_{CO} = 2 - \frac{F_{O_2 in} - F_{O_2 out}}{F_{CH_4 in} - F_{CH_4 out}} \tag{8}$$

Obviously, equation 8 suggests that the ideal synthesis gas reactor should operate at a $CH_4:O_2$ mole ratio of 2:1 (29.6% CH_4 in air) if total conversion of the reactants occurs. At high enough temperatures, this mixture is completely converted to H_2 and CO at thermodynamic equilibrium. Below about 1270 K, the equilibrium mole fraction of carbon becomes significant. Hence, a reactor must be maintained above 1270 K to avoid carbon formation.

Results

Role of Temperature and Catalyst. We first show results for typical experiment using a Pt-coated (11.6% by weight), 50 ppi alumina foam monolith 7 mm in length and a Rh-coated (9.83% by weight), 80 ppi alumina foam monolith 10 mm in length. In Figure 1, the measured product gas CH_4 composition and H_2 and CO selectivities are shown as a function of feed composition and preheat temperature. The mass flow rate was maintained at 4 slpm (standard liters per minute), which corresponds to a superficial velocity of ~80 cm/s (assuming a gas temperature of 1270 K and a void fraction of 0.85) and thus a residence time of ~10 ms. At the flow velocities in these experiments, mass transfer effects (such as those described in the next section) were not important for either of the monoliths, so the observed differences can be attributed to differences between Pt and Rh.

For a given feed temperature, CH_4 conversion is essentially complete for feed CH_4 compositions sufficiently lean in CH_4. Near this breakthrough point, the H_2 and CO concentrations are at their maxima. Above this, the CH_4 breakthrough increases dramatically, while the O_2 conversion remains essentially complete for all compositions. The CH_4 breakthrough occurs because all of the O_2 has been consumed early in the monolith. This is because some of the CH_4 is converted to

CO_2 and H_2O instead of forming only H_2 and CO, leaving no O_2 to react with the remaining CH_4. As the ideal feed composition (29.6% CH_4) is approached from leaner compositions, the CH_4 conversion decreases drastically. In fact, for feed compositions richer than the CH_4 breakthrough point, the O_2/CH_4 conversion ratio actually increases as the O_2/CH_4 feed ratio decreases. Thus, the H_2 and CO selectivities are optimal near the breakthrough point.

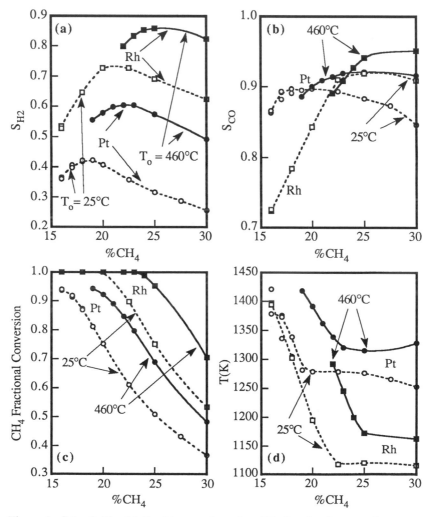

Figure 1. Selectivities (defined in equations 6 and 7), fractional conversion of CH_4, and product gas temperatures for CH_4 oxidation over a 7 mm long, 12 wt.% Pt, 50 ppi foam monolith (circles) and a 10 mm long, 9.83 wt.% Rh, 80 ppi foam monolith (squares) with 4 slpm of feed gases and 2 different feed gas temperatures (open symbols = 25°C and filled symbols = 460°C).

As the feed CH_4 concentration increases from the lean side of the breakthrough point, the increase in the H_2 and CO selectivities results in decreasing product gas temperatures (for both catalysts) since the formation of these species is much less exothermic than the formation of CO_2 and H_2O. On the fuel rich side of the breakthrough point, the product gas temperature levels off just above 1270 K for Pt as the heat generated by the formation of H_2O and CO_2 becomes relatively independent of the feed composition. The Rh catalyst gives qualitatively similar adiabatic temperatures, but because the Rh catalyst is much more selective, less heat is generated by total oxidation reactions, resulting in adiabatic reaction temperatures typically 100 to 150 K cooler than for the same feed conditions with a Pt catalyst.

Increasing the temperature of the feed gases by preheating shifts the CH_4 breakthrough point and the maxima in H_2 and CO selectivities to richer CH_4 compositions, as shown in Figure 1. In addition, the optimal H_2 selectivity increases from 42% to 53% for Pt and from 73% to 86% for Rh when the feed is preheated to about 460°C, while the optimal CO selectivity increases only slightly for both metals. A difference in feed gas temperatures of 435 K results in a *maximum* difference in product gas temperature of only 180 K for a given catalyst since preheating increases the selectivity of H_2 and CO formation, resulting in a lower adiabatic temperature rise for a given feed composition.

In additional experiments, a second catalytic monolith was added immediately after the first monolith. Although the residence time was doubled in these experiments, neither the water-gas shift reaction (2) or the steam reforming reaction (1) was found to significantly improve the reaction conversion and selectivity. From these data, it is apparent that the primary hurdle to achieving the perfect reactor operation involves the selective oxidation of CH_4 to H_2 and CO only. If CO_2 and H_2O are formed, the amount of available O_2 is obviously reduced accordingly. From stoichiometry, this results in unreacted CH_4 in the product gases since the reforming reaction is too slow to consume this methane at these short residence times. Thus, the only way to improve S_{H2} and S_{CO} at these short residence times is to maximize the partial oxidation reaction selectivity.

From the data shown in Figure 1, two key points are evident: (1) increasing the adiabatic reaction temperature increases the selectivity of synthesis gas formation, especially S_{H2}, and (2) Rh is a better catalyst than Pt for producing H_2 by direct oxidation of CH_4.

Role of Mass Transfer and Gas Phase Reactions. Figure 2 illustrates the effect of gas flow rate on selectivity and conversion. For these experiments, 10 layers of Pt-10% Rh gauze were used and the surface temperature of the first layer of gauze was kept constant by heating or cooling the reactor tube in order to decouple the effects of temperature and velocity on selectivity. As shown, although increasing the gas velocity decreases the residence time of the gases, the fractional conversion of CH_4, $\frac{F_{CH4in}-F_{CH4out}}{F_{CH4in}}$, actually increases with increasing flow rates. In addition, the selectivities of H_2 and CO formation increase significantly with increasing flow rates. At all flow rates used, all of the available O_2 is consumed. This change in reaction selectivity and conversion can be explained by a simple model based on a series-parallel reaction scheme with boundary layer mass transfer effects (17).

At the relatively high temperature (1227°C) and pressure (1.4 atm) used in this experiment, one might expect that gas phase reactions may play a significant role in affecting the product gas composition. Other studies with empty quartz tube reactors have shown that significant conversions of rich methane-oxygen mixtures to CO, CO_2, C_2H_4, and C_2H_6 are obtained at temperatures as low as 650°C (18). Thus, one might expect that reactions in the boundary layer near the catalyst surface might be important. In our experiments, however, we only observe at most small quantities (< 0.5 mole %) of ethane and ethylene in the product gases.

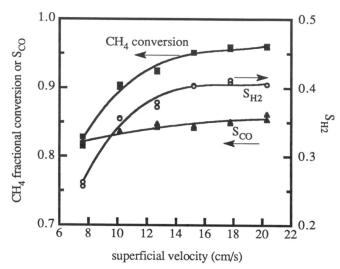

Figure 2. Effect of flow velocity on conversion and selectivity for CH_4 oxidation over 10 layers of 40 mesh Pt-10% Rh gauze. The feed contained 16% CH_4 in air and the front layer of gauze was maintained at 1227±5°C.

Nonetheless, the data in Figure 2 suggests that even if gas phase reactions are significant, the primary initial products of the surface reactions are CO and H_2. Thus, the effect of the mass transfer rate would be similar to that described above. If gas phase production of CO_2, H_2O, and hydrocarbons were competing with the heterogeneous reactions, increasing the flow velocity at a fixed temperature would decrease the residence time of reactant species in the boundary layer, reducing the rate of formation of CO_2, H_2O, and hydrocarbons.

Variation of Number of Gauze Layers. The selectivities and conversions were also studied as a function of the gauze pack thickness by using a gauze catalyst in which the number of gauzes was varied from 1 to 5 at a constant gas flow rate. In these experiments, essentially all of the O_2 was consumed with even one layer of gauze at a total flow rate of 5 slpm with 17% CH_4 in air. However, as the number of layers was increased, the selectivities and CH_4 conversion increased until additional layers result in no significant change in the product composition. These experiments demonstrated that *essentially all of the reactions occurred on the first three layers of gauze.* The rate of methane steam reforming was not significant at these short residence times since the additional fourth and fifth layers did not increase the CH_4 conversion. In addition, product distributions were far from water-gas shift or reforming equilibrium.

These experiments demonstrated that hydrogen must be a *primary product* of the direct oxidation of methane. With only three layers of gauze (giving a contact time of about 10^{-4} seconds), S_{H2} was ~40% with ~90% conversion of CH_4.

Effect of Catalyst Loading. In another set of experiments, two extruded monoliths of identical size and geometry were coated with Pt. One sample was coated with a high loading of Pt (11.6 wt.%) and the second was coated with a very low loading of Pt (about 0.1 wt.%). Methane oxidation experiments over these monoliths gave a similar maximum CO selectivity (~82%) for gases at room temperature, but the sample with a low Pt loading gave a much lower maximum H_2 selectivity than the monolith with a high Pt loading (10% vs. 45%).

Similarly, a Rh foam monolith with 0.56 wt.% Rh gave a lower optimal H_2 selectivity than a Rh foam monolith with 9.83 wt.% Rh (75% vs. 87%). In both the Pt and the Rh experiments, the samples with the lower metal loadings had significantly higher adiabatic reaction temperatures because of the heat generated by the formation of H_2O. As demonstrated by these experiments, the formation of H_2 occurs on the noble metal surface, not in the gas phase or on the catalyst support.

Comparison of Monoliths. By comparing the various samples with different geometries, thermal conductivities, catalysts, and catalyst loadings, we have identified several important factors which affect the methane conversion and product selectivity. For example, the catalyst support geometry strongly influences the mass transfer rate, which influences selectivity as discussed earlier. In addition, materials with a relatively low thermal conductivity will support a higher surface temperature gradient in the axial (flow) direction. This confirms that most of the exothermic oxidation reactions are occurring on or near the front surface of the catalyst, resulting in large temperature gradients in the ceramic samples, with the hottest point localized at the upstream end of the monolith. With such high surface temperatures and, thus, significant rates of heat loss from the reactor, the observed axial temperature drops should be expected.

In tests of all of the different types of samples, catalyst activation and deactivation were generally not observed over periods of several hours. However, SEM micrographs of gauzes revealed that, as in HCN synthesis, the gauze wires undergo surface roughening during the first few hours of use. Unlike the HCN synthesis catalyst, *the surface roughening of the gauze was not accompanied by an increase in selectivity.* Some gauze catalysts did show a drastic decrease in activity and selectivity to synthesis gas after being exposed to very high temperatures (> 1300°C) for a time of several hours or after being tested over a wide range of feed conditions. This decrease in activity was accompanied by a change from a faceted surface morphology to smoother, rounded surface structures and pits.

The supported catalyst samples showed no significant change in selectivity after operation for many hours at variable feeds and temperatures as high as 1300°C. As with the gauze catalysts, SEM micrographs of new and used Pt foam catalysts showed a striking contrast in microstructure. The fresh catalysts contained polycrystalline structures with well-defined facets and characteristic dimensions on the order of 1 μm, while the used catalysts had rounded surface structures of similar characteristic dimensions. Nevertheless, the catalytic selectivities did not change significantly during this transformation in microstructure, suggesting that the activity of these catalysts is that of the pure metals.

As discussed earlier, the most interesting result is that Rh catalysts give a better H_2 selectivity than Pt catalysts. This point will be the primary focus of the following section.

Discussion

From these experiments, one can see that the direct partial oxidation of CH_4 to synthesis gas over catalytic monoliths is governed by a combination of transport and kinetic effects, with the transport of gas phase species governed by the catalyst geometry and flow velocity and the kinetics determined by the nature of the catalyst and the reactor temperature. Under the conditions utilized here, the direct oxidation reaction has been studied independent of the steam reforming and shift reactions. Thermodynamic calculations demonstrate that the product gas mixture is far from the equilibrium composition for either of these two reactions.

Because H_2 and CO are produced with such high selectivities and conversions at these short residence times, the primary mechanism of formation of these products must be methane pyrolysis. The surface reactions which produce H_2 and CO occur in an oxygen-depleted environment, and the major surface species are probably adsorbed C or CH_x and H. The C reacts with oxygen to produce CO, which desorbs before being further oxidized to CO_2. Adsorbed H atoms may either combine to form H_2 which desorbs or react with oxygen to make adsorbed OH species, which then combine with additional adsorbed H atoms to form H_2O. Thus, H_2 and CO can be formed in the following reaction sequence:

$$CH_{4g} \rightarrow CH_{xs} + (4\text{-}x)H_s \tag{9a}$$

$$2\ H_s \rightarrow H_{2g} \tag{9b}$$

$$O_{2g} \rightarrow 2\ O_s \tag{9c}$$

$$C_s + O_s \rightarrow CO_s \rightarrow CO_g. \tag{9d}$$

However, any reactions involving OH, such as

$$CH_{xs} + O_s \rightarrow CH_{x\text{-}1,s} + OH_s \tag{10a}$$

$$H_s + O_s \rightarrow OH_s \tag{10b}$$

$$OH_s + H_s \rightarrow H_2O_g \tag{10c}$$

$$CO_s + O_s \rightarrow CO_{2g}, \tag{10d}$$

should all lead *almost inevitably* to H_2O and CO_2 because reverse reactions of these products are slow or thermodynamically unfavorable.

A CH_4 pyrolysis mechanism appears to be consistent with our observation that preheating improves partial oxidation selectivity. First, higher feed temperatures increase the adiabatic surface temperature and consequently decrease the surface coverage of O adatoms, thus decreasing reactions 10a-d. Second, high surface temperatures also increase the rate of H atom recombination and desorption of H_2, reaction 9b. Third, methane adsorption on Pt and Rh is known to be an activated process. From molecular beam experiments which examined methane chemisorption on Pt and Rh (19-21), it is known that CH_4 must overcome an activation energy barrier for chemisorption to occur.   Thus, the rate of reaction 9a is accelerated exponentially by higher temperatures, which is consistent with the data in Figure 1.

The difference in H_2 selectivity between Pt and Rh can be explained by the relative instability of the OH species on Rh surfaces. For the H_2-O_2-H_2O reaction system on both Pt and Rh, the elementary reaction steps have been identified and reaction rate parameters have been determined using laser induced fluorescence (LIF) to monitor the formation of OH radicals during hydrogen oxidation and water decomposition at high surface temperatures. These results have been fit to a model based on the mechanism (22). From these LIF experiments, it has been demonstrated that the formation of OH by reaction 10b is much less favorable on Rh than on Pt. This explains why Rh catalysts give significantly higher H_2 selectivities than Pt catalysts in our methane oxidation experiments.

By combining rate parameters from these O-H studies with rate parameters from the literature for the various steps in CO oxidation over Pt and Rh catalysts, we have developed a model based on the surface reaction mechanism outlined in reactions (9-10) which accurately predicts the experimentally observed selectivities and conversions. The accuracy of this model suggests that the proposed mechanism includes all of the major reactions steps during methane oxidation on Pt and Rh at these high temperatures. Furthermore, this model, which is described in detail in another publication (23), confirms that the very different selectivities obtained over Rh and Pt catalysts are the result of OH_S formation, step 10b, occurring at significantly slower rates on Rh than on Pt.

In addition to importance of the catalyst composition and temperature, we have shown that methane partial oxidation selectivity is strongly affected by the mass transfer rate. Our experiments show that increasing the linear velocity of the gases or choosing a catalyst geometry that gives thinner boundary layers enhances the selectivity of formation of H_2 and CO. Since H_2 and CO are essentially intermediate products in the total oxidation of CH_4, and the reactions $H_2 + O_2 \rightarrow H_2O$ and CO + $\frac{1}{2} O_2 \rightarrow CO_2$ are extremely fast, it is necessary to minimize the residence time of reactant species in the boundary layer near the catalyst surface. The existence of thinner boundary layers essentially means that the concentration of the partial oxidation products in the gas near the surface is reduced, decreasing the rate of total oxidation of these species.

As the mass transfer rate is increased, reactions producing H_2O and CO_2, whether in the boundary layer or on the catalyst surface, are reduced. Thus, because the direct oxidation of CH_4 is so fast, the mass transfer rate must be high or H_2 and CO will react with O_2 to form the total oxidation products, reducing the partial oxidation selectivity and decreasing the amount of O_2 available to react with CH_4.

Summary

These results provide several insights into the partial oxidation of CH_4 to synthesis gas. It is evident that this reaction system is governed by a combination of kinetic and transport effects. The reaction kinetics depend on the nature of the catalyst and the surface temperature, while transport of the gas species to the catalytic surface is a function of the catalyst geometry and flow velocity.

We have shown that CH_4 can be oxidized directly to synthesis gas over a Pt or Rh monolith catalyst with surprisingly high selectivities for contact times of 10^{-2} to 10^{-4} s. Although these experiments do not give the equilibrium yields of H_2 and CO reported in other work at much longer residence times, they do show that both H_2 and CO are *primary products* of the direct oxidation of CH_4 over a noble metal catalyst.

We also note that, even though mass transfer effects are significant, the kinetics of individual steps strongly influence selectivities. On Rh surfaces, the H_2 selectivity is much higher than on Pt, ($S_{H2} \sim 0.9$ vs. 0.7), and this is due primarily to the slower rate of hydrogen oxidation on Rh than on Pt.

The short residence times used in these tests have allowed us to study this direct oxidation independent of the reforming and shift reactions. By optimizing the selectivity of direct formation of the partial oxidation products, the relatively long residence times required for steam reforming of the unreacted CH_4 can be reduced, requiring less catalyst and a much smaller reactor.

Acknowledgments

The authors would like to thank Dr. L. Campbell of Advanced Catalyst Systems Inc for preparing many of the catalyst samples used in these experiments.

Literature Cited

1. Korchnak, J. D.; Dunster, M.; English, A. *World Intellectual Property Organization* **1990**, WO 90/06282 and WO 90/06297.
2. Blanks, R. F.; Wittrig, T. S.; Peterson, D. A. *Chem. Eng. Sci.* **1990**, *45*, 2407.
3. Dissanayake, D.; Rosynek, M. P.; Kharas, K. C. C.; Lunsford, J. H. *J. Catal.* **1991**, *132*, 117.
4. Ashcroft, A. T.; Cheetham, A. K.; Foord, J. S.; Green, M. L. H.; Grey, C. P.; Murrell, A. J.; Vernon, P. D. F. *Nature* **1990**, *344*, 319.
5. Vernon, P. D. F.; Green, M. L. H.; Cheetham, A. K.; and Ashcroft, A. T. *Catal. Lett.* **1990**, *6*, 181.
6. Hasenberg, D.; Schmidt, L. D. *J. Catal.* **1987**, *104*, 441.
7. Trimm, D. L.; Lam, C. *Chem. Eng. Sci.* **1980**, *35*, 1405.
8. Firth, J. G.; Holland, H. B. *Faraday Soc. Trans.* **1969**, *65*, 1121.
9. Otto, K. *Langmuir* **1989**, *5*, 1364.
10. Griffin, T. A.; Pfefferle, L. D. *AIChE J.* **1990**, *36*, 861.
11. *Catalyst Handbook;* M. V. Twigg, Ed.; Wolfe Publishing Ltd: London, 1989.
12. Satterfield, C. N. *Heterogeneous Catalysis in Practice;* McGraw-Hill: New York, 1980.
13. Blieszner, J. W., Ph.D. Thesis, University of Minnesota, 1979.
14. Schwartz, S. B.; Schmidt, L. D.; Fisher, G. B. *J. Phys. Chem.* **1986**, *90*, 6194.
15. Schwartz, S. B., Ph.D. Thesis, University of Minnesota, 1986.
16. Hickman, D. A.; Schmidt, L. D. *J. Catal.*, submitted 1992.
17. Hickman, D. A.; Schmidt, L. D., *J. Catal.*, in press.
18. Lane, G. S.; Wolf, E. E. *J. Catal.* **1988**, *113*, 144.
19. Schoofs, G. R.; Arumainayagam, C. R.; McMaster, M. C.; Madix, R. J. *Surf. Sci.* **1989**, *215*, 1.
20. Stewart, C. N.; Ehrlich, G. *Chem. Phys. Lett.* **1972**, *16*, 203.
21. Stewart, C. N.; Ehrlich, G. *J. Chem. Phys.* **1975**, *62*, 4672.
22. Williams, W. R.; Marks, C. M.; Schmidt, L. D. *J. Phys. Chem.*, in press.
23. Hickman, D. A.; Schmidt, L. D. *AIChE J.*, submitted 1992.

RECEIVED October 30, 1992

Chapter 33

Partial Oxidation Using Membrane Reactors

Lewis A. Bernstein, Sunil Agarwalla, and Carl R. F. Lund

Department of Chemical Engineering, State University of New York, Buffalo, NY 14260

The viability of one particular use of a membrane reactor for partial oxidation reactions has been studied through mathematical modeling. The partial oxidation of methane has been used as a model selective oxidation reaction, where the intermediate product is much more reactive than the reactant. Kinetic data for V_2O_5/SiO_2 catalysts for methane partial oxidation are available in the literature and have been used in the modeling. Values have been selected for the other key parameters which appear in the dimensionless form of the reactor design equations based upon the physical properties of commercially available membrane materials. This parametric study has identified which parameters are most important, and what the values of these parameters must be to realize a performance enhancement over a plug-flow reactor.

Membrane reactors can offer an improvement in performance over conventional reactor configurations for many types of reactions. Heterogeneous catalytic reactions in membrane reactors [1] and the membranes used in them [2,3] have been reviewed recently. One well studied application in this area is to remove a product from the reaction zone of an equilibrium limited reaction to obtain an increase in conversion [4-10]. The present study involves heterogeneous

0097–6156/93/0523–0427$06.00/0

catalytic partial oxidations of hydrocarbons, where conversion generally is not limited by thermodynamics. Instead, the key to successful partial oxidation is stopping the reaction short of complete conversion (to CO_2 and H_2O). In one approach to doing this, a membrane can be used to supply one of the reactants, e.g. oxygen, at a controlled rate or level throughout the reactor [1, 11-14]. It is also possible to use a solid oxygen anion electrolyte with an applied potential in this approach [1].

The present study investigates a different approach. The membrane is used to allow the desired intermediate product to escape from the reaction zone before it is consumed by further reaction. This use of a membrane reactor was first suggested by Michaels [15]. The partial oxidation of methane, which is a challenging reaction of the type proposed for this application of membrane reactors, has been analyzed herein. There is no thermodynamic limitation for the production of carbon dioxide and water; actually these products are favored. It is desired to remove any partial oxidation product, for example formaldehyde, before it has a chance to be further oxidized.

A schematic diagram of a concentric-tube membrane reactor configuration for doing this is given in Figure 1. In this configuration, reactants are fed to the tube side, which is a packed bed reactor whose walls are constructed of a membrane material. Inert gas is fed to the shell side to sweep away the products that have diffused to that side. Due to the temperatures required for partial oxidations, choices of membrane materials are limited to ceramic or porous glass membranes such as Vycor. For such membranes, reactants, products, and intermediates normally diffuse through the membrane according to the Knudsen mechanism. Relative diffusion rates are then fixed by the molecular weight (or size) of the diffusing species.

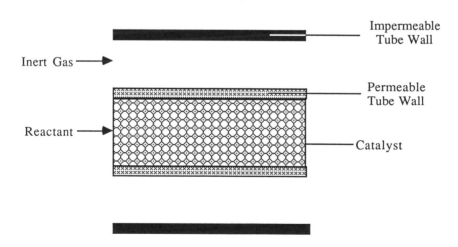

Figure 1. A schematic representation of a membrane reactor for improving yields of intermediate products in a partial oxidation reaction.

Agarwalla and Lund [16] analyzed the use of a membrane reactor for a generalized catalytic series reaction of the form,

$$A \rightarrow B$$
$$B \rightarrow C$$

where B is the desired intermediate product. They concluded that in order for a membrane reactor to show improvement over a conventional plug flow reactor the membrane must be permselective for the intermediate product, B. That is, B must pass through the membrane faster than A. However, the methane partial oxidation reaction is not purely a series reaction, but a series-parallel reaction. One of the reactants is reacted in series, the other in parallel as below:

$$A + B \rightarrow R + T$$
$$A + R \rightarrow S + T$$

This is seen in the kinetic sequence proposed by Spencer and Periera [17].

$$CH_4 + O_2 \rightarrow HCHO + H_2O$$
$$HCHO + \frac{1}{2} O_2 \rightarrow CO + H_2O$$
$$CO + \frac{1}{2} O_2 \rightarrow CO_2 + H_2O$$

It is expected that the conclusions reached in the analysis of the series reaction will also be valid for methane partial oxidation. The first objective of this study was to verify this expectation. The second objective of the study was to determine how much faster than methane formaldehyde must permeate for the membrane reactor to begin to outperform a plug-flow reactor.

Methods

The reactor was evaluated through mathematical modeling. Mole balances for all species on both the shell and tube sides were solved. The assumptions used in writing and solving the equations are as follows: The reactor operates at steady state, plug flow conditions exist on both sides of the membrane, boundary layer effects between the bulk fluids and membrane surface are negligible (that is the bulk and surface concentrations are equal), and radial diffusion that obeys Fick's Law is the only mechanism of transport within the membrane. All of these assumptions are made routinely in modeling inorganic membrane reactors of ceramic or Vycor glass construction [19-23].

Kinetic expressions for the three step pathway given above for the partial oxidation of methane to formaldehyde over a vanadium oxide-silica catalyst were determined by Spencer and Periera [17]. The kinetic parameters were measured in the temperature range from 773 to 873 K, at one atmosphere pressure, and with a nine to one ratio of methane to oxygen in the feed stream. Since these conditions were used in determining the kinetic data, the modeling studies were constrained to these values of temperature, pressure and feed composition.

Mole balance expressions were developed for a general series reaction by Agarwalla and Lund [16], and the same procedures were used here to develop the species balance equations shown in Table I. Boundary conditions and parameter definitions are presented in Tables II and III. Note that the boundary conditions are given only for co-current flow of reactants and inert, which is the only configuration studied. Previous work [16], has shown that counter-current operation is less effective than co-current operation.

From Table III it is evident that many parameters exist. The Damköhler number, Da, is a ratio of the time scale for reactor residency to the time scale for chemical reaction. The Peclet number, Pe, as defined for this problem is a ratio of the time scale for reactor residency to the rate of permeation through the membrane. The product, DaPe, is a ratio of the rate of reaction to the rate of permeation. Choosing a temperature fixes κ_2, κ_3, and all of the Knudsen diffusivities, S_i. Four main parameters remain once these values are determined. These are the Damköhler number, the Damköhler-Peclet product, DaPe, the ratio of the shell side to tube side pressure, ϕ, and the inert gas sweep rate ratio, Y_I, all of which can be varied independently. In all simulations the Damköhler number was allowed to vary over whatever range was required to generate conversions from zero to ca. ninety-nine percent of the limiting reagent, oxygen. The Damköhler-Peclet number varies with the thickness or the pore diameter of the membrane. Realistic values of membrane thickness and pore diameter (based on commercially available materials) were used when determining the range in which to vary the Damköhler-Peclet number. The ratio of inert gas to reactant gas flow rate, Y_I, and the ratio of pressure on the shell side to pressure on the tube side, ϕ, were studied over reasonable ranges. The relative permeability of the intermediate product, S_{HCHO}, was varied in some cases as well.

Once all of the conditions were determined and parameters chosen, the equations were solved by an implicit Euler method. The program was written with a self adjusting step size and analytic Jacobian to reduce error and run time.

Results

In all cases studied, the membrane reactor offered a lower yield of formaldehyde than a plug flow reactor if all species were constrained to Knudsen diffusivities. Thus the conclusion reached by Agarwalla and Lund for a series reaction network appears to be true for series-parallel networks, too. That is, the membrane reactor will outperform a plug flow reactor only when the membrane offers enhanced permeability of the desired intermediate product. Therefore, the relative permeability of HCHO was varied to determine how much enhancement of permeability is needed. From Figure 2 it is evident that a large permselectivity is not needed, usually on the order of two to four times as permeable as the methane. An asymptotically approached upper limit of

Table I. Mole balances on all species in the reactor

$$\frac{dF^t_{CH_4}}{dZ} = Da\left[-\frac{F^t_{CH_4}}{F^t_{total}} - \frac{1}{DaPe}\left(\frac{F^t_{CH_4}}{F^t_{total}} - \phi\frac{F^s_{CH_4}}{F^s_{total}}\right)\right]$$

$$\frac{dF^t_{O_2}}{dZ} = Da\left[-\frac{F^t_{CH_4}}{F^t_{total}} - \frac{\kappa_2}{2}\frac{F^t_{HCHO}}{F^t_{total}} - \frac{\kappa_3}{2}\frac{F^t_{CO}}{F^t_{total}} - \frac{1}{DaPe\,S_{O_2}}\left(\frac{F^t_{O_2}}{F^t_{total}} - \phi\frac{F^s_{O_2}}{F^s_{total}}\right)\right]$$

$$\frac{dF^t_{HCHO}}{dZ} = Da\left[\frac{F^t_{CH_4}}{F^t_{total}} - \kappa_2\frac{F^t_{HCHO}}{F^t_{total}} - \frac{1}{DaPe\,S_{HCHO}}\left(\frac{F^t_{HCHO}}{F^t_{total}} - \phi\frac{F^s_{HCHO}}{F^s_{total}}\right)\right]$$

$$\frac{dF^t_{H_2O}}{dZ} = Da\left[\frac{F^t_{CH_4}}{F^t_{total}} + \kappa_2\frac{F^t_{HCHO}}{F^t_{total}} - \frac{1}{DaPe\,S_{H_2O}}\left(\frac{F^t_{H_2O}}{F^t_{total}} - \phi\frac{F^s_{H_2O}}{F^s_{total}}\right)\right]$$

$$\frac{dF^t_{CO}}{dZ} = Da\left[\kappa_2\frac{F^t_{HCHO}}{F^t_{total}} - \kappa_3\frac{F^t_{CO}}{F^t_{total}} - \frac{1}{DaPe\,S_{CO}}\left(\frac{F^t_{CO}}{F^t_{total}} - \phi\frac{F^s_{CO}}{F^s_{total}}\right)\right]$$

$$\frac{dF^t_{CO_2}}{dZ} = Da\left[\kappa_3\frac{F^t_{CO_2}}{F^t_{total}} - \frac{1}{DaPe\,S_{CO_2}}\left(\frac{F^t_{CO_2}}{F^t_{total}} - \phi\frac{F^s_{CO_2}}{F^s_{total}}\right)\right]$$

$$\frac{dF^t_{Inert}}{dZ} = -\frac{Da}{DaPe\,S_{Inert}}\left(\frac{F^t_{Inert}}{F^t_{total}} - \phi\frac{F^s_{Inert}}{F^s_{total}}\right)$$

$$\frac{dF^s_{CH_4}}{dZ} = \frac{Da}{DaPe}\left(\frac{F^t_{CH_4}}{F^t_{total}} - \phi\frac{F^s_{CH_4}}{F^s_{total}}\right)$$

$$\frac{dF^s_{O_2}}{dZ} = \frac{Da}{DaPe\,S_{O_2}}\left(\frac{F^t_{O_2}}{F^t_{total}} - \phi\frac{F^s_{O_2}}{F^s_{total}}\right)$$

$$\frac{dF^s_{HCHO}}{dZ} = \frac{Da}{DaPe\,S_{HCHO}}\left(\frac{F^t_{HCHO}}{F^t_{total}} - \phi\frac{F^s_{HCHO}}{F^s_{total}}\right)$$

$$\frac{dF^s_{H_2O}}{dZ} = \frac{Da}{DaPe\,S_{H_2O}}\left(\frac{F^t_{H_2O}}{F^t_{total}} - \phi\frac{F^s_{H_2O}}{F^s_{total}}\right)$$

$$\frac{dF^s_{CO}}{dZ} = \frac{Da}{DaPe\,S_{CO}}\left(\frac{F^t_{CO}}{F^t_{total}} - \phi\frac{F^s_{CO}}{F^s_{total}}\right)$$

$$\frac{dF^s_{CO_2}}{dZ} = \frac{Da}{DaPe\,S_{CO_2}}\left(\frac{F^t_{CO_2}}{F^t_{total}} - \phi\frac{F^s_{CO_2}}{F^s_{total}}\right)$$

$$\frac{dF^s_{Inert}}{dZ} = \frac{Da}{DaPe\,S_{Inert}}\left(\frac{F^t_{Inert}}{F^t_{total}} - \phi\frac{F^s_{Inert}}{F^s_{total}}\right)$$

Table II. Boundary conditions for the membrane reactor

Variable	Boundary	Value
$F^t_{CH_4}$	$Z = 0$	1
$F^t_{O_2}$	$Z = 0$	1/9
F^t_{HCHO}	$Z = 0$	0
$F^t_{H_2O}$	$Z = 0$	0
F^t_{CO}	$Z = 0$	0
$F^t_{CO_2}$	$Z = 0$	0
F^t_{Inert}	$Z = 0$	0
$F^s_{CH_4}$	$Z = 0$	0
$F^s_{O_2}$	$Z = 0$	0
F^s_{HCHO}	$Z = 0$	0
$F^s_{H_2O}$	$Z = 0$	0
F^s_{CO}	$Z = 0$	0
$F^s_{CO_2}$	$Z = 0$	0
F^s_{Inert}	$Z = 0$	Y_I

Table III. Critical dimensionless parameters for series reactions in a membrane reactor

Ratio Characteristic of	Definition
Reaction vs. Flow Rate	$Da = \dfrac{\pi d^2 L k_1 P^t}{4 \left(\dot{n}^t_{CH_4}\right)^0 RT}$
Reaction vs. Permeation Rate	$DaPe = -\dfrac{d^2 k_1 \ln\left(\dfrac{d}{d+2\,t}\right)}{8\, D_{CH_4}}$
Shell Side Flow vs. Tube Side Flow	$Y_I = \dfrac{\left(\dot{n}^s_{inert}\right)^0}{\left(\dot{n}^t_{CH_4}\right)^0}$
Shell Side Pressure vs. Tube Side Pressure	$\phi = \dfrac{P^s}{P^t}$
Membrane Permeability of i vs. CH_4	$S_i = \dfrac{D_i}{D_{CH_4}}$
Relative Reaction Rate of Reaction 2	$\kappa_2 = \dfrac{k_2}{k_1}$
Relative Reaction Rate of Reaction 3	$\kappa_3 = \dfrac{k_3}{k_1}$
Moles CH_4 of Reacted/Mole CH_4 fed	$Conv = \dfrac{\left(\dot{n}^t_{CH_4}\right)^0 - \dot{n}^t_{CH_4} - \dot{n}^s_{CH_4}}{\left(\dot{n}^t_{CH_4}\right)^0}$
Methane Reacted to form HCHO	$Sel = \dfrac{\dot{n}^t_{HCHO} + \dot{n}^s_{HCHO}}{\left(\dot{n}^t_{CH_4}\right)^0 - \dot{n}^t_{CH_4} - \dot{n}^s_{CH_4}}$

$i = CH_4, O_2, HCHO, H_2O, CO, CO_2,$ or Inert

enhanced performance from the effect of the permselectivity of formaldehyde is seen as well.

The Damköhler-Peclet product also had an impact on performance; the optimal value ranged from is 1.0×10^{-2} at 773K, to 1.0×10^{-1} at 873K. Little or no improvement was observed when the pressure in the tube was larger than the pressure in the shell, and no improvement was seen when the shell pressure exceeded the tube pressure. When the inert gas sweep rate was increased, the membrane reactor improved until the amount of sweep gas to reactant gas was approximately one hundred as seen in Figure 3. Once again there was an asymptotic limit to the amount of enhancement seen. There was no improvement when the permeabilities of any other component were increased over the permeability of methane.

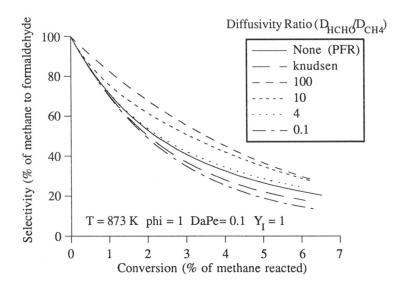

Figure 2. Effect of permselectivity of HCHO, S_{HCHO}, upon membrane reactor performance at 873K.

Discussion

In order for a membrane reactor to produce yields of HCHO greater than in a plug flow reactor, the membrane must be permselective for this species. The more permselective the membrane is to formaldehyde the better the membrane reactor performs until the formaldehyde is approximately one thousand times more permeable than methane. At this limit, the concentration of HCHO is essentially equal on both sides of the membrane at all times. No further improvement is possible by increasing the diffusivity of the formaldehyde further because there is no longer any concentration gradient. Existing membranes separate only on the basis of size and hence to not offer the requisite permselectivity. New materials are under development which use pore surface affinity for the permeants to alter permselectivities. The present study shows that if such materials can be developed, membrane reactors may represent a viable reactor alternative.

At each temperature an optimal range of DaPe exists. These represent a reactor system where the time scale for reaction and permeation are similar. At higher values of DaPe, corresponding to thick membranes and/or narrow pores, there is virtually no permeation. The performance of the membrane reactor then approaches that of a plug flow reactor consisting of just the inner tube. For low values of DaPe, corresponding to thin membranes and/or large diameter pores, there is virtually no resistance to permeation for any species. Here the performance of the reactor approaches that of a plug flow reactor with some bypass. The bypass is due to a certain amount of methane never coming into contact with catalyst because it has diffused to the shell side where no catalyst is present and therefore never reacted. That is, there is near perfect mixing across the membrane.

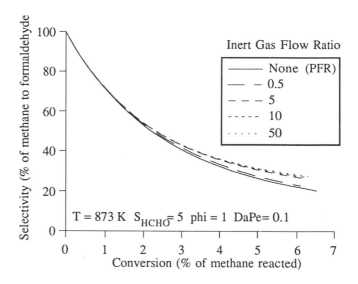

Figure 3. Effect of sweep gas flow rate on reactor performance at 873K.

As sweep gas flow rate is increased, the performance of the reactor improves until the flow rate is about one thousand times the reactant flow rate. The concentration of all species, but most importantly formaldehyde decreases in the shell side of the reactor as this happens. This increases the driving force for permeation of all species. After increasing this flow rate to a certain point further increases in inert gas flow rate do not change the concentration gradient of any species along the reactor because the shell concentrations of all species is essentially zero. However, this decrease in concentration is also a negative from a separations viewpoint. The products must then be recovered from both the tube and shell sides of the reactor, where they are present at very low concentrations.

Conclusions

Partial oxidation of methane in the membrane reactor configuration shown in Figure 1 will not lead to higher yields of desired products than a plug flow reactor unless the diffusivity of the intermediate product, formaldehyde, is approximately four times that of methane. Presently available membranes that can withstand partial oxidation temperatures do not satisfy this criterion.

This suggests areas for further study: developing high temperature molecular affinity based membranes or developing a low temperature methane partial oxidation catalysts to take advantage of existing permselective polymer materials. For example, Kapton polyimide films with a glass transition temperature of about 625K possess the necessary permselectivity [23].

Acknowledgments

This work is based upon work supported by the National Science Foundation under Grant No. CBT-8857100. The Government has certain rights in this material. The authors would also like to acknowledge an equipment grant from Sun Microsystems, Inc. for the 3/260 workstation used in the calculations.

Literature Cited

[1]   Eng, D., and Stoukides, M., *Proc. 9th Int. Congr. Catal.*, Calgary, Vol 2, p. 974. The Chem. Soc. of Canada, Ottawa, 1988.
[2]   Hsieh, H. P., *AIChE Symp. Ser.* $\underline{85}$ (268), 53 (1989).
[3]   Hsieh, H. P., *Catal. Rev.-Sci. Eng.* $\underline{33}$, 1 (1991).
[4]   Mohan, K., and Govind, R., *AIChE J.*, $\underline{32}$, 2083 (1986).
[5]   Oertel, M., Schmitz, J., Weirich, W., Jendryssek-Nuemann, D,. and Schuleten, R., *Chem. Eng. Tech.* $\underline{10}$, 248 (1987).
[6]   Raymont, M. E. D., *Hydrocarbon Proc.* $\underline{54}$, 139 (1975).
[7]   Uemiya, S., Sato, N., Ando, H., and Kikuchi, E., *Ind. Eng. Chem. Res.* $\underline{30}$, 585 (1991).
[8]   Uemiya, S., Sato, N., Ando, H., and Kikuchi, E., *Appl. Catal.* $\underline{67}$, 223 (1991).
[9]   Champagnie, A. M., Tsotsis, T. T., Minet, R.G., and Webster, I. A., *Chem. Eng. Sci.* $\underline{45}$, 2423 (1990).
[10]   Kameyama, T., Dokiya, M., Fujishige, M., Yokokawa, H., and Fukuda, K., *Ind. Eng. Chem. Fundam.* $\underline{20}$, 97 (1981).
[11]   Gryaznov, V. M. *Platinum Metals Rev.* $\underline{30}$, 68 (1986).
[12]   Gryaznov, V. M., Smirnov, V. S., and Slinko, M. G., *Proc. 5th Int. Congr. Catal.*, p. 80 - 1139. J. W. Hightower, ed. Elsevier, New York, 1973.
[13]   Nagamoto, H., and Inoue, H., *Chem. Eng. Comm.* $\underline{34}$, 315 (1985).
[14]   Dicosimo, R., Burrington, J. D., and Grasselli, R. K., U. S. Patent 4,571,443 (1986).

[15] Michaels, *Chem. Eng. Prog.* 64, 31 (1968).

[16] Agarwalla, S., and Lund, C. R. F., *J. Mem. Sci.*, 70 (1992) 129-141

[17] Spencer, N. D., and Pereira, C. J., *J. Catal.* 116, 399 (1989).

[18] N. Itoh, A Membrane Reactor Using Palladium, *AIChE J.*, 33 (1987) 1576.

[19] Y.-M. Sun and S.-J. Khang, *Ind. Eng. Chem. Res.*, 27 (1988) 1136.

[20] N. Itoh, Y. Shindo, K. Haraya and T. Hakuta, *J. Chem. Eng. Jpn.*, (1988) 399.

[21] Y.-M. Sun and S.-J. Khang, *Ind. Eng. Chem. Res.*, 29 (1990) 232.

[22] S. Uemiya, N. Sato, H. Ando, and E. Kikuchi, *Ind. Eng. Chem. Res.*, 30 (1991) 585.

[23] M. C. Hausladen, M.S. Thesis, SUNY-Buffalo, 1992.

RECEIVED October 30, 1992

Chapter 34

Selective Oxidation of Hydrocarbons by Supercritical Wet Oxidation

Fouad O. Azzam and Sunggyu Lee

Process Research Center, Department of Chemical Engineering,
The University of Akron, Akron, OH 44325–3906

Wet oxidation is a process in which partial or complete oxidation reactions occur in a supercritical water medium. Supercritical water exhibits drastically enhanced solvent power toward both oxygen and hydrocarbons, therefore permitting oxidation reactions to be carried out in a nearly homogeneous phase. The selectivity of this partial destructive oxidation can be controlled by varying the process operating parameters, thereby creating a favorable environment for desired byproducts while simultaneously reducing unwanted products. The advantages of these types of processes include; no use of classical conventional catalyst, milder temperature conditions for certain applications, very short residence time, high selectivity, and in most cases a "near zero" discharge. The selective partial oxidation of chlorinated hydrocarbons (CHC) is discussed in this paper, with particular emphasis on the recovery of reaction byproducts for reuse or further processing.

BACKGROUND

Wet Oxidation

Wet oxidation is a chemical process in which oxidation reactions take place under a water blanket. This process utilizes water in its supercritical state as the reaction medium, and high pressure oxygen as the oxidizing agent. Wet oxidation makes use of the increased solubility of oxygen in the supercritical water phase. While only 9.2 mg of oxygen at room temperature dissolve into a liter of water, the level of dissolved oxygen in the supercritical medium is practically infinite, which results in a homogeneous mixture of oxygen and supercritical water (1,2). This permits the process to utilize a lesser amount of excess oxygen in order to achieve the maximum destruction of the waste.

Supercritical water also exhibits a very strong solvent power toward most chemical species. This dramatically increased solvating power is due to the sharp increase of the fluid density as well as the polar nature of the fluid. Also, since many organics are completely miscible in supercritical water, the problem of mass transport

0097–6156/93/0523–0438$06.00/0

resistances can be eliminated, thereby achieving a complete and unhindered oxidation reaction.

In the wet oxidation process, materials partially or completely dissolve into a homogeneous, condensed-phase mixture of oxygen and water, and chemical reactions between the material and oxygen take place in the bulk water phase. This condensed-phase makes wet oxidation an ideal process to transform materials which would otherwise be non-soluble in water to a harmless mixture of carbon dioxide and water. Since oxidation reactions are also exothermic, the high thermal mass of supercritical water makes this reaction medium better suited for thermal control, reactor stability, and heat dissipation. The purpose of this research was to establish a new method for selectively oxidizing waste hydrocarbons into new and reusable products.

Wet Oxidation vs. Catalytic Oxidation

Partial wet oxidation or controlled wet oxidation is, in a sense, similar to that of catalytic oxidation. Catalytic oxidation provides a conventional catalyst in order to boost and control the oxidative reaction, whereas wet oxidation provides a favored atmosphere for the reaction to occur. More accurately, catalytic oxidation provides a surface upon which intimate contact between the reactants takes place compared to the thermodynamic (or fluid dynamic) contact provided by wet oxidation. In wet oxidation, it can be said that the supercritical water phase acts as the "catalyst" for the reaction.

Generally, the time required for catalytic oxidation is greater than that for wet oxidation. This is commonly due to the mass transport resistances which are encountered by the bulk phase reactants as surface reactions proceed. This is particularly true in commercial applications of catalytic oxidation. On the other hand, wet oxidation occurs at a molecular scale where only a controlled contact time is required. Since wet oxidation reactions occur in a homogeneous or semi-homogeneous phase, mass transport resistances are practically eliminated thereby permitting a more uniform distribution of reactants at any point in the reactor. This lack of resistance allows for more throughput per pass for wet oxidation which translates to a smaller recycling stream and therefore a more efficient process. Furthermore, by eliminating the use of conventional catalyst, the regeneration time associated with catalytic oxidation is eliminated. In wet oxidation, water can easily be recycled back into the oxidation vessel for further reuse.

PILOT PLANT LAYOUT

Figure 1 is a schematic representation of the wet oxidation micro-pilot plant. The system consists of three sections, an electrically heated oxidation vessel, a high pressure solvent delivery system, and a water cooled depressurization and collection chamber. A more detailed description of the pilot plant can be found in previous publications (*3,4*).

Primary Oxidation Vessel

The heart of the system is the oxidation vessel depicted in Figure 1. This vessel is a high pressure, 1000 cc, Hastelloy C-276, bolted closure reactor manufactured by Autoclave Engineers Inc. Hastelloy C-276 was chosen as the material of construction due to its excellent corrosion resistance to a wide variety of chemical process environments, which include processes utilizing strong oxidizers (*5,6,7*). The unit is fitted with 1/8" and 1/4" Hastelloy C-276 feed delivery and product outlet lines

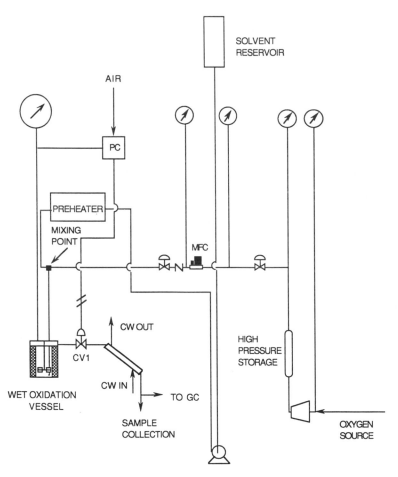

Figure 1. Schematic process flow diagram of wet oxidation mini-pilot plant.

respectively. The reactor is equipped with a thermowell, cooling coil, and a top mounted air driven agitator (Magnedrive).

Solvent Delivery System

Solvent (water) is delivered to the reactor via a high pressure micro-metering pump. This pump is capable of precisely delivering the solvent against a 5000 psi back pressure. The flow capacity of this unit is 6000 cc (STP)/hr at a motor speed of 85 RPM.

Oxygen is delivered to the reactor via a high pressure oxygen compressor (Haskel AGT 30/75). This air driven unit is capable of pressurizing pure oxygen to a maximum pressure of 5000 psig. The compressor is also equipped with a variable pressure safety relief valve and an automated air pilot switch. Both of these safety features make it practically impossible to overpressurize the oxygen storage cylinder, thereby decreasing the chances of failure due to spontaneous oxidation caused by overpressurization of the components (*8,9,10*).

Effluent Depressurization

Depressurization of the oxidation unit is achieved with the use of a high pressure control valve (Annin Wee Willie). This control valve can be set to operate in any one of three control actions: proportional, reset, or derivative. Following the control valve, the hot products are directed to a water cooled high pressure condenser. Here they are cooled to ambient conditions before being sent to a holding vessel for GC analysis.

WET OXIDATION OF CHLORINATED HYDROCARBONS

Chemistry

In the absence of oxygen, the thermal degradation of poly(vinyl chloride) involves dehydrochlorination, which gives polyene sequences followed by crosslinking. The dehydrochlorination also takes place even in the presence of oxygen or in other oxidative environments. When oxygen is present, chain scissions involving C-C bond breakages as well as dehydrochlorination take place. The relative rates of these two modes of reactions depend upon the concentration of oxygen, the temperature, the pressure in the case of supercritical wet oxidation, etc.

In the supercritical wet oxidation, the concentration of oxygen in the reactive system is high and the contact between the reactants is more intimate, thus making chain scission reactions much more active. This appears to be a major reason for more rapid degradation of PVC in an oxygen environment and production of monomers and dimers.

The following reaction mechanisms are proposed (efforts are currently being made to experimentally confirm these mechanisms):

I. Dehydrochlorination and Oxidation

$$\sim CH_2\text{-}CHCl\text{-}CH_2\text{-}CHCl\sim \xrightarrow{\;ROO\bullet\;} \sim \overset{\bullet}{C}H\text{-}CHCl\text{-}CH_2\text{-}CHCl\sim \xrightarrow[-Cl\bullet]{}$$

$$\sim CH=CH\text{-}CH_2\text{-}CHCl \sim \xrightarrow{+O_2} 2CO_2 + H_2O + CH_2=CHCl \qquad (1)$$

The dehydrochlorination rate increases substantially in a wet oxidation environment and this acceleration is believed to be due to peroxy radicals, which are formed by the straight oxidation of a hydrocarbon or a fraction of polymer. However, this also is yet to be confirmed experimentally.

II. Dehydrochlorination and Chain Scissions

$$\sim CH_2\text{-}CHCl\text{-}CH_2\text{-}CHCl\sim \longrightarrow CH_2\text{=}CH\text{-}CH\text{=}CHCl + HCl \qquad (2)$$

$$\sim CH_2\text{-}CHCl\text{-}CH_2\text{-}CHCl\sim \longrightarrow 2CH_2\text{=}CHCl \qquad (3)$$

$$\sim CH_2\text{-}CHCl\text{-}CH_2\text{-}CHCl\sim \longrightarrow \overset{\overset{\displaystyle Cl}{|}}{CH_2}\text{=}C\text{-}CH\text{=}CHCl + H_2 \qquad (4)$$

III. Oxidation (Free-Radical Initiated)

$$\sim CH_2\text{-}CHCl\text{-}CH_2\text{-}CHCl\sim \xrightarrow{ROO\bullet} \sim \overset{\bullet}{C}H\text{-}CHCl\text{-}CH_2\text{-}CHCl\sim \xrightarrow[-Cl\bullet]{}$$

$$\sim CH\text{=}CH\text{-}CH_2\text{-}CHCl \sim \xrightarrow[-HCl,\, -ROOCl]{} \sim CH\text{=}CH\text{-}CH\text{=}CH\sim$$

$$\xrightarrow{+O_2} CO_2 + H_2O \qquad (5)$$

$$CH_2\text{=}CHCl + 5/2\, O_2 \longrightarrow 2CO_2 + H_2O + HCl \qquad (6)$$

The final products of Route III are CO_2, H_2O, and HCl in their stoichiometric amounts.

IV. Hydrochlorination

$$CH_2\text{=}CHCl + HCl \longrightarrow CH_2Cl\text{-}CH_2Cl \qquad (7)$$

$$CH_2\text{=}CHCl + HCl \longrightarrow CH_3\text{-}CHCl_2 \qquad (8)$$

$$CH_2\text{=}CH\text{-}CH\text{=}CH_2 + 2HCl \longrightarrow CH_3\text{-}CHCl\text{-}CHCl\text{-}CH_3 \qquad (9)$$

It is conceivable that all these reactions, included in Routes I, II, II, and IV, take place competitively in the system, even though their relative kinetic rates depend on various operating parameters, in particular the oxygen concentration, the reactor residence time, and the pressure and temperature. For example, if the reaction mixture is left too long in a wet oxidation environment, the reaction would proceed to completion, resulting in producing only H_2O, CO_2, and HCl. Therefore, in order to maximize the production of vinyl chloride monomer or dimers, an optimal process condition must be sought, especially in terms of the residence time, percent excess of oxygen, temperature, and pressure.

Experimental

Selective partial wet oxidation of poly(vinyl chloride) (PVC) was performed using the wet oxidation pilot plant described in earlier sections. Experiments were initiated by charging the primary oxidation vessel with a preweighed amount of PVC resin. A typical oxidation utilized 0.5 to 2 g of PVC. Once charged, the oxidation vessel was brought up to the desired extraction/reaction temperature and pressure by heating and the constant addition of preheated supercritical water. Oxidation temperatures ranged between 390 and 440 °C with pressures ranging from 230 to 275 atm. Once the operating parameters were established, the injection of high pressure oxygen was initiated. Oxygen flow rates ranged between 50 and 200 scc/min. Reactor effluents were then collected at preset intervals and were directed to a gas chromatograph for analysis.

Results and Discussion

The results of partial wet oxidation of PVC were quite encouraging. The results of these experiments are summarized in Figure 2. For this particular case, PVC was oxidized to give monomers, dimers and lighter hydrocarbons. The percent recovery of select hydrocarbons have been plotted with respect to reactor residence time. It should be noted that the percent recovery is the amount of a particular hydrocarbon which is present in the reactor at the time of sampling. This amount is based on a gas chromatograph signal which has been precalibrated and is coupled with a mass balance which accounts for the theoretical amount of carbon which has been combusted. From the figure, it can be shown that the recovery of 1,2-dichloroethane increases as the reaction time progresses. This trend continues until a recovery of 2.5% is achieved. The same is also true for 1,1-dichloroethane, however, the change in the rate of recovery is not as dramatic until approximately 1.5 minutes into the reaction. At this point, the percentage of recovery increases at a much more significant pace until a maximum of 9 percent is achieved at 1.75 minutes. Perhaps a more interesting compound is vinyl chloride. With this compound, the percent recovery goes through several changes before reaching a maximum. These changes are attributed to several different reactions which take place during the course of partial oxidation. Initially, as PVC is thermally degraded, there is a buildup of hydrochloric acid and vinyl chloride monomer. As the wet oxidation process continues, the existing monomer is destroyed by the incoming oxygen stream which causes the recovery of vinyl chloride to decrease. This is clearly illustrated in the figure between 0.75 and 1.25 minutes. This explains the decreasing nature of the recovery curve. Once the majority of this monomer is destroyed, the oxygen begins to attack the polymer and smaller molecular weight fractions thereby releasing more vinyl chloride. Since the oxygen concentration in the reactor is not sufficient to destroy this newly formed monomer, a rise in the recovery percentage is noticed. This occurs between 1.25 and 1.75 minutes until a maximum recovery of 8 percent is achieved. It should be noted that increasing the reaction time past 1.75 minutes results in a complete wet oxidation and the destruction of the hydrocarbons.

Safety

Safety must be an integral part of any oxidation process. Safety begins with sound equipment, such as ASME code vessels and continues through the operating and shut

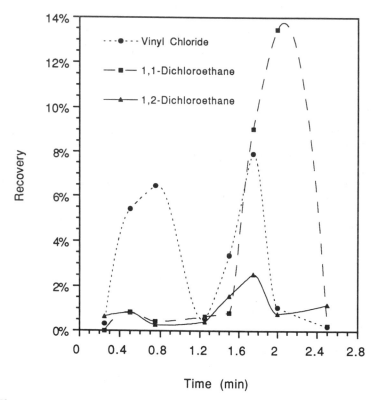

Figure 2. Recovery of select hydrocarbons from wet oxidation of poly(vinyl chloride).

down procedures for the equipment. This is particularly true for processes utilizing unusually oxidative materials coupled with high temperature and pressure operations. For this reason, a brief background into the safety considerations which were utilized in the undertaking of this research are included here.

In the construction of the wet oxidation unit, several areas of safety were considered. Of utmost importance was that of personal safety. Since this type of operation demands the use of high pressures and temperatures, operator contact with the high pressure vessels had to be limited. To accommodate this criterion, a barrier was constructed to shield the operator from any unforeseen releases from the reactor. This barrier was constructed from 1/4 inch steel and is designed in a manner that will fully contain any releases. This barrier is also equipped with two explosion vents to direct the force of any explosions away from the main walls and into a safe area. To further maximize personnel safety, all operator assisted controls are mounted on the outside of the unit.

A policy of a three level safety design was also adopted. This includes the use of utilities and safety devices. An orthogonal design was used such that interruption of any part of these services would not affect the remaining sections. Manual bypasses were also provided in order to permit operator intervention at any point in the process. Finally, any releases of material from any of the relief devices were directed into secondary holding vessels in order to prevent releases into the atmosphere.

A brief but important point must also be made regarding the use of oxygen as the primary oxidant. Oxygen in itself is an oxidizer. However, high pressure oxygen (greater than 2800 psig) becomes extremely oxidative and the possibility of spontaneous detonations increases significantly should it come in contact with a hydrocarbon. For this reason, special handling techniques must be employed. All oxygen process lines, fittings, holding vessels and any other part which is in direct contact with the oxygen must be specially cleaned for oxygen service. This can be done either commercially or in house, however, the final result must be a hydrocarbon free system. Any lubrication needed in the assembly of the oxygen systems must be hydrocarbon free and specially rated for this particular application. With these types of precautions, supercritical wet oxidation reactions can be made safer, and spontaneous detonations can be eliminated.

Further Potential Applications

The principle of supercritical wet oxidation can be applied to several areas, including municipal waste treatment, chemical waste treatment, polymeric waste treatment, and the treatment of mildly radioactive waste. Since the basic principle of wet oxidation involves the rapid oxidation of organic material, any substance which is mildly oxidative can be subjected to this process. With such high temperature operation, the remaining fraction of inorganic material can be simply precipitated as a salt and then easily collected from the bottom of the oxidation vessel.

Conclusions

Several conclusions can be drawn from this experimental work.

1. The operability of the wet oxidation pilot plant was successfully demonstrated, particularly from a standpoint of reaction controllability.
2. The system was successfully operated for both complete and partial oxidation of a chlorinated hydrocarbon (PVC).
3. Selectivity can be controlled by variation of process parameters and reactor residence time.

4. The research confirms the viability of wet oxidation as a new method of selectively oxidizing hydrocarbons into fresh and usable products.

ACKNOWLEDGMENTS

The authors are grateful for a generous fellowship support provided by BF Goodrich Co.

LITERATURE CITED

1. Pray, H. A., Schweickert, C. E., and Minnich, B. H., "Solubility of Hydrogen, Oxygen, and Helium in Water at Elevated Temperatures", *Industrial and Engineering Chemistry*, *44*:5, pp. 1146-1151, 1951.
2. Himmelblau, D. M., "Solubilities of Inert Gases in Water 0 C to Near the Critical Point of Water", *Journal of Chemical and Engineering Data*, *5*, pp. 10-15, 1959.
3. Azzam, F. 0., Fullerton, K. F., Vamosi, J. E. and Lee, S. "Application of Wet Oxidation for Waste Treatment", invited paper for Symposium on The Environmental Issues of Energy Conversion Technology, paper No 87G, AIChE Summer National Meeting, Pittsburgh, PA, Aug 18-21, 1991.
4. Azzam, F. 0. and Lee, S., "Design and Operation of Wet Oxidation Mini-Pilot Plant for Complete and Partial Combustion", invited paper for Symposium on The Role of Pilot Plants in Commercialization of Processes, paper No 36E, AIChE Spring National Meeting, New Orleans, LA, Mar 29 - Apr 2, 1992.
5. Corrosion Resistance of Hastelloy Alloys, *Properties Data Booklet*, Cabot Corporation, Stellite Division, 1978.
6. Corrosion Resistance of Tantalum and Niobium Metals, *NRC Inc.*, Bulletin No. 3000.
7. Metals Handbook, 8th Ed., Vol. 1, *Properties Selection of Metals*, American Society of Metals, Novelty, Ohio, 1961.
8. Fryer, D. M., "High Pressure Safety System Analysis- A Proposed Method", *Design, Inspection, and Operation of High Pressure Vessels and Piping Systems*, PVP, Vol. 48.
9. Loving, F. A., "Barricading Hazardous Reactions", *Industrial and Engineering Chemistry*, Vol. 49, October 1957.
10. Pohto, H. A., "Energy Release from Rupturing High-Pressure Vessels - A Possible Code Consideration", *Journal of Pressure Vessel Technology*, Vol 101, May 1979.

RECEIVED October 30, 1992

INDEXES

Author Index

Affiliation Index

Subject Index

Production: Peggy D. Smith and C. Buzzell-Martin
Indexing: Deborah H. Steiner
Acquisition: Anne Wilson
Cover design: Sarah Chung

Printed and bound by Maple Press, York, PA

Bestsellers from ACS Books

The ACS Style Guide: A Manual for Authors and Editors
Edited by Janet S. Dodd
264 pp; clothbound ISBN 0–8412–0917–0; paperback ISBN 0–8412–0943–X

The Basics of Technical Communicating
By B. Edward Cain
ACS Professional Reference Book; 198 pp;
clothbound ISBN 0–8412–1451–4; paperback ISBN 0–8412–1452–2

Chemical Activities (student and teacher editions)
By Christie L. Borgford and Lee R. Summerlin
330 pp; spiralbound ISBN 0–8412–1417–4; teacher ed. ISBN 0–8412–1416–6

Chemical Demonstrations: A Sourcebook for Teachers,
Volumes 1 and 2, Second Edition
Volume 1 by Lee R. Summerlin and James L. Ealy, Jr.;
Vol. 1, 198 pp; spiralbound ISBN 0–8412–1481–6;
Volume 2 by Lee R. Summerlin, Christie L. Borgford, and Julie B. Ealy
Vol. 2, 234 pp; spiralbound ISBN 0–8412–1535–9

Chemistry and Crime: From Sherlock Holmes to Today's Courtroom
Edited by Samuel M. Gerber
135 pp; clothbound ISBN 0–8412–0784–4; paperback ISBN 0–8412–0785–2

Writing the Laboratory Notebook
By Howard M. Kanare
145 pp; clothbound ISBN 0–8412–0906–5; paperback ISBN 0–8412–0933–2

Developing a Chemical Hygiene Plan
By Jay A. Young, Warren K. Kingsley, and George H. Wahl, Jr.
paperback ISBN 0–8412–1876–5

Introduction to Microwave Sample Preparation: Theory and Practice
Edited by H. M. Kingston and Lois B. Jassie
263 pp; clothbound ISBN 0–8412–1450–6

Principles of Environmental Sampling
Edited by Lawrence H. Keith
ACS Professional Reference Book; 458 pp;
clothbound ISBN 0–8412–1173–6; paperback ISBN 0–8412–1437–9

Biotechnology and Materials Science: Chemistry for the Future
Edited by Mary L. Good (Jacqueline K. Barton, Associate Editor)
135 pp; clothbound ISBN 0–8412–1472–7; paperback ISBN 0–8412–1473–5

For further information and a free catalog of ACS books, contact:
American Chemical Society
Distribution Office, Department 225
1155 16th Street, NW, Washington, DC 20036
Telephone 800–227–5558